Progress in Mathematical Ecology

Progress in Mathematical Ecology

Special Issue Editor

Sergei Petrovskii

MDPI • Basel • Beijing • Wuhan • Barcelona • Belgrade

MDPI

Special Issue Editor
Sergei Petrovskii
University of Leicester
UK

Editorial Office
MDPI
St. Alban-Anlage 66
Basel, Switzerland

This is a reprint of articles from the Special Issue published online in the open access journal *Mathematics* (ISSN 2227-7390) from 2017 to 2018 (available at: https://www.mdpi.com/journal/mathematics/special_issues/Mathematical_Ecology)

For citation purposes, cite each article independently as indicated on the article page online and as indicated below:

LastName, A.A.; LastName, B.B.; LastName, C.C. Article Title. *Journal Name* **Year**, *Article Number*, Page Range.

ISBN 978-3-03897-312-6 (Pbk)
ISBN 978-3-03897-313-3 (PDF)

Contents

About the Special Issue Editor

Sergei Petrovskii, is an applied mathematician with about thirty years of research experience in mathematical ecology and ecological modelling. His research spans a broad variety of problems of ecology and population dynamics, with a particular emphasis on modelling complex multiscale environmental and ecological systems. He has published four books and 100+ papers in peer-reviewed journals. He presently holds the position of Chair in Applied Mathematics at the University of Leicester (UK). He is also the Editor-in-Chief of *Ecological Complexity* (a journal on theoretical and mathematical ecology published by Elsevier) and a member of the editorial board of three other journals. He has been an invited or keynote speaker at 20+ international conferences and workshops. He is the founder and scientific coordinator of the MPDE (Models in Population Dynamics and Ecology) conference series.

mathematics

MDPI

Editorial

Progress in Mathematical Ecology

Sergei Petrovskii

Department of Mathematics, University of Leicester, Leicester LE1 7RH, UK; sp237@le.ac.uk

Received: 11 September 2018; Accepted: 11 September 2018; Published: 13 September 2018

Mathematical modelling plays a special role in ecology. Although traditional ecology is a largely empirical science, replicated experiments are not often possible because of the high complexity of ecological interactions and the impossibility to reproduce the weather conditions. Moreover, large-scale field experiments (where the consequences are usually not fully known) can be damaging for the ecological communities and costly or even dangerous for humans. Mathematical modelling provides an efficient supplement and sometimes even a substitute to an empirical study; it creates a virtual laboratory where different hypotheses can be tested safely, and at relatively low cost.

Application of mathematics to problems arising in ecology and population dynamics—to which we broadly refer as mathematical ecology—has a long and glorious history. It was mutually beneficial for both of the disciplines involved and there are many examples of that. Prey–predator population cycles are one of the core concepts of contemporary ecological theory, but the route of this idea goes back to the works of Vito Volterra. Food web analysis is a fruitful and fast developing area of ecology, but its progress would hardly be the same without the insights and tools from the mathematical theory of networks. Ecological research on the spatial spread of traits and species was the starting point for the mathematical theory of diffusion-reaction waves, and Alan Turing's study on morphogenesis led to the development of the theory of self-organized pattern formation.

New times have brought new challenges and stimulated further developments in mathematical ecology. This volume of Mathematics presents a carefully selected collection of papers that, taken together, provide a rather comprehensive account of the state of the art in this science. A broad range of problems is addressed. The first two papers [1,2] consider some new aspects of the classical problem of pattern formation. The dynamics of infectious diseases is another old problem where many aspects are still poorly understood; this is addressed [3,4]. Comparisons between different patterns of animal movement are made, [5] and the effect of dispersal and migration on the population dynamics in fragmented habitats are considered [6,7]. Biological invasion of an alien population of "ecosystem engineers" is considered [8], where it is shown how nonlinear feedbacks may lead to a self-organized regime shift resulting in the collapse of the invading population. The structural complexity of ecosystems is revisited in a review paper [9], where the meaning and value of various biodiversity indices is reassessed. The final paper [10] provides a new mathematical insight into a classical problem of genealogy.

We hope that, along with, obviously, shedding new light on several important problems, this volume also brings new questions and hence will stimulate future advances in mathematical ecology.

References

1. Köhnke, M.C.; Malchow, H. Impact of Parameter Variability and Environmental Noise on the Klausmeier Model of Vegetation Pattern Formation. *Mathematics* **2017**, *5*, 69. [CrossRef]
2. Banerjee, M.; Mukherjee, N.; Volpert, V. Prey-Predator Model with a Nonlocal Bistable Dynamics of Prey. *Mathematics* **2018**, *6*, 41. [CrossRef]
3. Barbara, F.; La Morgia, V.; Parodi, V.; Toscano, G.; Venturino, E. Analysis of the Incidence of Poxvirus on the Dynamics between Red and Grey Squirrels. *Mathematics* **2018**, *6*, 113. [CrossRef]

4. Fatehi, F.; Kyrychko, Y.N.; Blyuss, K.B. Effects of Viral and Cytokine Delays on Dynamics of Autoimmunity. *Mathematics* **2018**, *6*, 66. [CrossRef]

5. Ahmed, D.A.; Petrovskii, S.V.; Tilles, P.F.C. The "Lévy or Diffusion" Controversy: How Important Is the Movement Pattern in the Context of Trapping? *Mathematics* **2018**, *6*, 77. [CrossRef]

6. Alharbi, W.; Petrovskii, S.V. Critical Domain Problem for the Reaction—Telegraph Equation Model of Population Dynamics. *Mathematics* **2018**, *6*, 59. [CrossRef]

7. Pal, N.; Samanta, S.; Martcheva, M.; Chattopadhyay, J. Role of Bi-Directional Migration in Two Similar Types of Ecosystems. *Mathematics* **2018**, *6*, 36. [CrossRef]

8. Fontanari, J.F. The Collapse of Ecosystem Engineer Populations. *Mathematics* **2018**, *6*, 9. [CrossRef]

9. Daly, A.J.; Baetens, J.M.; De Baets, B. Ecological Diversity: Measuring the Unmeasurable. *Mathematics* **2018**, *6*, 119. [CrossRef]

10. Slade, P.F. Linearization of the Kingman Coalescent. *Mathematics* **2018**, *6*, 82. [CrossRef]

![Σ] *mathematics*

MDPI

Article

Analysis of the Incidence of Poxvirus on the Dynamics between Red and Grey Squirrels

Fadi Barbara [1,†], Valentina La Morgia [2,†], Valerio Parodi [1,†], Giuseppe Toscano [1,†] and Ezio Venturino [1,3,*,†]

1   Department of Mathematics "Giuseppe Peano", University of Turin, via Carlo Alberto 10, 10123 Torino, Italy; fadi.barbara@gmail.com (F.B.); parodi.valerio@gmail.com (V.P.); giuseppe.toscano@gmail.com (G.T.)
2   Institute for Environmental Protection and Research (ISPRA), via Ca'Fornacetta 9, 40064 Ozzano Emilia (BO), Italy; valentina.lamorgia@isprambiente.it
3   Member of the Research Group GNCS of INdAM
*   Correspondence: ezio.venturino@unito.it; Tel.: +39-011-670-2833
†   These authors contributed equally to this work.

Received: 23 March 2018; Accepted: 18 June 2018; Published: 1 July 2018

Abstract: A model for the interactions of the invasive grey squirrel species as asymptomatic carriers of the poxvirus with the native red squirrel is presented and analyzed. Equilibria of the dynamical system are assessed, and their sensitivity in terms of the ecosystem parameters is investigated through numerical simulations. The findings are in line with both field and theoretical research. The results indicate that mainly the reproduction rate of the alien population should be drastically reduced to repel the invasion, and to achieve disease eradication, actions must be performed to reduce the intraspecific transmission rate; also, the native species mortality plays a role: if grey squirrels are controlled, increasing it may help in the red squirrel preservation, while the invaders vanish; on the contrary, decreasing it in favorable situations, the coexistence of the two species may occur. Preservation or restoration of the native red squirrel requires removal of the grey squirrels or keeping them at low values. Wildlife managers should exert a constant effort to achieve a harsh reduction of the grey squirrel growth rate and to protect the remnant red squirrel population.

Keywords: competition; dynamical systems; invasive species; squirrel poxvirus (SQPV); transmissible diseases

MSC: primary 92D25; secondary 92D30; 92D40

1. Introduction

The Eurasian red squirrel (*Sciurus vulgaris* Linnaeus, 1758) is an arboreal mammal with a large Palearctic range, and in Europe, it is a native and widespread species in most areas, with a few exceptions. It is absent from the southwest of Spain and from the majority of Mediterranean islands, and it occurs only sporadically in the Balkans [1]. The species is known to locally disappear because of the competition with the Eastern grey squirrel (*Sciurus carolinensis* Gmelin, 1788), which has been introduced from North America into the British Isles, mainly from 1890 onwards [2,3], and in Italy [4], where the species was first reported in 1948. Because of its severe impacts on the red squirrel and on forest biodiversity [5], the grey squirrel was previously reported by the International Union for Conservation of Nature (IUCN) in the list of 100 worst invasive species [6]. More recently, it has been included in the list of invasive alien species of European concern (EU Regulation No. 1143/2014; EU Implementing Regulation No. 1141/2016).

The grey squirrel out-competes the smaller red squirrel in the use of trophic resources. Both species feed on tree seeds, and they largely overlap in their use of habitat [7–9]. The work in [10] suggested that interspecific competition for scatterhoarded seeds, with grey squirrels pilfering red squirrels' food caches, can play a role in the replacement of red by grey squirrels. In sympatric conditions, when the core area of red squirrel home ranges largely overlaps with grey squirrel ranges, these authors observed a reduction in the daily energy intake by red squirrels. The spring body mass of red squirrels was also negatively correlated with the percentage of interspecific core area overlap. In a high-quality mixed deciduous woodland in Italy, [9] estimated a large tree-species niche overlap (about 70%), and taking into account both activity and foraging behavior in time and space, there was no evidence of interspecific niche partitioning. Thus, red squirrels seemed unable to avoid competition with grey squirrels. Furthermore, the invasive species often occurs at higher densities than the native one, and it has higher breeding rates. When red squirrels tend to avoid good quality habitats with high grey densities, their demography may be negatively affected, with a decrease in juvenile recruitment [11]. Available data thus suggest that competition between red and grey squirrels for habitat and food is very likely and that it may affect red squirrel population dynamics, finally playing a role in the observed replacement of the red species by the grey squirrel.

In the United Kingdom, the interspecific competition between squirrels is also mediated by a squirrel poxvirus (SQPV), the *Chordopoxviridae* genus [12,13]. The virus causes exudative erythematous dermatitis [14], characterized by ulcerated and hemorrhagic scabs affecting the skin around the eyes, nose and lips first, then spreading to the ventral thorax, inguinal area and feet [15,16]. The disease was unrecorded before the grey squirrel introduction [16]; it was then reported in England since 1930 [2] and later confirmed by electron microscopy in several regions [14,16]. Recently, the virus was also detected in Scotland [17]. Serological evidence supported the theory that the virus was introduced to the U.K. with the grey squirrel [13] and [18] proved that the parapoxvirus can cause a debilitating disease in red squirrels, whilst having no apparent impact on greys. According to the experimental tests of these authors, in red squirrels, the disease effects would include ulceration and infection of the skin lesions, along with lethargy, almost certainly causing mortality in the wild within 10–20 days. Only a few red squirrels could survive the infection [19] or at least they could experience a prolonged disease course, so that ultimately the disease may also be transmitted between red squirrels in the wild [18]. Transmission of the poxviruses differs between genera, being spread either by aerosol, contaminated fomites, direct contact or arthropods [13]. In particular, for the SQPV, current data support a direct transmission mechanism. According to [20], in England and Wales, cases of SQPV disease occurred in red squirrels only in geographic areas with seropositive grey squirrels, and a critical community size of squirrels may be required for disease epidemics in red squirrels to occur. This is also suggested by the delay observed between the establishment of invading grey squirrels in new areas and the cases of the disease in the red squirrels. Grey squirrel must probably reach a threshold density or number before the virus is transmitted to reds. Anyway, no evidence was found that the density or number of both squirrel species was controlling the onset of disease outbreaks, as expected when micro-parasites have a reservoir host, and the even spread of cases of disease across months also suggested that a direct rather than vector-borne transmission route was more likely.

Consistent with the hypothesis of an endemic infection of low pathogenicity in most grey squirrel populations, in several United Kingdom populations [19], it was found that 61% of grey squirrels appeared to have been exposed to SQPV, whereas only four (about 3%) of the red squirrels tested had antibodies to parapoxvirus. Out of these four animals, three also had parapoxvirus-associated disease. Overall, these data clearly supported the hypothesis that parapoxvirus spill-over from grey squirrel reservoirs of infection may significantly affect red squirrel populations. As a consequence, in England, the decline of the latter has been attributed not only to direct competition, but it has also been associated with the epidemic outbreaks of the disease [15,21]. Indeed, clinical signs observed in red squirrels by [18] were the same observed on squirrels recovered dead in the wild [16,20–23], and detailed studies demonstrated that the competitive interactions alone could not account for the observed rate and pattern of replacement observed in some areas of the United Kingdom [24,25].

Since this viral disease had such a significant impact on the decline of the red squirrel in the U.K., it is particularly interesting to focus on the epidemiological dynamics of the red-grey squirrel system. In this respect, deterministic approaches such as the one developed by [26] are particularly useful to provide a clear understanding of the interaction mechanisms. Here, in particular, we develop a mathematical model for a three-population system, including grey squirrels and both healthy and infected red squirrels, in order to highlight the potential effects of the SQPV transmission on the long-term population dynamics of the native species. Even though, so far, the virus has been detected only in the United Kingdom, the development of general models is fundamental, and because grey squirrels have been mainly introduced outside North America as pets, multiple releases in the wild have occurred and are still expected. As a consequence, it is not possible to exclude that the virus will appear in new areas in the future. In this respect, our model provides insights into the dynamics of the two-species system, identifying the parameter space allowing for species coexistence and for disease eradication. Management alternatives to achieve red squirrel conservation are also discussed.

In particular, this paper attempts a different approach than other current research in the field [25,27]. While these models include space in the model formulation and tend to be more specific (e.g., [27] for the Anglesey peninsula in U.K.), here we try to concentrate on a more ample view. The main aim is indeed to obtain more general results and to possibly provide an indication of how to approach the invasion problem in different scenarios. Furthermore, in setting up our model, we consider a frequency-dependent term for disease transmission, in epidemiology also known as standard incidence. This assumption is justified in the discussion on the model formulation, but here, we stress that this approach is rather different from previous model setups that have appeared in the grey squirrel invasion literature. In this way, red squirrels can be wiped out also by the epidemics. This remark prompts the field ecologists to perform new investigations, to ascertain actual disease-related extinction, thus absolving one of the tasks of theoretical models, i.e., suggesting the possible appearance of ecological phenomena. Finally, although recent investigations suggest that squirrelpox prevalence may fluctuate in grey populations, thereby implying also a change in the disease transmission, our basic assumption removes the possibility that some of the grey squirrels are virus-free. This may very well not be the case, but apart from the mathematical simplification that ensues from having a three-dimensional system rather than dealing with four differential equations, we are actually looking at the worst possible invasive scenario, in order to preserve the utmost conservative viewpoint. The analysis of a worst scenario is also justified if we consider that, where SQPV has been present, the decline in red squirrel distribution can be 17–25 times faster than in areas where SQPV is known not to be present [28].

2. Materials and Methods

2.1. The Model

We introduce at first the model using the grey squirrel G, healthy red squirrel R and infected red squirrel I populations. All the populations are counted by numbers. As a consequence, essentially all the parameters denoting rates are frequencies, measured in t^{-1}. The main point is that the red squirrels are partitioned into two subpopulations, to investigate possibly the epidemic effects on them. The grey squirrels are instead assumed to be immune from the disease, although asymptomatic carriers of the virus. Therefore, interactions between healthy red squirrels and grey squirrels may lead to new cases of the disease and move the red individual to the class of infected.

Note that all squirrels are local creatures. Indeed, they usually establish home ranges whose extension varies as a function of habitat type, (e.g., from 2 to about 30 hectares for red squirrels; [29–31]), and they hardly travel long distances. Dispersal occurs usually once or twice a year, and movements are confined within the range of a few kilometers (e.g., for the grey squirrel, see [32,33]). Basically then, for the interaction terms between populations due to possible competition for resources, we assumed a functional response similar to the classical Holling Type II, thereby modeling the fact that if a population is large, because these are essentially sedentary animals, only the influence of the closest neighbors will most directly determine the dynamics of an individual of the other population. The disease transmission is assumed to be modeled by a standard incidence function, for the same reasons. Both of these assumptions, as in standard epidemic models, lead to a singularity in the cases respectively that both populations or the disease-affected one vanish. Mathematically, this means that the origin and the case $G = N = 0$ must be excluded from the domain of the dynamical system (1). Assuming further that all the parameters are nonnegative, the model reads as follows:

$$
\begin{aligned}
\frac{dG}{dt} &= rG\left(1 - \frac{G}{E}\right) - \frac{kRG}{G + c(R + I)}, \\
\frac{dR}{dt} &= sR\left(1 - \frac{R}{K}\right) - \frac{aRG}{G + c(R + I)} - R\frac{\lambda I + \beta G}{R + I + G}, \\
\frac{dI}{dt} &= R\frac{\lambda I + \beta G}{R + I + G} - \mu I.
\end{aligned}
\tag{1}
$$

Note that we can take the initial condition $I(0) = 0$ to indicate that initially the disease may very well be absent.

The first equation describes the grey squirrel dynamics, while the remaining two illustrate respectively the healthy and infected red squirrel populations' evolution. For the grey squirrel, logistic growth is assumed, and then, competition only with the healthy red squirrels is taken into account. Indeed, the point is that the disease is highly virulent for the latter, weakens and kills them in just a short time, so that any of their possible interference for sharing resources with the grey squirrels can be easily ruled out. Note that we take the competition term to be a saturating function both in terms of G and R because the squirrels are mainly residing in a fixed place, for which their interactions with possible neighbors are limited. Thus, even if either one of the populations grows at high values, there is a saturation in the contact rate. The direct competition with infected squirrels instead is disregarded because the latter are too weak, but we take them into account in the denominator, because if there is a large number of them, the interactions between the susceptible red squirrels and the grey ones will diminish, as more infected red squirrels will be the neighbors of the grey, and therefore, the competition damage to the grey population will be reduced.

The healthy red squirrels also follow a logistic growth in which, for the very same reasons just outlined, the intraspecific competition is confined only to their healthy similars. Note that the two carrying capacities of the grey and red squirrels are assumed to be different, respectively E and K, because in suitable habitats, the grey squirrel can live at higher densities (e.g., 1.45–2.99 individuals /ha, up to 7–10 or even more than 10 ind./ha [34–37]) than the red squirrel, whose densities are usually lower than 1–2 ind./ha [7,27,30,31,38–42]. Furthermore, their net reproductive rates differ, r and s, respectively [43]. The competition has the same form as in the first equation because similar considerations as for the grey squirrel competition term hold for this corresponding red squirrel term in this equation. In addition, the red squirrels suffer from the competition of the grey ones, third term of the equation, at a rate a that once more differs from the corresponding one holding for the grey squirrels, k, because of the latter having a larger body size and being more efficient in the use of trophic resources [10,44]. Further, we need to consider the disease effects. Note that there are two ways that a susceptible red squirrel can contract the disease. This can occur if it comes in contact with an infected consimilar at rate λ, which is modeled by the fourth term, or also at rate β by the interaction with a grey individual, which is always an asymptomatic carrier, as stated earlier, the last term of the second equation. Rates of viral shedding from infected grey and red squirrels may be different [45], but a transmission rate of 3.27 between and within each species was obtained by matching model and field data on the seroprevalence in the U.K. [27,46]. Note that we take the interspecific transmission term to be in a rational form because the squirrels are mainly residing in a fixed place, for which their interactions with possible neighbors are limited. Thus, even if either one of the populations grows at high values, there is a saturation in the contact rate.

Infected red squirrels are recruited via the two infection mechanisms just described for the healthy red ones and appear as the first two terms of the last equation. In addition, they experience natural plus disease-related mortality at rate μ. In view of their weakness, the fast replication process of the virus in their bodies and high lethality of the disease, no other vital dynamics is possible for this class of individuals. Indeed, pregnancy lasts longer than the disease itself: on average, they indeed respectively last 38–40 [47] and 10–20 days [18]. Table 1 summarizes the model parameters.

Table 1. A summary of the model parameters and their interpretation.

Name	Interpretation	Unit
r	grey squirrels' net growth rate	t^{-1}
E	grey squirrels' carrying capacity	pure number
s	red squirrels' net growth rate	t^{-1}
k	grey squirrels' damage due to interspecific competition	t^{-1}
a	red squirrels' damage due to interspecific competition	t^{-1}
K	red squirrels' carrying capacity	pure number
λ	intraspecific disease transmission rate for red squirrels	t^{-1}
β	interspecific disease transmission rate for red squirrels	t^{-1}
μ	red squirrels' natural plus disease-related mortality	t^{-1}
c	competitivity weight among different squirrel species	t^{-1}

2.2. Model Reparametrization

We now reformulate Model (1) in terms of the total red squirrel population $N = R + I$, the disease prevalence $i = IN^{-1}$ among the latter, while keeping the grey squirrels' population unchanged. This transformation is rather common in epidemic [48] and ecoepidemic models [49,50], in spite of the fact that it breaks down when $N = 0$, because it allows distinguishing the two different vanishing modes of this population, namely when the disease remains endemic in it, affecting the whole vanishing population, or when some of the red squirrels remain still healthy in the process. Two equilibria arise for the situation in which the local population is eradicated by the successful replacement of the invaders, both corresponding to red squirrels' disappearance in the original Model (1). They are not artificial because they allows us to understand better the role played by the epidemic in the native species disappearance. Thus, they represent a valuable piece of information that, without this transformation, we would not be able to gather. The reformulation leads to the following system:

$$\frac{dG}{dt} = G\left[r\left(1 - \frac{G}{E}\right) - \frac{kN(1-i)}{G+cN}\right], \tag{2}$$
$$\frac{dN}{dt} = N\left[s(1-i)\left(1 - \frac{N(1-i)}{K}\right) - \frac{a(1-i)G}{G+cN} - \mu i\right],$$
$$\frac{di}{dt} = (1-i)\left[\frac{\lambda iN + \beta G}{G+N} - \mu i - si\left(1 - \frac{N(1-i)}{K}\right) + \frac{aiG}{G+cN}\right],$$

once again initially assuming no disease in the red squirrel population, $i(0) = 0$.

2.3. Study of the Ecosystem Behavior

Taking into account the described model, we analyze the existence and stability of equilibrium solutions, and we also perform numerical simulations to better investigate the current situation. For the latter, we use a set of reference parameter values:

$$r = 1.2, \quad E = 1.25, \quad k = 0.61, \quad s = 1, \quad K = 0.5, \quad a = 1.65, \quad \lambda = 3.27, \quad \beta = 3.27. \tag{3}$$

These values are obtained from the literature, mainly from [27,51]. Instead, the natural plus disease-related mortality for the red squirrels and the weight of the two populations are arbitrarily assumed:

$$\mu = 2, \quad c = 1. \tag{4}$$

Note that the natural plus disease-related mortality μ is chosen at a high value in view of the high virulence and lethality of the epidemics, for the red squirrels. In addition, note that the literature [27,43,51] reports that some of them vary in the following suitable ranges:

$$r \in [1.2, 1.3], \quad E \in [0.6, 2.5], \quad K \in [0.35, 0.65]. \tag{5}$$

Thus, we also investigate the sensitivity of the system with respect to some of the ecosystem parameters. In all simulations the initial conditions are chosen as follows, with no initial infectives, as this was the pristine situation before the introduction of the invasive grey population:

$$G(0) = 0.07, \quad G(0) = 2.80, \quad G(0) = 0.$$

3. Results

3.1. Equilibria

As noted earlier, the origin P_0 and the point at which the ecosystem collapses, but with the disease remaining endemic among the red squirrels, $P_2 = (0, 0, 1)$, are outside the dynamical system domain. There are only four possible remaining equilibria: the healthy-red-squirrels-only point $P_1 = (0, K, 0)$, the best possible ecosystem outcome, then an equilibrium in which the red population collapses, but the disease remains endemic, while grey squirrels thrive, for which, however, a pair of points is detected, $P_3 = (E, 0, 1)$ and $P_4 = (E, 0, i_4)$; the grey-squirrels-free point $P_5 = (0, N_5, i_5)$, which is another desirable ecosystem outcome, implying the invasive grey squirrel extinction, although the disease remains endemic among the red ones, and the coexistence of the three populations $P_* = (G_*, N_*, i_*)$, which would also be less welcome, but still acceptable, as the local individuals would thrive anyway. Note that the pair of points P_3 and P_4 denotes two really different situations: although the native squirrel population disappears, in the former case, the disease affects the whole vanishing population, while in the latter, it is still endemic, but part of the red squirrels are still healthy.

3.1.1. Feasibility

Specifically, for the red-squirrels-free equilibrium, observe that there are two cases depending on the status of the disease in the population, while the species itself is disappearing. In both, the grey squirrels attain carrying capacity. In the former, P_3, the totality of red squirrels is infected, while the population declines to zero, while for the second, equilibrium P_4, for the disease prevalence among the vanishing red squirrels, we find the value:

$$i_4 = \frac{\beta}{\mu + s - a}$$

indicating that if $i_4 < 1$, a fraction of the healthy red squirrels is still present in the vanishing population. This restriction implies:

$$a < \mu + s \tag{6}$$

for feasibility, while:

$$a + \beta \leq \mu + s \tag{7}$$

to have really a different outcome than P_3.

For P_5, we find:

$$N_5 = \frac{K\lambda(s + \mu - \lambda)}{s\mu}, \quad i_5 = 1 - \frac{\mu}{\lambda}.$$

Feasibility requires then:

$$s + \mu \geq \lambda \geq \mu. \tag{8}$$

The possible coexistence equilibrium is investigated numerically, both for feasibility as well as for stability.

3.1.2. Stability

The Jacobian $J = (J_{k\ell})$ has the following entries:

$$J_{11} = r\left(1 - \frac{2G}{E}\right) - \frac{ckN^2(1-i)}{(G+cN)^2}, \quad J_{12} = -\frac{kG^2(1-i)}{(G+cN)^2}, \quad J_{13} = \frac{kGN}{G+cN}, \quad J_{21} = -\frac{aN(1-i)}{(G+cN)^2},$$

$$J_{22} = s(1-i)\left(1 - \frac{2N(1-i)}{K}\right) - \frac{aG^2(1-i)}{(G+cN)^2} - \mu i, \quad J_{23} = -sN\left(1 - \frac{2N(1-i)}{K}\right) + \frac{aGN}{G+cN} - \mu N,$$

$$J_{31} = (1-i)N\left[\frac{\beta - \lambda i}{(G+N)^2} + \frac{aic}{(G+cN)^2}\right], \quad J_{32} = (1-i)\left[\frac{(\lambda i - \beta)G}{(G+N)^2} - \frac{iacG}{(G+cN)^2} + \frac{is}{K}(1-i)\right],$$

$$J_{33} = i\left[\mu + s\left(1 - \frac{N(1-i)}{K}\right) - \frac{aG}{G+cN}\right] - \frac{\lambda Ni + \beta G}{G+N} + (1-i)\left[\frac{\lambda N}{G+N} - (\mu + s) + \frac{sN}{K}(1-2i) + \frac{aG}{G+cN}\right].$$

Again, although the points P_0 and P_2 are not equilibria, in their neighborhood, the trajectories are repulsed away from them, because the dominant terms are the linear ones, and they depend on the positive reproduction rates $r, s > 0$, so that the populations will rebound if reduced to very low terms.

We now evaluate J at each equilibrium, to assess their local stability. P_1 is conditionally stable; one eigenvalue is $-s < 0$, and the remaining ones provide its stability conditions:

$$\lambda < \mu, \quad cr < k. \tag{9}$$

P_3 has two negative eigenvalues, $-r$ and $-\mu$, while the third one provides the stability condition:

$$\mu + s < \beta + a. \tag{10}$$

There is an obvious transcritical bifurcation between P_3 and P_4; compare (7) and (10).

Furthermore, for P_4, the eigenvalues are explicit, $-r$, $s - a - \beta$ and $\beta - (\mu + s - a)$, leading to the following stability conditions:

$$s < a + \beta < \mu + s. \tag{11}$$

At P_5, again, one eigenvalue is explicit, giving:

$$cr\lambda < k\mu \tag{12}$$

for stability. Other inequalities come from the Routh–Hurwitz conditions on the remaining minor $\hat{J}(P_5)$. The trace condition is satisfied, since the diagonal entries are negative in view of the feasibility conditions (8):

$$J_{22} = (s + \mu - \lambda)\left(\frac{\mu}{\lambda} - 2\right) < 0, \quad J_{33} = \frac{\mu^2}{\lambda^2}(\lambda - s - \mu) < 0$$

while the determinant ultimately gives the stability condition:

$$4\lambda\mu^2 + 2s\lambda\mu + 2\lambda^3 < 5\lambda^2\mu + \mu^3 + s\mu^2 + s\lambda^2 \tag{13}$$

9

Remark 1. *The equilibria P_1, P_3 and P_4 have real eigenvalues, so that no Hopf bifurcation is possible in their neighborhood.*

Proposition 1. *Taking λ as the bifurcation parameter, there exists a Hopf bifurcation at P_5, for the critical threshold:*

$$\lambda_\dagger = s + \mu.$$

Proof. Note that the trace is:

$$\mathrm{tr}(\widehat{J}(P_5)) = (s + \mu - \lambda)\phi(\lambda), \quad \phi(\lambda) = \frac{1}{\lambda^2}\left(\lambda^2 - \mu\lambda + \mu^2\right) > 0, \quad \forall \lambda.$$

Now, for $\lambda_\dagger$, the trace vanishes, while for the determinant, we find:

$$\det(\widehat{J}(P_5)) = -(s + \mu - \lambda_\dagger)^2 \left(\frac{\mu}{\lambda} - 2\right)\frac{\mu^2}{\lambda_\dagger^2} - N_5(s + \mu - 2\lambda_\dagger)\frac{si_5}{K}(1 - i_5)^2 = N_5\lambda_\dagger\frac{si_5}{K}(1 - i_5)^2 > 0.$$

Hence, the eigenvalues are pure imaginary. □

3.2. Simulations

In this section, the ecosystem behavior is explored by means of simulations carried out by standard MATLAB routines for the integration of ordinary differential equations.

The first results are shown in Figure 1. It is clear that the red squirrels disappear with all individuals contracting the epidemics. The role of the mortality appears only in the speed of disappearance; the faster it is, the higher the mortality rate. For this, compare the two frames, where the right one shows the simulations for a value of the mortality, which is one tenth of the one used in the left frame. The results shown in the left frame of Figure 2 indicate nevertheless that even a one thousand-fold reduction in this mortality does not prevent extinction, but as can also be seen by a close look at the prevalence graphs in Figure 1, favor a faster epidemics propagation, because the disease affects the whole population in a much shorter time. A ten-fold increase in this mortality rate prevents the disease from reaching every individual of the red squirrel population, but it clearly favors the vanishing of this population; see the right frame of Figure 2. However, this can hardly be of any help because the native squirrel population cannot be preserved anyway. This situation is known in communities, where the optimum for an (infected) individual, namely its life span, becomes a danger for the community, because more individuals would be infected by its presence. Conversely, the worst for the single individual is in this case optimal from the community point of view: by being removed, it cannot further spread the epidemics.

In all these situations, the grey squirrels are unaffected, as should be expected. Indeed, they might suffer from the red squirrels' competition, but as the latter disappear, after a short transient, they reach their natural carrying capacity.

We then investigate the sensitivity of the system to the choice of the ecosystem parameters, keeping all the parameters that do not vary as given in (3), (4) and in (5) when not otherwise specifically stated.

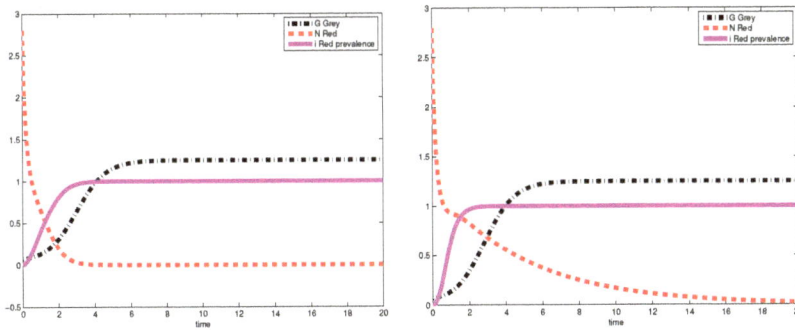

Figure 1. The three populations are plotted versus time: G grey squirrels, dash-dotted; N total red squirrels, dashed; i disease prevalence among red squirrels, continuous line. (**Left frame**) This is the current situation obtained with the basic parameter values (3); red squirrels are wiped out. (**Right frame**) This is the situation with the basic parameter values (3), but taking a much lower value of the natural plus disease-related mortality $\mu = 0.2$, one tenth of the previous one. In both cases, it is apparent that the red squirrel population is doomed, disappearing while all individuals contract the epidemics. Only the speed of its disappearance varies, being lower for lower mortality rates.

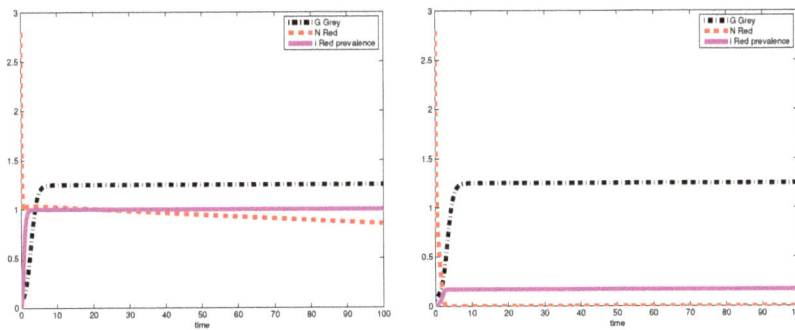

Figure 2. The three populations are plotted versus time: G grey squirrels, dash-dotted; N total red squirrels, dashed; i disease prevalence among red squirrels, continuous line. (**Left frame**) This is the current situation obtained with the basic parameter values (3), but $\mu = 0.002$. The red squirrels would eventually disappear, but in a much longer time span. (**Right frame**) This is the situation with the basic parameter values (3), but taking a ten-times higher value of the natural plus disease-related mortality, $\mu = 20.0$. The red squirrel population still vanishes, but now, not all its individuals will be affected by the epidemics. This of course is no help in the preservation cause of the native squirrel population.

For the arbitrary parameters c and μ, we use a range that encompasses the conditions (4). Figure 3 shows that if the red squirrels have much more weight in the interspecific competition (small c), for moderate to large values of the natural plus disease-related mortality, they could outcompete the invading species, at the same time eradicating the disease, as well. However, this occurs for a value of c perhaps too small. The next question is to evaluate how the system responds to changes in the reproductive rate of grey squirrels, so we compare again the range of c with a suitable range of r around the value given in (5). However, the red squirrels vanish altogether with the disease affecting the whole population, over the entire range of the parameter space. The results are not shown. Combining c with the disease intraspecific transmission rate λ, again, a small stripe of low parameter values for c and moderate for λ ensure that the native species thrives with low endemicity and invading species almost vanishing, the results being shown in Figure 4.

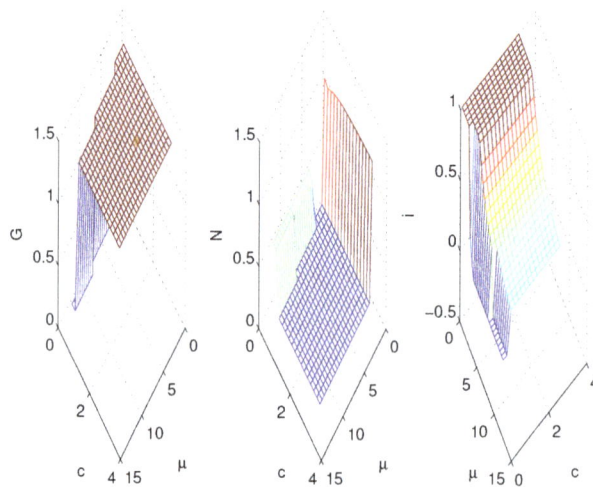

Figure 3. The ecosystem behavior as a function of c and μ; in a very narrow stripe for moderate values of the mortality and small values of c, meaning that the native squirrels can better compete with the invaders, the red squirrels could thrive, with the disease being eradicated, as well as the invaders vanishing.

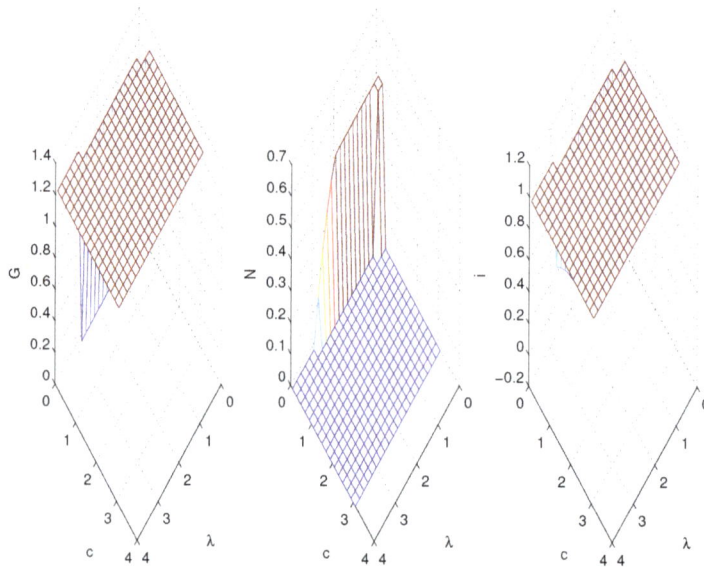

Figure 4. In the c and λ parameter space, similar results hold as for Figure 3.

The next question is to evaluate how the system responds to changes in the reproductive rate of grey squirrels. The simulations show that no change in the behavior occurs; letting r vary as indicated in (5) and taking $\mu \in [0.1, 10.0]$, no substantial changes occur in the ecosystem behavior. To better explore this situation, we explore also what happens if a reduction in the reproductive rates is implemented in the wild. Figure 5 shows the continuation plot when $r \in [0.01, 1.01]$, for the same range of μ. If artificially, the growth rate of the grey squirrel is controlled, there might be hope to restore the pristine situation, for a suitably medium value of the natural-plus-disease-related mortality.

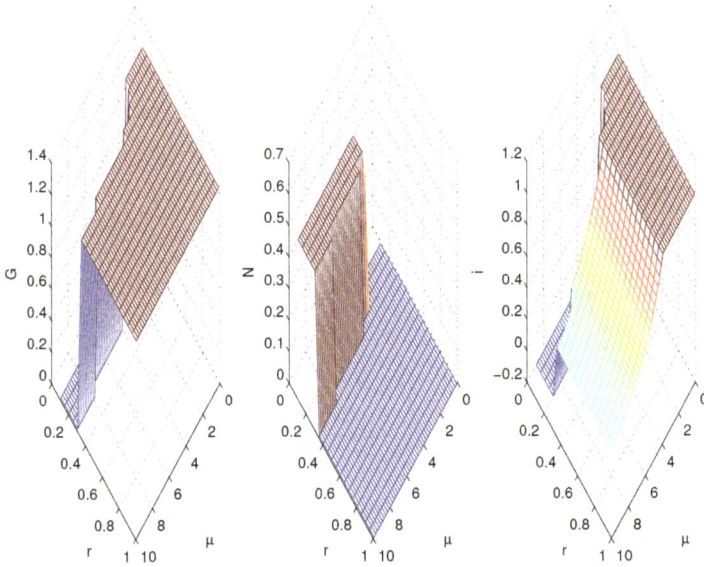

Figure 5. The ecosystem behavior as function of r and μ. The grey squirrel reproduction rate must fall really to low values to allow the survival of the native population together with the disappearance of both the disease and the invading species. Higher grey squirrel reproduction rates are allowed if the red squirrels' natural plus disease-related mortality becomes higher.

Simultaneous perturbations of the two carrying capacities do not lead to any change whatsoever. No changes are observed either in the $K - \lambda$ parameter space; native squirrels are wiped out, all carrying the disease in the process. Essentially, the same occurs in the $K - c$ and in the $K - \mu$ parameter space; in the latter, only an extremely narrow stripe, surprisingly for low K, ensures red squirrel survival, but with the disease infecting the whole population; otherwise, the red squirrels disappear anyway even by increasing their own carrying capacity; see Figure 6.

In the $E - r$ parameter space, no changes in the ecosystem behavior are observed; for the other parameters given in (3), the invasion of the grey squirrels is successful.

Letting the grey squirrel carrying capacity E vary as suggested in (5), coupling it once again with μ leads to almost a direct linear relationship with the grey squirrel population size, this time independent of μ. The results (not shown) indicate that the red squirrel population is wiped out, and the prevalence is heavily affected by μ and almost independent of changes in E. Furthermore, in the $\beta - r$ parameter space, independent of the grey squirrel reproduction rate r, a small transmission rate induces a reduction in the prevalence, but the red squirrels are wiped out anyway. Combining the interspecific transmission rate β with K, the same effect occurs: for low values of β, the disease prevalence drops, independent of the value of the red squirrels' carrying capacity, but the latter do not survive anyway.

We now investigate the combined changes of both reproductive rate and carrying capacity of grey squirrels, at first with a low value of the natural plus disease-related mortality, $\mu = 0.2$. In this case, the red squirrels are wiped out with an epidemic that in the process affects the whole population. The grey squirrel population instead grows linearly with its carrying capacity, almost independently of the reproduction rate (result not shown).

13

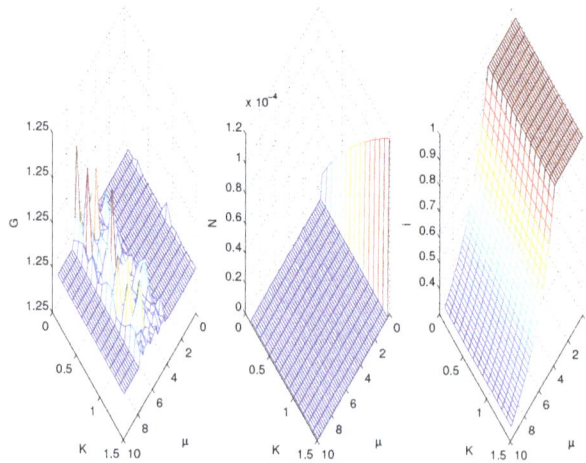

Figure 6. For low values of K independent of the interspecific transmission rate β, an apparent, but very low recovery of the red squirrels occurs, although they are all infected. The oscillations observed in the grey squirrel population are due to noise, as they attain the carrying capacity throughout the whole parameter space.

Taking a larger intermediate value, $\mu = 5.0$, instead shows that there is again a narrow stripe in the parameter space in which the red squirrels thrive. It is obtained for low values of the grey squirrel reproduction rate, $r \leq 0.2$ approximately, while the grey squirrel population increases linearly with E; see Figure 7. We finally take $\mu = 10.0$. The behavior is similar as for $\mu = 5.0$, but now for small values of E and moderate values of r, the red squirrels survive, at a similar level as before; compare Figures 7 and 8, the latter corresponding to this simulation. In this case, the disease is also heavily affected, since it is eradicated for very low r, but also when it is endemic, its prevalence is very much reduced; note indeed that the vertical scale for i in Figure 8 is less than half of the corresponding one in Figure 7. The grey squirrel population, when surviving, is hardly influenced by the change of r.

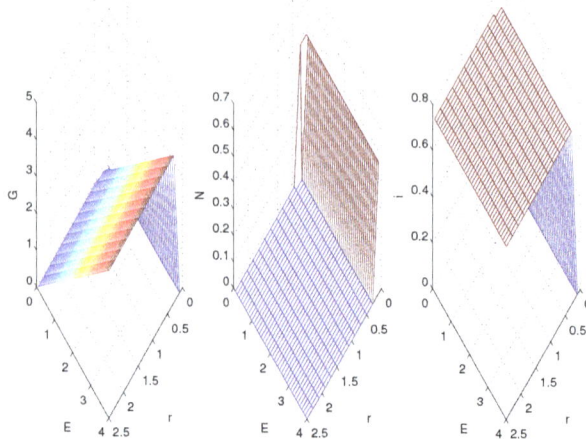

Figure 7. Simulation with $\mu = 5.0$. The red squirrels thrive for low r, independent of E. For small E, they also thrive, but at low population values. The grey squirrel population increases linearly with E.

14

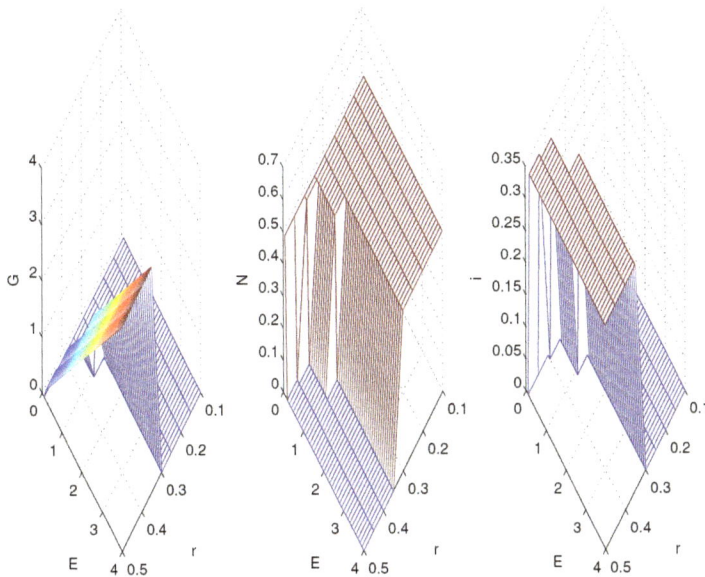

Figure 8. Simulation with $\mu = 10.0$. Now, the red squirrels survive also for small E, at higher values. Disease prevalence is very much reduced; compare the right plot's vertical scale of each frame.

We consider also changes in the respective competitive rates, a and k, but in the present conditions (3) and (4), no significant changes are observed; the red squirrels are wiped out, although the disease does not affect the whole population.

We finally tried to see how the influence of the disease transmission is felt by the ecosystem. By comparing the respective influences of the two transmission rates, trying to lessen the interspecific transmission rate β appears to induce more important changes in the ecosystem than reducing the intraspecific one, λ.

Furthermore, a drastic reduction of the interspecific transmission rate β combined with higher natural-plus-disease-related mortality μ cannot help in preserving the native population. The analysis of the combination of λ and μ does not lead to native squirrels' survival, as should be expected from the previous above results on the intraspecific disease transmission rate, but their prevalence decreases as the mortality increases.

Decreasing r, E and β, while increasing μ from the reference values (3) and (4), namely taking $r = 0.3$, $E = 0.5$, $\mu = 10.0$, $\beta = 0.6$, the red squirrel population is preserved with the epidemics disappearing, while the grey squirrels are also eliminated; see the left frame of Figure 9.

Increasing r alone, in these same conditions, however, does not lead to coexistence, as past a value of the invasive species reproduction rate that lies in the range $[0.57, 0.58]$, the grey squirrels instead eliminate the native species; see the right frame of Figure 9.

We attempt at last multiple changes of the parameters, taking new values as follows:

$$r = 0.5, \quad a = 0.4, \quad \mu = 1.5, \quad \beta = 0.4. \tag{14}$$

For these values, the coexistence of the three populations occurs with the disease not affecting the whole native population; see the left frame of Figure 10. Therefore, for our in silico experiments, from now on, we use the parameters (14). In this four-dimensional parameter space, we study once more the sensitivity. At first, we use the pair $E \in [0.1, 0.8]$ and $\beta \in [0.1, 1.0]$ that has not yet been considered; the results are not shown. When both parameters attain their lowest value in these respective ranges, the red squirrel population survives, with low endemicity, while the grey squirrels' competitors are sensibly kept in check. We also once more compare the coexistence as a function of $E \in [0.1, 0.8]$ and the mortality $\mu \in [8, 12]$; see Figure 11. Coexistence occurs for lower values of E with a thriving native population at moderate levels, low prevalence and the grey squirrels kept in check at reasonably low values. Furthermore, this result appears to be almost independent of the values of the mortality. The most relevant parameter for obtaining this outcome is the interspecific disease transmission rate β, because if it is low, the native population thrives also for higher values of the grey squirrel carrying capacity.

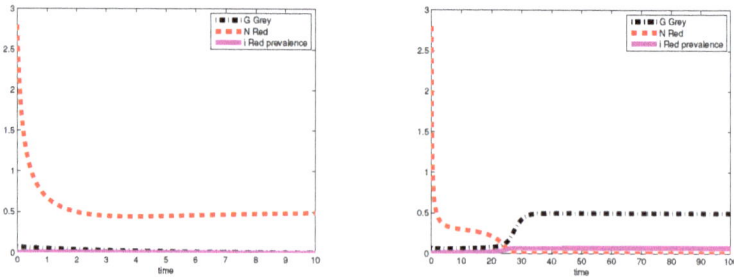

Figure 9. (**Left frame**) Native squirrels are preserved, and disease, as well as invaders are eradicated with very low prevalence for $r = 0.3$, $E = 0.5$, $\mu = 10.0$, $\beta = 0.6$. (**Right frame**) Invader squirrels thrive while the natives disappear, with the disease remaining endemic, through $r = 0.58$ and the remaining values as before.

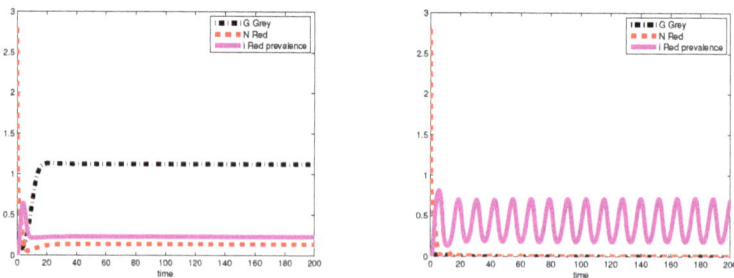

Figure 10. (**Left frame**) The coexistence equilibrium obtained with the basic parameters (3) coupled with (14). When both values of the parameters in their ranges are low, native squirrels can thrive, and prevalence is low; while the grey ones are reduced, and disease prevalence also drops. This ecosystem behavior depends more on the values of E than on those of β. (**Right frame**) Persistent oscillations in the prevalence, obtained for $r = 0.1$, $\beta = 0.3$, and the other parameters as given in (3) and (14).

Next, we compare the ecosystem outcome as a function once more of the grey squirrel reproduction rate r and the competition a suffered by the native population. The native population survives less for increasing competition, but better if r is very small, just in which case, the invaders suffer greatly; the disease prevalence is confined to reasonably low values for intermediate values of the parameters, in the range explored (results not shown).

As functions of the grey squirrels reproduction rate and interspecific disease transmission rate, the populations survive in an intermediate range of this parameter space, with declining native population as the incidence rate grows. The grey squirrels suffer from a very own low reproduction rate, as should be expected, but in this range, while the red squirrel population is also much reduced or vanishes altogether, persistent prevalence oscillations are observed, indicating that the disease remains endemic while the population disappears; see the right frame of Figure 10.

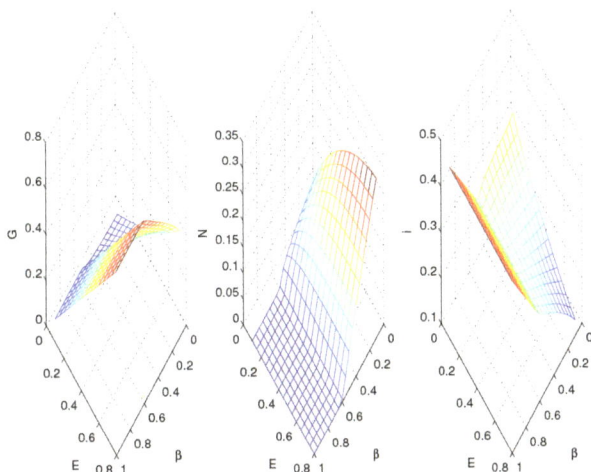

Figure 11. Coexistence sensitivity in terms of E and β. When both values of the parameters in their ranges are low, native squirrels can thrive, and prevalence is low; while the grey ones are reduced. Increasing the disease transmission rate soon leads to the native squirrel disappearance. This ecosystem behavior depends more on the values of β than on those of E.

The ecosystem behavior in terms of μ and β shows that the role of mortality is less important; the red squirrels are preserved independently of it if the transmission rate is low enough. Prevalence drops with higher mortality though, the invasive population picks up instead with higher transmission rates.

Low values of r and moderate values of the mortality ensure the disappearance of the invaders and the survival of the red squirrels. Some oscillations for the prevalence are detected in the same range. A simultaneous reduction in both a and β leads to coexistence with a somewhat reduced grey squirrel population, while the native species thrives at good levels, and the disease is sensibly reduced.

4. Discussion

Our analyses revealed that the three-populations system has different equilibria. From a conservation and wildlife management perspective, the most desirable one is the healthy-red-squirrels-only point $P_1 = (0, K, 0)$, followed by $P_5 = (0, N_5, i_5)$. Apart from the origin, P_0, these are the only equilibria leading to the grey squirrel extinction, and it is thus particularly interesting to evaluate their conditions. Equilibrium P_1 is always feasible, and stability can be attained only if conditions (9) are satisfied, while P_5 is feasible and stable if (8), (12) and (13) hold (see Table 2). For both points, the conditions often involve the intraspecific disease transmission rate for red squirrels (λ) and the overall red squirrel mortality (μ), whose combination did not lead to native squirrel survival according to our simulations. Using the reference parameter values (3), the system is rather likely to evolve from a two-population ecosystem towards a grey-squirrel-only situation, with the disappearance of the red squirrel population in a time span that depends on the value of the red squirrel overall mortality (Figures 1 and 2). Indeed, excluding the unstable P_2, all other equilibria imply the persistence of the grey squirrel population, with the native species vanishing while the disease remains endemic in it. This is consistent with the results obtained through other models. For instance in [52], it was also found that, in the absence of control of the invading species, native populations are driven to extinction and that this occurs both in the absence of disease (through competition only; see also [25,43]) and more rapidly when the disease is included [28].

17

Table 2. Summary of the feasibility and stability conditions for the equilibria with at least one vanishing population.

Equilibrium	Feasibility	Stability
$P_1 = (0, N_1, 0)$	always	$\lambda < \mu,\ cr < k$
$P_3 = (G_3, 0, i_3)$	$i_3 = 1$	$\mu + s < \beta + a$
$P_4 = (G_4, 0, i_4)$	$a < \mu + s$	$s < a + \beta < \mu + s$
$i_4 \leq 1$	$a + \beta \leq \mu + s$	
$P_5 = (0, N_5, i_5),\ i_5 < 1$	$s + \mu \geq \lambda \geq \mu$	$cr\lambda < k\mu$
		$4\lambda\mu^2 + 2s\lambda\mu + 3\lambda^3 < 5\lambda^2\mu + \mu^3 + s\mu^2 + s\lambda^2$

Investigating the sensitivity of the system with respect to the parameter values, we highlighted that the red squirrel could outcompete the alien species and at the same time eradicate the disease if it had much more weight in the interspecific competition (small c, Figures 3 and 4). This is unlikely to happen in most of the habitat types where the grey squirrel successfully settles (e.g., where its carrying capacity is the highest, as in mixed and broadleaved woodlands), but it could locally occur in coniferous forests [51].

The system may also respond to changes in the growth rate (r) of the grey squirrel population. The latter is certainly the parameter that can be manipulated most easily in natural conditions. As shown in Figures 5 and 6, if artificially, the growth rate of the grey squirrel is controlled, the alien species may disappear from the system, with cascade effects on the transmission of the disease to reds. The native species can thus recover and return to its natural carrying capacity. However, to obtain such a result, the reduction in r must be very harsh: the parameter value should be smaller than 0.2. This is required for all values of E (Figure 7), i.e., independent of the habitat type and of its carrying capacity for the grey squirrel. Interestingly, decreasing the growth rate (r), the grey squirrel carrying capacity (E) and the intraspecific transmission rate (β) at the same time and increasing the red mortality (μ), the red squirrel population can be preserved even for $r = 0.3$ (Figure 9). In this case, the grey squirrels are eliminated, and the epidemics disappears. Again, [51,52] also found that grey squirrel control can protect red populations in strongholds, although in their specific case study, periodic outbreaks of the disease could not be prevented because of occasional recolonization of the stronghold by grey squirrels from neighboring areas. The work in [52] also highlighted that there is a threshold level of control (expressed as the proportion of grey squirrels to be removed), above which the invading population can be prevented from establishing and the native species can be protected. Another study [28] also investigated what level of population control of grey squirrels would be necessary to suppress the disease-induced decline in red squirrels. They found that the kill rate required to avoid the spread of the invasive species should be in excess of 60% grey squirrels. For any level of control below this threshold, the grey squirrels were predicted to expand with a concomitant decline in red squirrel populations.

Since achieving $r \leq 0.3$ could still be difficult, managers could also aim at a larger 0.5 value. Assuming that the reduction in the growth rate is achieved through the removal of the grey squirrels, the red squirrel damage due to interspecific competition (a) and the intraspecific transmission rate (β) could also be smaller (e.g., 0.4 as in (14)), and the overall red mortality (μ) should also be reduced. Under these conditions, the coexistence of the three populations is ensured, and the grey squirrels are kept in check at reasonably low values. We also note that the reduction in the interspecific transmission rate (β) appears more important in the ecosystem than the intraspecific one, λ, and more important than E (Figure 11). In this respect, [25] also indicated that the progress of the disease within individual populations is largely dependent on the probability of virus transmission between the two species.

Different combinations of the parameter values may thus allow wildlife managers to achieve different objectives. Simulations highlighted that the most desirable outcome could be achieved only in favorable conditions for the red squirrel (very small c) or for very small values of r (e.g., 0.2 or 0.3). How the growth rate of grey squirrels could be reduced, in practice, depends on the specific circumstances, including the stage of the invasion. Actually, two strategies have been proposed to eradicate or at least to control grey squirrels in the wild.

The first strategy involves the removal of squirrels through capture and subsequent euthanasia. Evidence for its effectiveness is provided by data from the U.K. [46], and other mathematical models highlighted its potential to protect red populations and to prevent the spread of SQPV [27,51,52]. According to our results, the removal control approach could ensure the conservation of the native red squirrel population if r is substantially reduced ($r \leq 0.5$). In this respect, it should be noted that the removal of the animals is advantageous, since it allows for a reduction in r, and in turn, it could affect other model parameters, such as a and β, in a favorable way for the conservation of red squirrels. Culling can indeed remove in parallel both the competitive and disease threats posed to red by grey squirrels [46]. Its application is also consistent with the provisions of EU Regulation No. 1143/2014. According to this regulation, rapid eradication should be now implemented when an alien species of European concern is detected in a new area. In the first phases of an invasion, the number of alien individuals should be limited, and their population should be in the lag phase that typically precedes rapid expansion and demographic growth. It should thus still be possible to affect the growth rate of the population by removing a few individuals. On the contrary, if the invasive population is large, reducing r would require a huge capture effort.

The second suggested strategy relies on the reduction of the reproductive rate, as achieved through surgical sterilization or inhibition of fertility, e.g., using immunocontraceptive vaccines. In particular, gonadotrophin-releasing hormone (GnRH) vaccines seem to offer great promise for the control of populations. They interfere with the regulation of reproductive hormones, and they have been successful at reducing fertility in several mammals [53]. Available data on their potential impact on the population dynamics of other squirrel species [54] suggest they could reduce populations, without eradicating them, if at least 70% of the females were treated [54–57]. At present, this requires the capture of the animals in order to inject the vaccine, and it could be necessary to capture and treat each animal more than once, since the long-term effects of the vaccine have not been demonstrated so far [58,59]. On the basis of a theoretical model, [56] concluded that 90% of females should be treated in order to eradicate a grey squirrel population within 7/8 years, ensuring a constant sterilization effort. The need to ensure a constant effort in reducing the reproductive rate is in agreement with the results of our model, since we found that, if the reproductive rate of the grey squirrel is artificially controlled, there might be hope to restore the pristine situation, for a suitably medium value of the natural-plus-disease-related mortality of the red squirrels. Anyway, sterilization also implies the subsequent release of the individuals in the wild, so that through this approach, we could not expect prompt effects on the model parameters, other than r. In particular, in this case, we would expect no immediate reduction in the occurrence of the disease and, thus, no major impact on the SQPV interspecific transmission rate.

In conclusion, our model confirms that the native red squirrel populations can be preserved or restored, under specific parameter conditions, if the grey squirrel population is removed or kept in check at low values. Independent of the adopted methodology, wildlife managers should aim at a harsh reduction of the grey squirrel growth rate, and a constant effort should be ensured to protect the remnant red squirrel populations, if the complete removal of the alien species is not achieved at the early stages of invasion. As concerns the SQPV, actions should mainly aim at the reduction of the interspecific transmission rate, one of the most important parameters in determining the ecosystem response to the disease, although how to effectively carry out this task in practice may still be an open question. At the moment, programs for the grey squirrel control, or even local eradication, are implemented both in the U.K. and in Italy. The insights provided by our general model could be considered by these programs to improve the planning of management actions. Although modeling outcomes have already been used by other authors to critically analyze the efficiency of removal activities (e.g., [27]), as implemented in specific contexts, future research could concentrate on the analysis of a more general system, including also the control strategy in its formulation, as is commonly used for instance in the case of epidemics [60–62]. This approach may provide useful indications to the ecologists that deal with this problem in practice.

Author Contributions: E.V. and V.L.M. cooperated in the model formulation. V.P., F.B. and G.T. worked on the mathematical part under E.V.'s supervision, V.L.M. contributed the ecological parts. The paper was jointly written by V.L.M., E.V. and V.P.

Acknowledgments: The research of this paper has been partially supported by the Dipartimento di Matematica "Giuseppe Peano" research project "Metodi numerici e computazionali per le scienze applicate". The paper was written while the last author was visiting the Fields Institute during the "Thematic Program on Emerging Challenges in Mathematical Biology"; he gratefully acknowledges I.Hambleton for his support and M. Lewis and H. Eberl for the invitation. This research has been carried out also within the framework of the LIFE U-SAVEREDSProject (LIFE13 BIO/IT/000204), funded by the LIFE+ Biodiversity Programme of the European Commission.

Conflicts of Interest: The authors declare no conflict of interest.

References

1. Shar, S.; Lkhagvasuren, D.; Bertolino, S.; Henttonen, H.; Kryštufek, B.; Meinig, H. *Sciurus vulgaris* (errata version published in 2017). The IUCN Red List of Threatened Species, 2017. Available online: http://www.iucnredlist.org/details/20025/0 (accessed on 27 June 2018).
2. Middleton, A.D. 38. The ecology of the American grey squirrel (*Sciurus carolinensis* Gmelin) in the British Isles. *Proc. Zool. Soc. Lond.* **1930**, *100*, 809–843. [CrossRef]
3. Shorten, M.; Elton, C. Some aspects of the biology of the grey squirrel (*Sciurus carolinensis*) in Great Britain. *Proc. Zool. Soc. Lond.* **1951**, *121*, 427–459. [CrossRef]
4. Martinoli, A.; Bertolino, S.; Preatoni, D.G.; Balduzzi, A.; Marsan, A.; Genovesi, P.; Tosi, G.; Wauters, L.A. Headcount 2010: The multiplication of the grey squirrel populations introduced to Italy. *Hystrix Ital. J. Mammal.* **2010**, *21*, 127–136.
5. Bertolino, S.; Martinoli, A.; Wauters, L.A. Risk Assessment for *Sciurus carolinensis* (grey squirrel). In *Invasive Alien Species—Framework for the Identification of Invasive Species of EU Concern*; Roy, H., Ed.; Natural Environment Research Council: Swindon, UK, 2014; p. 298.
6. Lowe, S.; Browne, M.; Boudjelas, S.; De Poorter, M. *100 of the World's Worst Invasive Alien Species*; The Invasive Species Specialist Group (ISSG), a Specialist Group of the Species Survival Commission (SSC) of the World Consevation Union (IUCN): Gland, Switzerland, 2000.
7. Gurnell, J. Squirrel numbers and the abundance of tree seeds. *Mamm. Rev.* **1983**, *13*, 133–148. [CrossRef]
8. Wauters, L.A.; Gurnell, J. The mechanism of replacement of red squirrels by grey squirrels: A test of the interference competition hypothesis. *Ethology* **1999**, *29*, 1053–1071. [CrossRef]
9. Wauters, L.A.; Gurnell, J.; Martinoli, A.; Tosi, G. Interspecific competition between native Eurasian red squirrels and alien grey squirrels: Does resource partitioning occur? *Behav. Ecol. Sociobiol.* **2002**, *52*, 332–341. [CrossRef]
10. Wauters, L.; Tosi, G.; Gurnell, J. Interspecific competition in tree squirrels: Do introduced grey squirrels (*Sciurus carolinensis*) deplete tree seeds hoarded by red squirrels (*S. vulgaris*)? *Behav. Ecol. Sociobiol.* **2002**, *51*, 360–367. [CrossRef]
11. Wauters, L.A.; Lurz, P.W.W.; Gurnell, J. Interspecific effects of grey squirrels (*Sciurus carolinensis*) on the space use and population demography of red squirrels (*Sciurus vulgaris*) in conifer plantations. *Ecol. Res.* **2000**, *15*, 271–284. [CrossRef]
12. Thomas, K.; Tompkins, D.M.; Sainsbury, A.W.; Wood, A.R.; Dalziel, R.; Nettleton, P.F.; McInnes, C.J. A novel poxvirus lethal to red squirrels (*Sciurus vulgaris*). *J. Gener. Virol.* **2003**, *84*, 3337–3341. [CrossRef] [PubMed]
13. McInnes, C.J.; Wood, A.R.; Thomas, K.; Sainsbury, A.W.; Gurnell, J.; Dein, F.J.; Nettleton, P.F. Genomic characterization of a novel poxvirus contributing to the decline of the red squirrel (*Sciurus vulgaris*) in the UK. *J. Gener. Virol.* **2006**, *87*, 2115–2125. [CrossRef] [PubMed]
14. Sainsbury, A.W.; Nettleton, P.; Gurnell, J. Recent developments in the study of parapoxvirus in red and grey squirrels. In *The Conservation of Red Squirrels, Sciurus vulgaris L.*; Gurnell, J., Lurz, P.W.W., Eds.; PTES (People's Trust for Endangered Species): London, UK, 1997; pp. 105–108.
15. Edwards, F.B. Red squirrel disease. *Vet. Rec.* **1962**, *74*, 739–741.

16. Sainsbury, A.; Gurnell, J. An investigation into the health and welfare of red squirrels, *Sciurus vulgaris*, involved in reintroduction studies. *Vet. Rec.* **1995**, *137*, 367–370. [CrossRef] [PubMed]

17. McInnes, C.J.; Coulter, L.; Dagleish, M.P.; Fiegna, C.; Gilray, J.; Willoughby, K.; Cole, M.; Milne, E.; Meredith, A.; Everest, D.J.; et al. First cases of squirrelpox in red squirrels (*Sciurus vulgaris*) in Scotland. *Vet. Rec.* **2009**, *164*, 528–531. [CrossRef] [PubMed]

18. Tompkins, D.M.; Sainsbury, A.W.; Nettleton, P.; Buxton, D.; Gurnell, J. Parapoxvirus causes a deleterious disease in red squirrels associated with UK population declines. *Proc. R. Soc. B Biol. Sci.* **2002**, *269*, 529–533. [CrossRef] [PubMed]

19. Sainsbury, A.W.; Nettleton, P.; Gilray, J.; Gurnell, J. Grey squirrels have high seroprevalence to a parapoxvirus associated with deaths in red squirrels. *Anim. Conserv.* **2000**, *3*, 229–233. [CrossRef]

20. Sainsbury, A.W.; Deaville, R.; Lawson, B.; Cooley, W.A.; Farelly, S.S.J.; Stack, M.J.; Duff, P.; McInnes, C.J.; Gurnell, J.; Russell, P.H.; et al. Poxviral disease in red squirrels *Sciurus vulgaris* in the UK: Spatial and temporal trends of an emerging threat. *EcoHealth* **2008**, *5*, 305–316. [CrossRef] [PubMed]

21. Sainsbury, T.; Ward, L. Parapoxvirus infection in red squirrels. *Vet. Rec.* **1996**, *138*, 400. [PubMed]

22. Scott, A.; Keymer, I.; Labram, J. Parapoxvirus infection of the red squirrel (*Sciurus vulgaris*). *Vet. Rec.* **1981**, *109*, 202–202. [CrossRef] [PubMed]

23. Duff, J.P.; Scott, A.; Keymer, I.F. Parapoxvirus infection of the grey squirrel. *Vet. Rec.* **1996**, *138*, 527. [PubMed]

24. Reynolds, J.C. Details of the geographic replacement of the red squirrel (*Sciurus vulgaris*) by the grey squirrel (*Sciurus carolinensis*) in Eastern England. *J. Anim. Ecol.* **1985**, *54*, 149–162. [CrossRef]

25. Rushton, S.P.; Lurz, P.W.W.; Gurnell, J.; Fuller, R. Modelling the spatial dynamics of parapoxvirus disease in red and grey squirrels: A possible cause of the decline in the red squirrel in the UK? *J. Appl. Ecol.* **2000**, *37*, 997–1012. [CrossRef]

26. Tompkins, D.M.; White, A.R.; Boots, M. Ecological replacement of native red squirrels by invasive greys driven by disease. *Ecol. Lett.* **2003**, *6*, 189–196. [CrossRef]

27. Jones, H.; White, A.; Lurz, P.; Shuttleworth, C. Mathematical models for invasive species management: Grey squirrel control on Anglesey. *Ecol. Model.* **2017**, *359*, 276–284. [CrossRef]

28. Rushton, S.P.; Lurz, P.W.W.; Gurnell, J.; Nettleton, P.; Bruemmer, C.; Shirley, M.D.F.; Sainsbury, A.W. Disease threats posed by alien species: The role of a poxvirus in the decline of the native red squirrel in Britain. *Epidemiol. Infect.* **2006**, *134*, 521–533. [CrossRef] [PubMed]

29. Halliwell, E.C. Red squirrel predation by pine martens in Scotland. In *The Conservation of Red Squirrels, Sciurus vulgaris L.*; PTES (People's Trust for Endangered Species): London, UK, 1997; pp. 39–48.

30. Lurz, P.W.W.; Garson, P.J.; Wauters, L.A. Effects of temporal and spatial variations in food supply on the space and habitat use of red squirrels (*Sciurus vulgaris* L.). *J. Zool.* **2000**, *251*, 167–178. [CrossRef]

31. Wauters, L.A.; Gurnell, J.; Preatoni, D.; Tosi, G. Effects of spatial variation in food availability on spacing behavior and demography of Eurasian red squirrels. *Ecography* **2001**, *24*, 525–538. [CrossRef]

32. Thompson, D.C. The social system of the grey squirrel. *Behaviour* **1978**, *64*, 305–328. [CrossRef]

33. Koprowski, J.L. *Sciurus carolinensis*. *Mamm. Species* **1994**, *480*, 1. [CrossRef]

34. Mosby, H.S. The influence of hunting on the population dynamics of a woodlot gray squirrel population. *J. Wildl. Manag.* **1969**, *33*, 709–717. [CrossRef]

35. Montgomery, S.D.; Whelan, J.B.; Mosby, H.S. Bioenergetics of a woodlot gray squirrel population. *J. Wildl. Manag.* **1975**, *39*, 709. [CrossRef]

36. Gurnell, J. The effects of food availability and winter weather on the dynamics of a grey squirrel population in Southern England. *J. Appl. Ecol.* **1996**, *33*, 325–338. [CrossRef]

37. Bertolino, S. *Attivazione di un Progetto di Monitoraggio Estansivo ed Intensivo dello Scoiattolo Grigio (Sciurus carolinensis) in Piemonte*; Technical Report; Regione Piemonte: Torino, Italy, 2004.

38. Lurz, P.W.W. The ecology of squirrels in spruce dominated plantations: Implications for forest management. *For. Ecol. Manag.* **1995**, *79*, 79–90. [CrossRef]

39. Wauters, L.A.; Lens, L. Effects of food availability and density on red squirrel (*Sciurus vulgaris*) reproduction. *Ecology* **1995**, *76*, 2460–2469. [CrossRef]

40. Lurz, P.; Garson, P.; Wauters, L. Effects of temporal and spatial variation in habitat quality on red squirrel dispersal behavior. *Anim. Behav.* **1997**, *54*, 427–435. [CrossRef] [PubMed]
41. Lurz, P.W.W.; Garson, P.J.; Ogilvie, J.F. Conifer species mixtures, cone crops and red squirrel conservation. *Forestry* **1998**, *71*, 67–71. [CrossRef]
42. Wauters, L.A.; Matthysen, E.; Adriaensen, F.; Tosi, G. Within-sex density dependence and population dynamics of red squirrels *Sciurus vulgaris*. *J. Anim. Ecol.* **2004**, *73*, 11–25. [CrossRef]
43. Okubo, A.; Maini, P.K.; Williamson, M.H.; Murray, J.D. On the spatial spread of the grey squirrel in Britain. *Proc. R. Soc. Lond. Ser. B Biol. Sci.* **1989**, *238*, 113–125. [CrossRef]
44. Bryce, J.M.; Speakman, J.R.; Johnson, P.J.; Macdonald, D.W. Competition between Eurasian red and introduced Eastern grey squirrels: The energetic significance of body-mass differences. *Proc. R. Soc. B Biol. Sci.* **2001**, *268*, 1731–1736. [CrossRef] [PubMed]
45. Atkin, J.W.; Radford, A.D.; Coyne, K.P.; Stavisky, J.; Chantrey, J. Detection of squirrel poxvirus by nested and real-time PCR from red (*Sciurus vulgaris*) and grey (*Sciurus carolinensis*) squirrels. *BCM Vet. Res.* **2010**, *6*, 33. [CrossRef] [PubMed]
46. Schuchert, P.; Shuttleworth, C.M.; McInnes, C.J.; Everest, D.J.; Rushton, S.P. Landscape scale impacts of culling upon a European grey squirrel population: Can trapping reduce population size and decrease the threat of squirrelpox virus infection for the native red squirrel? *Biol. Invasions* **2014**, *16*, 2381–2391. [CrossRef]
47. Gurnell, J. *The Natural History of Squirrels*; Helm: London, UK, 1987.
48. Roberts, M.G.; Heesterbeek, J.A.P. Cpidemiology. *J. Math. Biol.* **2013**, *66*, 1045–1064. [CrossRef] [PubMed]
49. Hilker, F.M.; Malchow, H. Strange Periodic Attractors in a Prey-Predator System with Infected Prey. *Math. Popul. Stud.* **2006**, *13*, 119–134. [CrossRef]
50. Oliveira, N.M.; Hilker, F.M. Modelling Disease Introduction as Biological Control of Invasive Predators to Preserve Endangered Prey. *Bull. Math. Biol.* **2010**, *72*, 444–468. [CrossRef] [PubMed]
51. White, A.; Lurz, P.W.W.; Jones, H.E.; Boots, M.; Bryce, J.; Tonkin, M.; Ramoo, K.; Bamforth, L.; Jarrott, A. The use of mathematical models in red squirrel conservation: Assessing the threat from grey invasion and disease to the Fleet basin stronghold. In *Red Squirrels Ecology, Conservation Management in Europe*; Shuttleworth, C., Lurz, P.W.W., Hayward, M.W., Eds.; European Squirrel Initiative (ESI): 2015; pp. 265–279.
52. White, A.; Bell, S.S.; Lurz, P.W.W.; Boots, M. Conservation management within strongholds in the face of disease-mediated invasions: Red and grey squirrels as a case study. *J. Appl. Ecol.* **2014**, *51*, 1631–1642. [CrossRef]
53. Massei, G.; Cowan, D.P.; Coats, J.; Gladwell, F.; Lane, J.E.; Miller, L.A. Effect of the GnRH vaccine GonaCon on the fertility, physiology and behavior of wild boar. *Wildl. Res.* **2008**, *35*, 540–547. [CrossRef]
54. Krause, S.K.; Kelt, D.A.; Van Vuren, D.H.; Gionfriddo, J.P. Regulation of tree squirrel populations with immunocontraception: A fox squirrel example. *Hum. Wildl. Interact.* **2014**, *8*, 1–12.
55. Moore, H.; Jenkins, N.M.; Wong, C. Immunocontraception in rodents: A review of the development of a sperm-based immunocontraceptive vaccine for the grey squirrel (*Sciurus carolinensis*). *Reprod. Fertil. Dev.* **1997**, *9*, 125–129. [CrossRef] [PubMed]
56. Cowan, D.P.; Massei, G. Wildlife contraception, individuals, and populations: How much fertility control is enough? In Proceedings of the 23rd Vertebrate Pest Conference, San Diego, CA, USA, 17–20 March 2008; pp. 220–228.
57. Massei, G.; Cowan, D. Fertility control to mitigate human–wildlife conflicts: A review. *Wildl. Res.* **2014**, *41*, 1–21. [CrossRef]
58. Pai, M. Field Evaluation of the Immunocontraceptive GonaCon™ in Reducing Eastern Gray Squirrel Fecundity in Urban Areas. Ph.D. Thesis, Clemson University, SC, USA, 2009.
59. Pai, M.; Bruner, R.; Schlafer, D.H.; Yarrow, G.K.; Yoder, C.A.; Miller, L.A. Immunocontraception in Eastern gray squirrels (*Sciurus carolinensis*): Morphologic changes in reproductive organs. *J. Zoo Wildl. Med.* **2011**, *42*, 718–722. [CrossRef] [PubMed]
60. Rodrigues, H.S.; Monteiro, M.T.T.; Torres, D.F. Dynamics of dengue epidemics when using optimal control. *Math. Comput. Model.* **2010**, *52*, 1667–1673. [CrossRef]

61.   Rodrigues, H.S.; Monteiro, M.T.T.; Torres, D.F.; Zinober, A. Dengue disease, basic reproduction number and control. *Int. J. Comput. Math.* **2012**, *89*, 334–346. [CrossRef]
62.   Rodrigues, H.S.; Monteiro, M.T.T.; Torres, D.F. Vaccination models and optimal control strategies to dengue. *Math. Biosci.* **2014**, *247*, 1–12. [CrossRef] [PubMed]

Σ **mathematics**   MDPI

Article

Linearization of the Kingman Coalescent

Paul F. Slade

Computational Biology and Bioinformatics Unit, Research School of Biology, R.N. Robertson Building 46, Australian National University, Canberra, ACT 0200, Australia; pfslade@gmail.com

Received: 5 February 2018; Accepted: 9 May 2018; Published: 14 May 2018

Abstract: Kingman's coalescent process is a mathematical model of genealogy in which only pairwise common ancestry may occur. Inter-arrival times between successive coalescence events have a negative exponential distribution whose rate equals the combinatorial term $\binom{n}{2}$ where n denotes the number of lineages present in the genealogy. These two standard constraints of Kingman's coalescent, obtained in the limit of a large population size, approximate the exact ancestral process of Wright-Fisher or Moran models under appropriate parameterization. Calculation of coalescence event probabilities with higher accuracy quantifies the dependence of sample and population sizes that adhere to Kingman's coalescent process. The convention that probabilities of leading order N^{-2} are negligible provided $n \ll N$ is examined at key stages of the mathematical derivation. Empirically, expected genealogical parity of the single-pair restricted Wright-Fisher haploid model exceeds 99% where $n \leq \frac{1}{2}\sqrt[3]{N}$; similarly, per expected interval where $n \leq \frac{1}{2}\sqrt{N/6}$. The fractional cubic root criterion is practicable, since although it corresponds to perfect parity and to an extent confounds identifiability it also accords with manageable conditional probabilities of multi-coalescence.

Keywords: Markov chain; multiple coalescence; transition probability; Wright-Fisher model

1. Introduction

Kingman's coalescent process is a mathematical model of ancestral lineages that inspired a paradigmatic era in population genetics [1–3]. Kingman's coalescent process [4–7] relies on negligibility of coalescence probabilities, and inter-arrival times, other than those of single pair-wise coalescence. Negligibility depends on terms of leading order N^{-2} or less that can be omitted from the process in the limit of a large population size. A comparative study of data generation simulators that implement Kingman's coalescent process demonstrates the utility of this conventional approximation to the exact ancestral process [8]. Phylogenetic trees in general contain a coalescent process of ancestral lineages from the corresponding sub-population within each branch of the phylogeny. The ancestral process within the branches of a phylogeny are often modeled using Kingman's coalescent [9] or theory of branching processes [10]. Statistical distribution theory of the Ewens' sampling formula is derived in population genetics by superimposing unique event mutations on the genealogical structure of Kingman's coalescent [11,12].

1.1. Coalescent Theory of Ancestral Processes

Kingman's coalescent process can be derived in a straightforward manner based on the genealogy of a Wright-Fisher model [13]. Consider a parent and an offspring generation, where the haploid population size N is kept fixed in each generation. The probability of zero coalescence events, such that none of the offspring are direct descendants of any parent in common, equals

$$\prod_{i=1}^{n-1}\left(1-\frac{i}{N}\right) = 1 - \left(\sum_{i=1}^{n-1}\frac{i}{N}\right) + \mathcal{O}\left(N^{-2}\right) = 1 - \frac{\binom{n}{2}}{N} + \mathcal{O}\left(N^{-2}\right) \tag{1}$$

with respect to n ancestral lines. This conventional approximation defines a geometric probability distribution for the number of generations that pass until a coalescence event,

$$\left\{1 - \frac{\binom{n}{2}}{N}\right\}^{j-1}\frac{\binom{n}{2}}{N} \tag{2}$$

where $j = 1, 2, 3, \dots$ denotes the generation in which at least one coalescence occurs. Recalibrated coalescent units of time $t = \frac{j}{N}$ generations in Equation (2) yields a negative exponential probability distribution, $\Pr(T > t) = e^{-\binom{n}{2}t}$, where T denotes the *waiting time* until a coalescence event in the limit of a large population size. Consider $\Pr(\check{T} \geq j) = (1-p)^j$, where $\check{T} \sim \text{Geom}(p)$. Take $p = \frac{\binom{n}{2}}{N}$ and $j = Nt$ to get an approximation of the geometric distribution relevant to Kingman's coalescent process. The binomial formula $(x+y)^n = \sum_{i=0}^{n}\binom{n}{i}x^i y^{n-i}$ thus yields an infinite series, in the limit of a large population size,

$$\Pr(\check{T} \geq j) = \left(1 - \frac{\binom{n}{2}}{N}\right)^{Nt} = 1 - \binom{n}{2}t + \binom{Nt}{2}\left(\frac{\binom{n}{2}}{N}\right)^2 - \binom{Nt}{3}\left(\frac{\binom{n}{2}}{N}\right)^3 + - \cdots$$

$$= 1 - \binom{n}{2}t + \left[\frac{\left(\binom{n}{2}t\right)^2}{2} - \frac{\binom{n}{2}^2 t}{2N}\right] - \left[\frac{\left(\binom{n}{2}t\right)^3}{3!} - \frac{\binom{n}{2}^3 t^2}{2N} + \frac{\binom{n}{2}^3 t}{3N^2}\right] + - \cdots \tag{3}$$

$$\approx 1 - \binom{n}{2}t + \frac{\left(\binom{n}{2}t\right)^2}{2} - \frac{\left(\binom{n}{2}t\right)^3}{3!} + - \cdots$$

Now, consider practical approximation, where $t = 1 \Rightarrow j = N$ and one unit of coalescent time equals N discrete generations in the geometric distribution. Thus, the negative exponential series in Equation (3) yields the conventional result, $\Pr(T > t) = e^{-\binom{n}{2}t}$, when the process is observed in this rewind coalescent time under the approximation of a large finite population size.

Simulation of the trade-off between n versus N had suggested that $n^2 < N$ should ensure Kingman's coalescent process ([14], pp. 5–6). Alternatively, a classic theoretical approximation due to R.A. Fisher yields a recursion of expected genealogical branch lengths to quantify single nucleotide polymorphisms as a function of sample size upon effective population size [15]. Further simulation study of the Kingman coalescent had suggested its validity threshold should be $n \approx \sqrt{2N}$ [16]. Evaluations in that work compared probabilities of pair-wise, multiple pair-wise and multi-coalescence events. Exploratory analysis concludes that Kingman's coalescent should be a robust approximation of the Wright-Fisher model in terms of genealogical timing, with external branch lengths likely to differ significantly. Another simulation study, under a similar approximation to the Kingman coalescent, calculates percentages of multi-coalescence events and statistics of mutational activity throughout a genealogy of high sample sizes with alternative demographics [17]. The results in Sections 2 and 3 herein clearly demonstrate the region of validity for the Kingman coalescent depends on population size. Furthermore, multi-coalescence events yield sensitivity in terms of fine-scale topological variation towards the tips. The negligibility of multiple coalescence events by which the Kingman coalescent should accurately approximate the exact Wright-Fisher ancestral process tends to be indirectly addressed in the literature of applied probability modeling and evolutionary biology on multi-coalescent processes.

1.2. Coalescent Theory of Branching Processes

An active research field on extension of discrete generations Wright-Fisher models, overlapping generations Moran models, and generalizations to the Cannings model, are based on their multinomial offspring distribution variance and moments to develop multi-coalescents [18–20]. Derivations of alternative coalescent processes usually retain the conventional proportionality to N^{-2} ([21], Theorem 3.2 via Equation (5); [22], Theorem 2.1 via Equation (4)). These generalizations are in turn based on the partition structures of equivalence classes described in terms of sampling distributions not originally connected to genealogy [23–25]. The corresponding convergence-to-coalescent results tend to rely upon fast continuous time scales rather than generational ancestral processes. Thus, multi-coalescent processes replace a multinomial offspring distribution with a variety of continuous population frequency distributions that yield non-negligible jump transitions of lineage decrements greater than one in continuous-time Markov chains. There are alternative approaches to the development of multi-coalescents: (i) branching process theory ([26–28], for an application see [29]); and (ii) measure-valued diffusion theory [30,31]. Both approaches model proliferation of lineages over time. Further examples include β-coalescent [32], Λ-coalescent [33,34], Ξ-coalescent [35,36], and Galton–Watson theory [37,38]. Technical mathematical treatments tend to assume the foundations of ancestral processes. The quantitative analysis of Sections 2 and 3 in this work clearly identifies regions of adherence and detraction from the Wright-Fisher ancestral process, in terms of transition probabilities and expected inter-arrival times, due to the linearization of Kingman's coalescent that neglects multi-coalescence events.

2. Ancestral Process, per Generation

Error threshold is the forefront of the issue for computationally-intensive methodologies and statistical models based on Kingman's coalescent. Six main points arise: (i) discrepancy between the exact and linearized non-coalescence probability in Equation (1); (ii) validity of the linearized coalescence probability in Equation (2); (iii) conditional probabilities of single-pair and multi-coalescences given at least one coalescence; (iv) parity of reduced ancestral processes that suppress multi-coalescences, when compared to the exact ancestral process; (v) genealogical topology; and (vi) subsequent inter-arrival times.

2.1. Zero Coalescence Events

The exact probability of k offspring genes that are descendants of k different parents, without shared ancestry in the parental generation, was given by Equation (1). The corresponding approximation derives from the product in Equation (1), where expansion yields

$$1 - N^{-1} \sum_{i=1}^{n-1} i + N^{-2} \sum_{i=1}^{n-2} i \sum_{j=i+1}^{n-1} j - N^{-3} \sum_{i=1}^{n-3} i \sum_{j=i+1}^{n-2} j \sum_{k=j+1}^{n-1} k +$$

$$N^{-4} \sum_{i=1}^{n-4} i \sum_{j=i+1}^{n-3} j \sum_{k=j+1}^{n-2} k \sum_{l=k+1}^{n-1} l + \cdots + (-1)^{n-2} N^{-(n-2)} (n-1)! \sum_{i=1}^{n-1} \frac{1}{i} + \tag{4}$$

$$(-1)^{n-1} N^{-(n-1)} (n-1)!.$$

In Equation (4), calculate the summation of the quadratic term, N^{-2}, to get a coefficient

$$[n(n-1)(n-2)(3n-1)]/24. \tag{5}$$

Similarly, the summation of the cubic term, N^{-3}, yields a coefficient

$$\left[n^2(n-1)^2(n-2)(n-3)\right]/48. \tag{6}$$

Derivation of Equations (5) and (6) are deferred to Appendix A.

The default population size in this work is set at $N = 2 \times 10^5$, unless otherwise stated, then the exponent increased and decreased by one or two to verify generality for criterion that are expressed as functions of N. Refer to Figure 1 that compares the first and third order approximation non-coalescence probabilities. The criterion $\sqrt{2N}$ [16] sets the error tolerance down to where the linearized non-coalescence probability, per generation, goes negative at $n = 633$; clearly, negativity must occur at $n(n-1) > 2N$. The criterion $\sqrt{N}$ [14] sets the error tolerance greater than 15%, and the corresponding proportion of the exact probability equals 0.825979 at $n = 447$. Reduction to precisely 1% error tolerance occurs at $n = 233$. Exact non-coalescence probability can be compared to its linearized, quadratic and cubic approximation; refer to Figures 2 and 3. The difference between the quadratic and cubic terms of Equation (4) determines the error of the linearization, since non-linear terms of higher degree do not significantly affect the exact value even with many lineages present in the genealogy; refer to Figure 4. Evaluation of the non-coalescence probability suggests a criterion of 1% proportional error after round-up be $\sqrt{N/3}$.

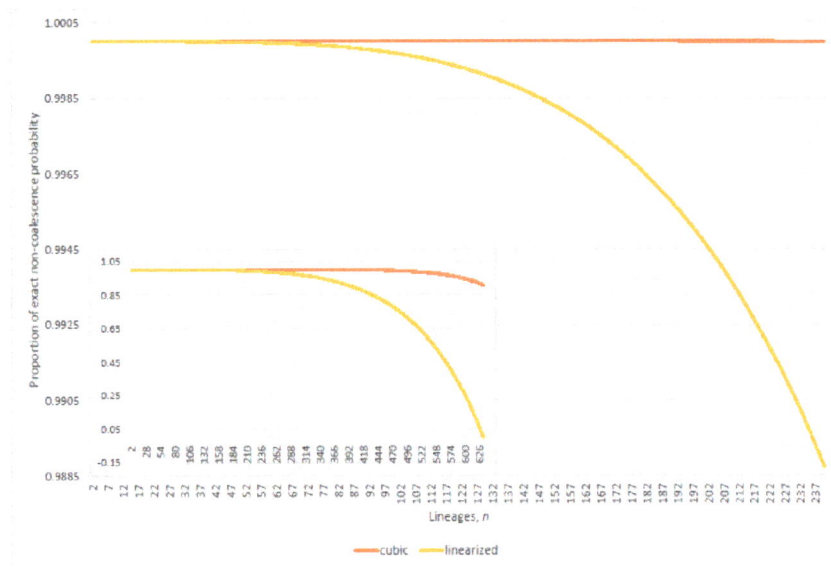

Figure 1. Proportions of the exact non-coalescence probability: quotients of the linearized $1 - \binom{n}{2}/N$ and (cubic) third order approximation of Equation (4) upon the exact non-coalescence probability of Equation (1), respectively. Population size $N = 2 \times 10^5$ and $n = 2, \ldots, 240$ (inset $n = 2, \ldots, 633$).

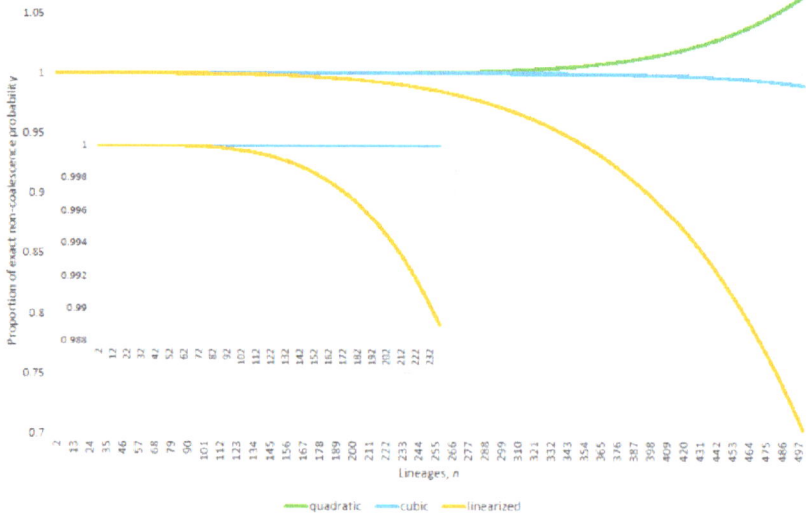

Figure 2. Proportion of exact non-coalescence probability: quotients of the linearized $1 - \binom{n}{2}/N$, (quadratic) second order and (cubic) third order approximation of Equation (4) upon the exact non-coalescence probability of Equation (1), respectively. Population size $N = 2 \times 10^5$ and $n = 2, \ldots, 500$ (inset $n = 2, \ldots, 240$).

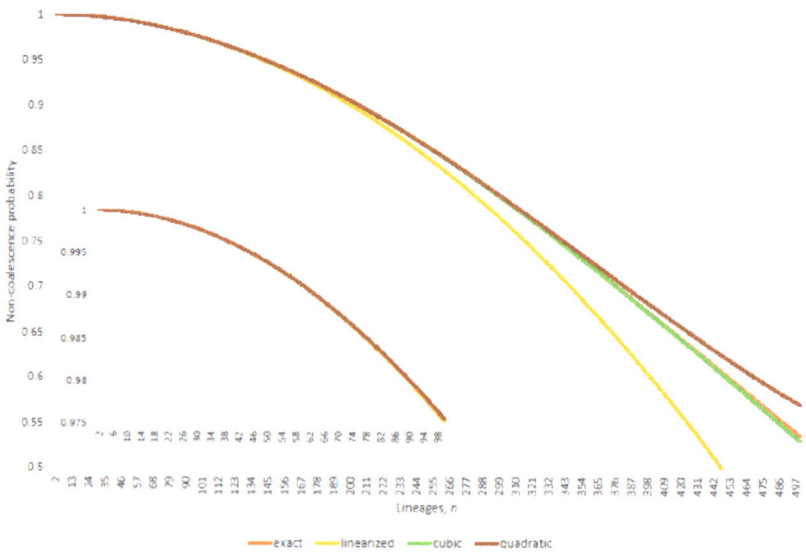

Figure 3. Non-coalescence probabilities: exact Equation (1), linearized $1 - \binom{n}{2}/N$, (quadratic) second order and (cubic) third order approximation of Equation (4), respectively. Population size $N = 2 \times 10^5$ and $n = 2, \ldots, 500$ (inset $n = 2, \ldots, 100$).

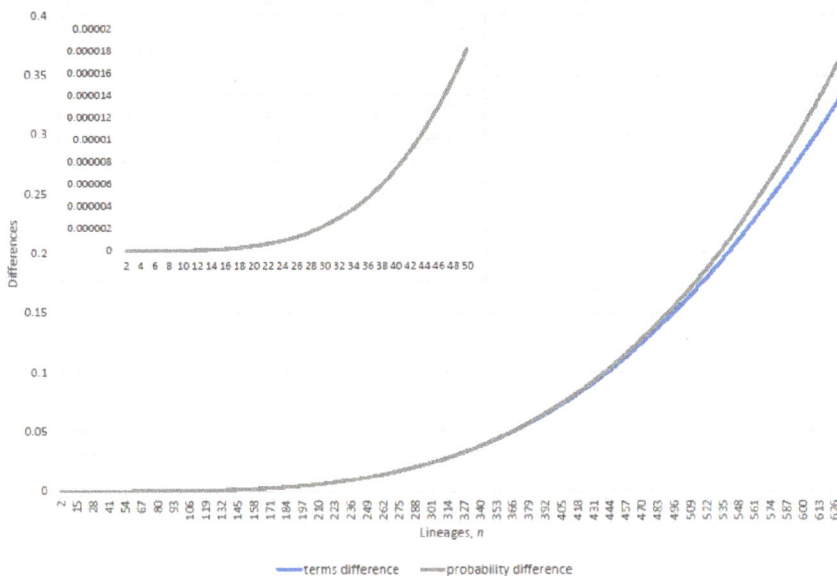

Figure 4. Quadratic term given by Equation (5) minus cubic term given by Equation (6) of Equation (4), and exact Equation (1) minus linearized $1 - \binom{n}{2}/N$ non-coalescence probabilities. Population size $N = 2 \times 10^5$ and $n = 2, \ldots, 633$ (inset $n = 2, \ldots, 50$).

Remark 1. *In this case, coalescence probability absolute error, linearized minus exact value, yields an empirical criterion for greater than (precisely) 99% expected genealogical parity; $n \leq 33$. The Wright-Fisher ancestral process restricted to single-pair coalescence thus yields $n \leq 34$. The total linearization error of the Kingman coalescent, which includes non-coalescence error, thus yields $n \leq 26$. Refer to the exposition of parity in Section 4 for the details of these criteria.*

2.2. Single Pair Coalescence Events

Identically to Equation (1), precisely two lineages with the same parent occurs with probability

$$\frac{\binom{n}{2}}{N} \prod_{i=1}^{n-2} \left(1 - \frac{i}{N}\right). \tag{7}$$

The form of Equation (7) can be explained by analogy to Equation (1). Common ancestry among two lineages occurs with probability $1 \cdot \frac{1}{N}$, since the same individual must be picked uniformly at random from the parent generation by two individuals from the offspring generation in a population of fixed size N. Exchangeability renders a combinatorial term $\binom{n}{2}$, since any single pair of the n lineages from the offspring generation participate in such a common ancestry event. There is no common ancestry among the remaining $n - 2$ lineages in the offspring generation, which yields the corresponding product of $(N - i)/N$ for $i = 1, 2, \ldots, n - 2$.

Compare the linearized probability of at least one coalescence $\frac{1}{2}n(n - 1)N^{-1}$ from Equation (2) and the exact pair-wise coalescence probability of Equation (7). Clearly, the linearization omits the corresponding non-coalescence probability product.

Remark 2. *The single-pair coalescence restriction is questionable prima facie with respect to the exact coalescence probabilities, since the complimentary event to non-coalescence in Equation (1) describes at least one coalescence. This includes combinations of single-pairs or multi-coalescence. The N^{-1} term of Equation (4) linearizes the probability of at least one coalescence, which is to be distinguished from the probability of single pair coalescence.*

The differences between the corresponding linearized and exact probabilities cancel out as equal and opposite, whereas the relative proportions yield asymmetric linearized substitutions; refer to Figure 5. Both substitutions equal the exact value at $n = 2$; as n increases, linearized non-coalescence probability underestimates and linearized coalescence probability overestimates. Although the absolute errors have zero sum, linearization exaggerates coalescence transition probabilities and by comparison slightly reduces non-coalescence transition probabilities; refer to Section 2.3. Thus, Kingman's coalescent detracts from the exact ancestral process.

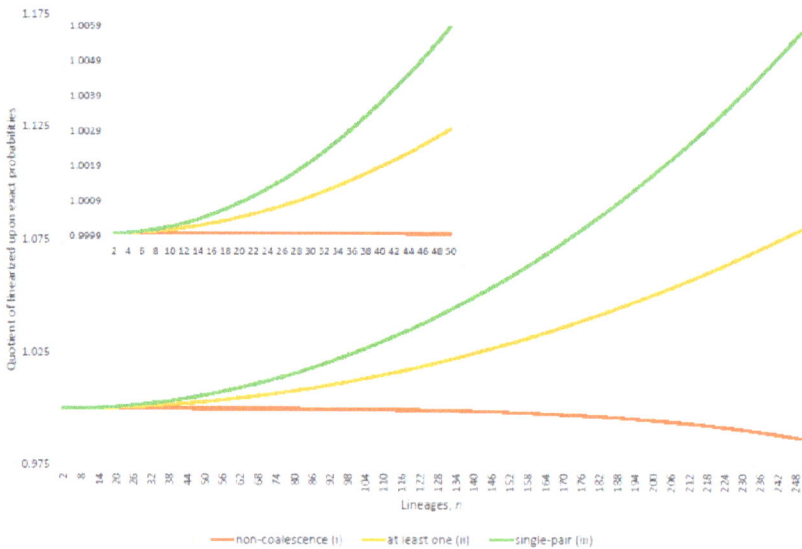

Figure 5. Quotients of linearized upon exact probabilities: (i) linearized $1 - \binom{n}{2}/N$ upon exact Equation (1) non-coalescence probability; (ii) linearized coalescence $\binom{n}{2}/N$ upon exact at least one coalescence probability (complimentary event of Equation (1)); and (iii) linearized coalescence $\binom{n}{2}/N$ upon the exact single-pair coalescence probability of Equation (7). Population size $N = 2 \times 10^5$ and $n = 2, \ldots, 250$ (inset $n = 2, \ldots, 50$).

Table 1 quantifies decreased accuracy of coalescence probability linearization, in Figure 5 (ii), for alternative population sizes, N.

Table 1. Percentage overestimation of linearized coalescence probability reached at n lineages.

N	1%	5%	10%	15%	20%	25%
2×10^4	30	64	90	109	124	138
2×10^6	284	629	882	1072	1229	1364

Remark 3. *Does the conventional substitution correspond to omission of the multi-coalescence probabilities, or constraint of emergent coalescence events by suppression of multi-coalescence and replacement with single pair coalescence? Answer: The latter, since the probability of at least one coalescence is linearized in Equation (1).*

Define the *absolute error (type I)* as the difference between linearized and exact single pair coalescence probabilities; $\binom{n}{2}/N$ minus Equation (7). The quotient of the absolute error (type I) upon the exact single pair coalescence probability defines the *relative error (type I)*. After cancellation, when n lineages remain, this equals the quotient of exact probabilities for at least one coalescence upon non-coalescence from $n-2$ lineages. Alternatively, define *absolute error (type II)* as the difference between linearized and exact at least one coalescence probabilities; $\binom{n}{2}/N$ minus the probability of the complimentary event to Equation (1). The quotient of the absolute error (type II) upon the exact at least one coalescence probability defines the *relative error (type II)*. Refer to Equations (14) and (16) in Section 4 for further explanation.

The absolute and relative errors heighten a probability structure that would be invisible otherwise; refer to Figures 6 and 7. Thus, the robustness of the Kingman coalescent gets a qualitative measure. The quotient of relative errors illustrates their comparative proportional growth as n increases; refer to Figure 8. In this case, a *minmax* transition occurs around $n = 20$ between two gradient phases that correspond to the quotients of relative error type I upon type II. Intuitively, the two types of relative errors follow maximum and minimum detraction, respectively; type I corresponds to suppression of multi-coalescence altogether, whereas type II corresponds to replacement of multi-coalescence events with a single-pair, which accords to the Kingman coalescent. The single-pair and at least one coalescence probabilities for small to moderate numbers of lineages look equivalent; refer to Figure 9 in Section 2.3.

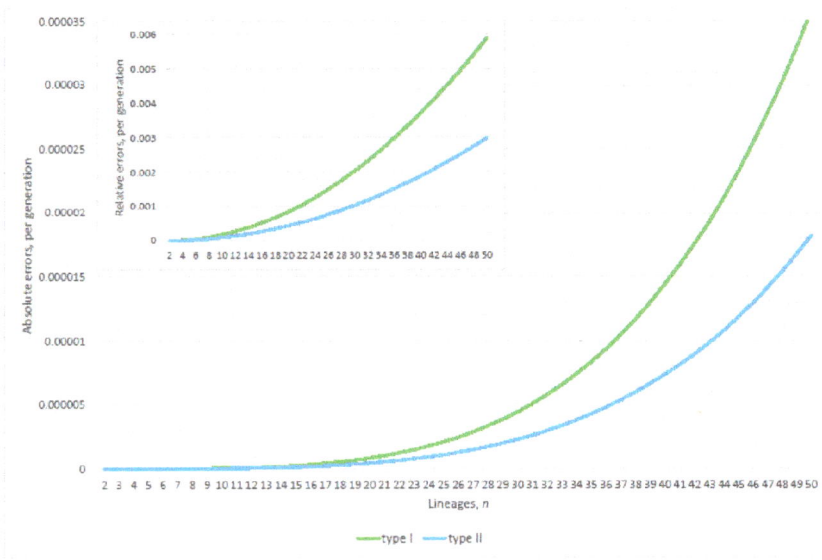

Figure 6. Absolute errors types I and II, inset relative errors types I and II; per generation, $n = 2, \ldots, 50$. Population size $N = 2 \times 10^5$.

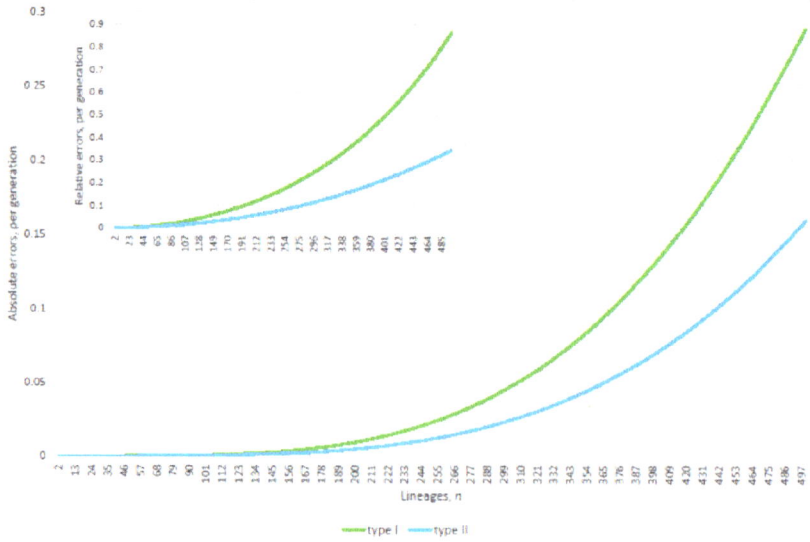

Figure 7. Absolute error types I and II, inset relative error types I and II; per generation, $n = 2, \ldots, 500$. Population size $N = 2 \times 10^5$.

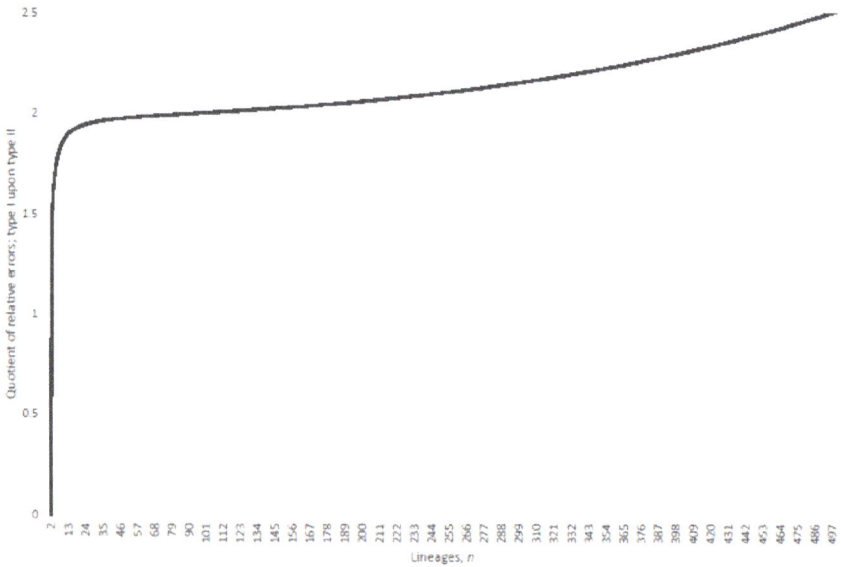

Figure 8. Minmax phase transition by comparison of relative errors; type I upon type II, per generation. Population size $N = 2 \times 10^5$ and $n = 2, \ldots, 500$.

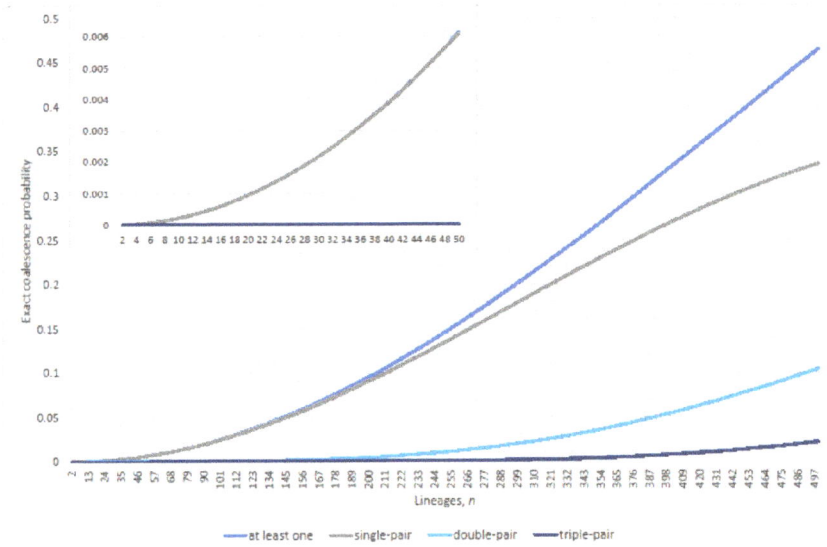

Figure 9. Exact coalescence probability: at least one coalescence (complimentary event of Equation (1)), single-pair (Equation (7)), double-pair (Equation (8)) and triple-pair (Equation (10)) coalescence probabilities. Population size $N = 2 \times 10^5$ and $n = 2, \ldots, 500$ (inset $n = 2, \ldots, 50$).

Relative error (type II), per generation, does not exceed precisely 1% where $n \leq 90$, in this case.

2.3. Multiple Coalescence Events

There is no implementation of multiple coalescence events in fastsimcoal, version 2.6 (fsc26), according to their online documentation [39–41]. Extension of an original SimCoal package [42] simulates genetic data serially sampled, Serial SimCoal [43], and implements a heuristic double-pair coalescence transition probability (software and documentation available online: http://web.stanford.edu/group/hadleylab/ssc/index.html).

Consider the ancestral process in which at coalescence a decrement of two lineages can occur; double-pair or triplet coalescence. Precisely two pairs of lineages, with a different parent in common for each pair, occurs with probability

$$\frac{1}{2} \frac{\binom{n}{2}\binom{n-2}{2}}{N^2} \prod_{i=1}^{n-3}\left(1 - \frac{i}{N}\right), \tag{8}$$

since discounting permutation of both pairs yields a factor one half. Similar with Equation (7), precisely one pair-wise common ancestry event occurs with probability $\binom{n}{2}/N$, since this event involves any two of the n lineages present in the offspring generation. The second pair-wise common ancestry event picks a different common parent to the first pair and this occurs with probability $\left(\frac{N-1}{N}\right)\frac{1}{N}\binom{n-2}{2}$. Permutation of the first and second pairs does not count due to the exchangeability of lineages in the ancestral process and requires the factor $\frac{1}{2}$. There is no common ancestry among the remaining $n - 4$ lineages in the offspring generation, which yields the corresponding product of $(N - i)/N$ for $i = 2, 3, \ldots, n - 3$.

Three lineages with the same parent occurs with probability

$$\frac{\binom{n}{3}}{N^2} \prod_{i=1}^{n-3} \left(1 - \frac{i}{N}\right). \tag{9}$$

Common ancestry among three lineages occurs with probability $1 \cdot \frac{1}{N} \cdot \frac{1}{N}$, since the same individual from the parent generation is picked uniformly at random by three individuals from the offspring generation in a population of fixed size N. Exchangeability renders a combinatorial term $\binom{n}{3}$, since any triplet of the n lineages from the offspring generation participate in such a common ancestry event. There is no common ancestry among the remaining $n - 3$ lineages in the offspring generation, which yields the corresponding product of $(N - i)/N$ for $i = 1, 2, \ldots, n - 3$.

Consider the ancestral process in which at coalescence a decrement of three lineages can occur; triple-pair, both a single-pair and a triplet, or quadruplet coalescence. Three pairs of lineages, with a different parent in common for each pair, occurs with probability

$$\frac{1}{3!} \frac{\binom{n}{2}\binom{n-2}{2}\binom{n-4}{2}}{N^3} \prod_{i=1}^{n-4} \left(1 - \frac{i}{N}\right), \tag{10}$$

since discounting permutation of the triple-pair yields a factor one sixth. Similar with Equation (8), the first pair-wise common ancestry event occurs with probability $\binom{n}{2}/N$. The second pair-wise common ancestry event picks a different common parent to the first pair and this occurs with probability $(\frac{N-1}{N})\binom{n-2}{2}/N$. The third pair-wise common ancestry event picks a different common parent to the first and second pairs and this occurs with probability $(\frac{N-2}{N})\binom{n-4}{2}/N$. Permutation of the first, second and third pairs does not count due to the exchangeability of lineages in the ancestral process and requires the factor $\frac{1}{6}$. There is no common ancestry among the remaining $n - 6$ lineages in the offspring generation, which yields the corresponding product of $(N - i)/N$ for $i = 3, 4, \ldots, n - 4$.

One single-pair and one triplet of lineages, with a different parent in common, occurs with probability

$$\frac{1}{2} \frac{\binom{n}{2}\binom{n-2}{3}}{N^3} \prod_{i=1}^{n-4} \left(1 - \frac{i}{N}\right), \tag{11}$$

since discounting permutation of the pair and the triplet yields a factor one half. The pair-wise common ancestry event occurs with probability $\binom{n}{2}/N$. Similar with Equation (9), the triplet common ancestry event now picks a different common parent to the pair and this occurs with probability $\left(\frac{N-1}{N}\right)\frac{1}{N}\frac{1}{N}\binom{n-2}{3}$. The alternative combinatorial product $\binom{n}{3}\binom{n-2}{2}$ yields the same function of n as in Equation (11). In this sense, the two alternatives cannot be distinguished. However, the usual permutation discount of simultaneous common ancestry events, one single pair and one triplet, applies with the factor $\frac{1}{2}$ due to exchangeability. There is no common ancestry among the remaining $n - 5$ lineages in the offspring generation, which yields the corresponding product of $(N - i)/N$ for $i = 2, 3, \ldots, n - 4$.

Precisely four lineages with the same parent occurs with probability

$$\frac{\binom{n}{4}}{N^3} \prod_{i=1}^{n-4} \left(1 - \frac{i}{N}\right). \tag{12}$$

Common ancestry among four lineages occurs with probability $1 \cdot \frac{1}{N} \cdot \frac{1}{N} \cdot \frac{1}{N}$, since the same individual from the parent generation is picked uniformly at random by four individuals from the offspring generation in a population of fixed size N. Exchangeability renders a combinatorial term $\binom{n}{4}$, since any quadruplet of the n lineages from the offspring generation participate in such a common ancestry event. There is no common ancestry among the remaining $n - 4$ lineages in the offspring generation, which yields the corresponding product of $(N - i)/N$ for $i = 1, 2, \ldots, n - 4$.

The probabilities of Equations (7)–(12) constitute a subset of all possible types of coalescences and therefore yield a restricted ancestral process. These probabilities correspond in every generation until a coalescence event occurs, with those of certain multi-coalescences equal to zero for small n. That is, such probabilities apply from one generation to the next among the offspring while n lineages remain. At coalescence, adjust n accordingly and continue the ancestral process, until eventually absorption occurs with a most recent common ancestor of the entire initial sample.

The exact at least one coalescence probability, compliment to Equation (1), and multiple exact coalescence probabilities of Equations (8) and (10) evaluated for small, moderate and larger numbers of lineages demonstrate their region of negligibility; refer to Figure 9.

The significance of coalescence probabilities of Equations (7)–(12) is of direct relevance to computer simulation and importance sampling methodology of the ancestral Markov chain, particularly as linearization errors accumulate. For the present purpose, quantitative analysis of conditional coalescence probability given the event of at least one coalescence, compliment to Equation (1), occupies Section 3.1.

3. Genealogical Topology and Expected Inter-Arrival Generations

Realization of the entire ancestral process yields one resultant genealogy. Statistical inference of genealogical time, for instance importance sampling methodologies, should be robust under a subset of ancestral transitions restricted to lineage decrements of one unless other genetic or exogenic processes act to emphasize the external branches.

3.1. Conditional Probabilities of Multi-Coalescence

The conditional probability of multi-coalescences given a coalescence event determine the genealogical topology in realization of the ancestral process. Refer to Figure 10, where conditional probability is given the event of at least one coalescence, either linearized or exact. Given exact coalescence: when $n = 10$, Pr(double-pair | coalescence) $< 1/14{,}286$, Pr(triplet | coalescence) $< 1/75{,}003$ and Pr(triple-pair | coalescence) $< 1/571{,}428{,}571$. When $n = 20$, $1/2615$, $1/33{,}344$ and $1/13{,}071{,}895$, respectively. Thus, in the region of most significance to timing, such multi-coalescence events rarely occur under genealogical stochastic reiteration.

Figures 11–13 illustrate the rapid decline of significant intervals for timing the genealogy and quantify the extent of multi-coalescence event rarity. Multi-coalescence event probabilities vary substantially within such regions, and negligibility becomes less extensive as population size decreases; refer to Figures 14–16.

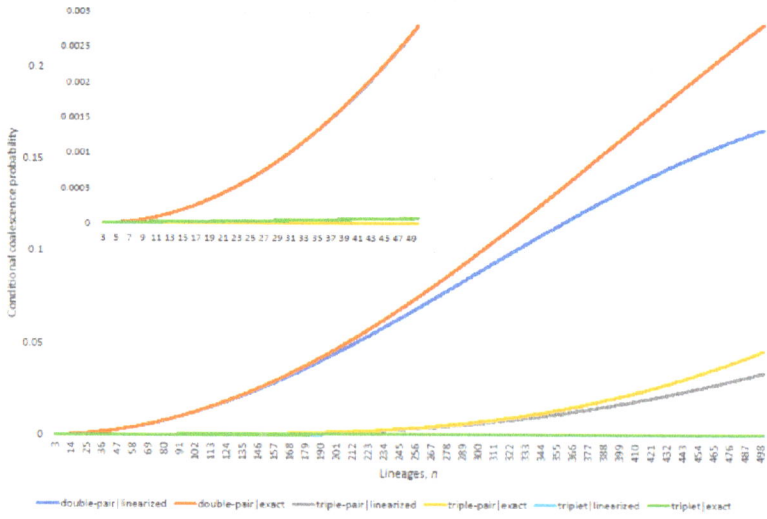

Figure 10. Conditional probabilities of double-pair (Equation (8)), triple-pair (Equation (10)), and triplet (Equation (9)) given either linearized coalescence $\binom{n}{2}/N$ or exact at least one coalescence (complimentary event of Equation (1)), respectively; $N = 2 \times 10^5$ and $n = 2, \dots, 500$ (inset $n = 2, \dots, 50$).

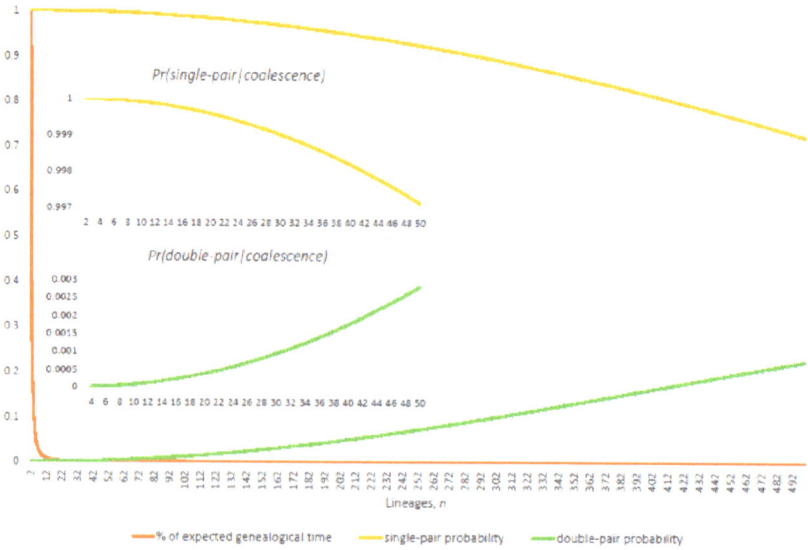

Figure 11. Exact conditional probabilities of single-pair (Equation (7)) and double-pair (Equation (8)) coalescence given the event of at least one coalescence (compliment of Equation (1)), respectively. Percentage of expected cumulative total genealogical inter-arrival generations shows significance of expected interval durations with n lineages present. Population size $N = 2 \times 10^5$ and $n = 2, \dots, 500$ (inset $n = 2, \dots, 50$).

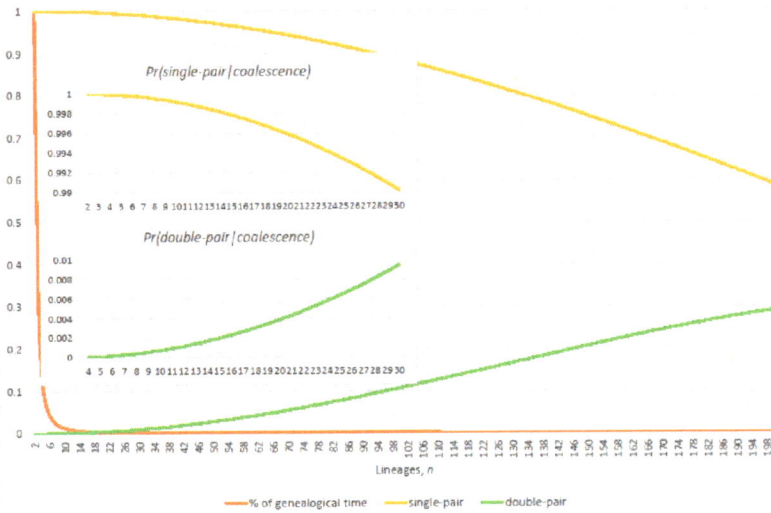

Figure 12. Exact conditional probabilities of single-pair (Equation (7)) and double-pair (Equation (8)) coalescence given the event of at least one coalescence (complimentary event of Equation (1)), respectively. Percentage of expected cumulative total genealogical inter-arrival generations shows the significance of expected interval durations with n lineages present. Population size $N = 2 \times 10^4$ and $n = 2, \dots, 200$ (inset $n = 2, \dots, 30$).

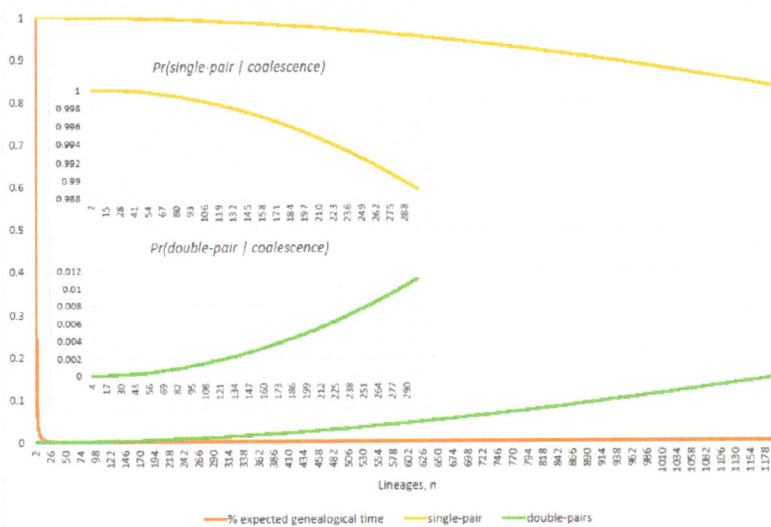

Figure 13. Exact conditional probabilities of single-pair (Equation (7)) and double-pair (Equation (8)) coalescence given the event of at least one coalescence (complimentary event to Equation (1)), respectively. Percentage of expected cumulative total genealogical inter-arrival generations shows significance of expected interval durations with n lineages present. Population size $N = 2 \times 10^6$ and $n = 2, 3. \dots, 1200$ (inset $n = 2, 3. \dots, 300$).

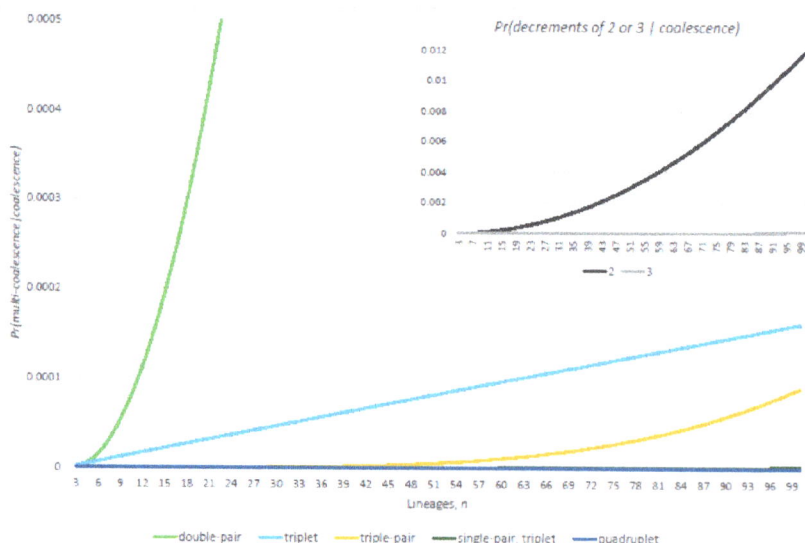

Figure 14. Exact conditional multi-coalescence probabilities given the event of at least one coalescence (complimentary event of Equation (1)); double-pair (Equation (8)), triplet (Equation (9)), triple-pair (Equation (10)), single-pair with triplet (Equation (11)), and quadruplet (Equation (12)) coalescence. Population size $N = 2 \times 10^5$ and $n = 3, \ldots, 100$.

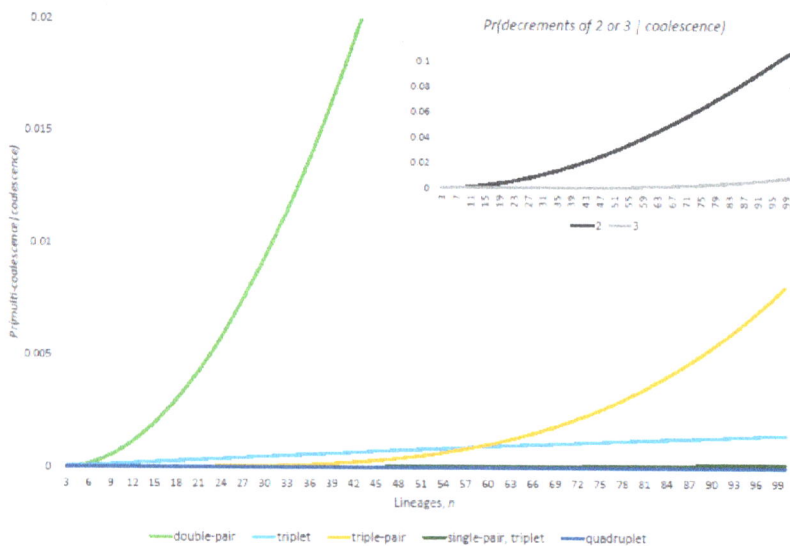

Figure 15. Exact conditional multi-coalescence probabilities given the event of at least one coalescence (complimentary event of Equation (1)); double-pair (Equation (8)), triplet (Equation (9)), triple-pair (Equation (10)), single-pair with triplet (Equation (11)), and quadruplet (Equation (12)) coalescence. Population size $N = 2 \times 10^4$ and $n = 3, \ldots, 100$.

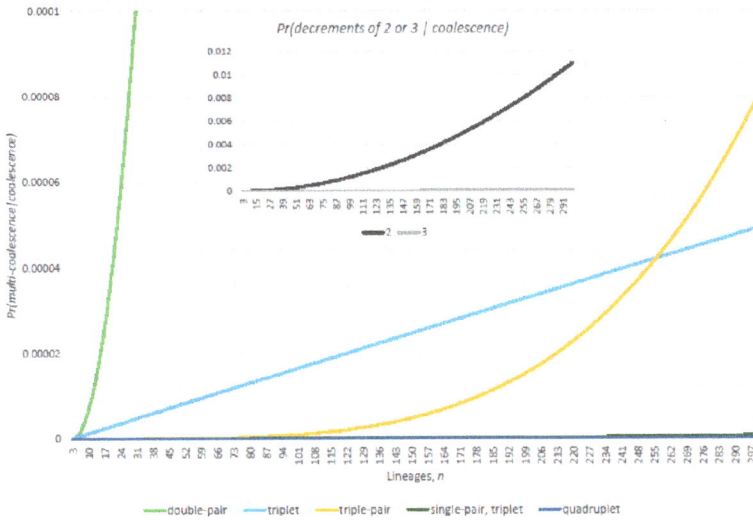

Figure 16. Exact conditional multi-coalescence probabilities given the event of at least one coalescence (complimentary event of Equation (1)); double-pair (Equation (8)), triplet (Equation (9)), triple-pair (Equation (10)), single-pair with triplet (Equation (11)), and quadruplet (Equation (12)) coalescence. Population size $N = 2 \times 10^6$ and $n = 3, \ldots, 300$.

In the first regions, these conditional probabilities of triplets exceed triple-pairs; this trend switches in the second regions and triple-pairs far exceed triplets. Thus, only the slightest relative contribution of multiple coalescence transition probabilities occurs in the ancestral process, per generation. Substantial replication of the ancestral process will be required before realizing a genealogy that contains multi-coalescence events. That is, unless the genealogy consists of many lineages or the population size is diminished substantially.

3.2. Single-Pairs Dominate Double-Pairs?

Consider the relative probabilities of double-pair and single-pair coalescence, namely the quotient of Equation (8) upon Equation (7),

$$\frac{(n-2)(n-3)}{4N - 4(n-2)}. \tag{13}$$

Equation (13) equals case $i = 1$ [16] (Equation (19)), which required correction since it should be $(n - 2i)(n - 2i - 1)/[2N(i+1) - 2(i+1)(n - i - 1)]$, where the denominator term $2N(i + 1)$ replaces $4N(i + 1)$. This expression equals the quotient of the $(i + 1)$st multiple and the ith multiple-pair coalescence probability. Thus, $i = 1$ corresponds to the quotient of double-pair upon single-pair coalescence probabilities.

The quotient of Equation (13) explains the dominance of expected inter-arrival times by single-pair coalescence. This is because the geometric distribution yields expectation equal to the reciprocal of the sum of Equation (7) plus Equation (8), when double and single-pair coalescences may occur in the ancestral process. Thus, double-pair coalescence is negligible in terms of the expected inter-arrival generations in the ancestral process due to Equation (13). Refer to Figure 17, the quotient of double-pair upon single-pair coalescence probabilities per generation has increased from nil at $n = 2$ to 1% (0.1%, $N = 2,000,000$) at $n = 92$, whereas the relative proportion of the total expected generations in the genealogy then equals 0.0121%. The expected inter-arrival generations determined by single-pair and double-pair coalescence probability, respectively, equals $1/p_s$ of Equation (7) and $1/p_d$ of Equation (8);

refer to Figure 18. The exact probability of avoiding a double-pair coalescence per expected interval, according to the geometric distribution with parameter p_d at $n = 92$ equals 0.990032.

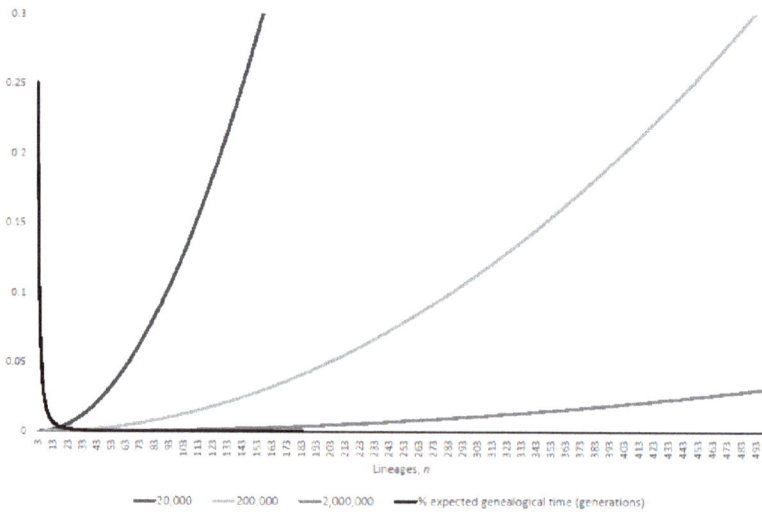

Figure 17. Quotient of single-pair coalescence probability upon double-pair coalescence probability. Evaluation of Equation (13) as lineages n vary; population sizes $N = 2 \times 10^4, 2 \times 10^5, 2 \times 10^6$ and $n = 3, 4, \ldots, 500$. Percentage of expected cumulative total genealogical inter-arrival generations shows significance of expected interval durations with n lineages present, the case $n = 2$ omitted (equals one).

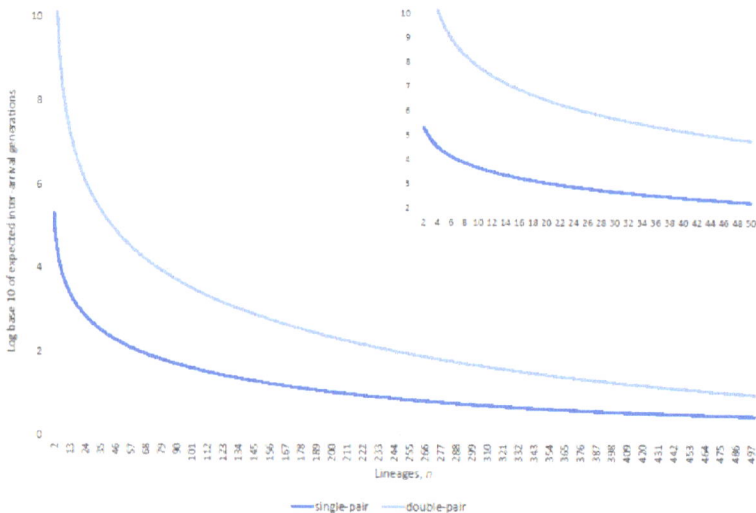

Figure 18. Logarithm base 10 of expected inter-arrival generations obtained from the reciprocals of single-pair (Equation (7)) and double-pair (Equation (8)) exact coalescence probabilities. Population size $N = 2 \times 10^5$ and $n = 2, 3, \ldots, 500$ (inset $n = 2, 3, \ldots, 50$). Due to a property of the Wright-Fisher model such that a geometric distribution determines the number of generations until a coalescence event occurs, the success probability of the distribution equals either Equation (7) or Equation (8).

In the next Section, calculation of multiple coalescence event probabilities per expected interval leads to a paradox of negligibility and its resolution obtained.

4. Parity of the Kingman Coalescent

Empirical calculations in this section yield a criterion of coalescence probability error, linearized minus exact value, such that expected genealogical parity be greater than 99% where $n \leq \frac{1}{2}\sqrt[3]{N}$. The Wright-Fisher ancestral process restricted to single pair coalescence empirically yields the same criterion as that just described. Total error of the Kingman coalescent that includes linearized non-coalescence probability thus yields $\frac{1}{2}\sqrt[3]{N/2}$.

In general, per generation, consider error to be the probability of a neglected coalescence; parity the probability of avoiding a neglected coalescence. The *parity, per expected interval*, is obtained by raising parity, per generation, to the power of an exponent given by $1/p$, where p equals the probability of coalescence, per generation. For instance, using the linearized coalescence probability yields the expected inter-arrival generations of Kingman's coalescent. The product of parity, per expected interval, across all intervals from the initial sample to its most recent common ancestor yields *expected genealogical parity*. Non-occurrence of neglected coalescence events anywhere in the expected genealogical realization represents *perfect* parity. This maximum stringency confounds observability, since the impact of neglected coalescence depends on position within the genealogy. Therefore, parity, per expected interval, is more directly informative.

4.1. Linearization Errors

The linearization of Kingman's coalescent yields error in both the non-coalescence and the coalescence probabilities, which cancel each other and sum to zero when the coalescence error is with respect to the exact probability of at least one coalescence. Consider n lineages to be present in the genealogy. Define the *linearization error (type I)* with respect to the exact probability of single-pair coalescence, per generation,

$$\left\{ \left\{ 1 - \frac{\binom{n}{2}}{N} \right\} - \prod_{i=1}^{n-1}\left(1 - \frac{i}{N}\right) \right\} + \left\{ \frac{\binom{n}{2}}{N} - \frac{\binom{n}{2}}{N}\prod_{i=1}^{n-2}\left(1 - \frac{i}{N}\right) \right\}. \tag{14}$$

Equation (14) simplifies as the exact multi-coalescence probability and is equivalent to the error of the Wright-Fisher ancestral process restricted to single-pair coalescence. Thus, one minus the linearization error (type I) defines *linearization parity (type I)*, per generation,

$$\left[1 + \frac{(n-1)(n-2)}{2N}\right]\prod_{i=1}^{n-2}\left(1 - \frac{i}{N}\right). \tag{15}$$

Note Equation (15) equals one plus the linearized coalescence probability then multiplied by the exact non-coalescence probability, while $n - 1$ lineages remain in the genealogy. Equation (15) is quantified as n varies, per expected interval and expected cumulative genealogy, according to reduced, mid-range and enlarged constant population sizes in Figures 19–24. These Figures also illustrate that inclusion of multi-coalescence transition probabilities of Equations (8)–(12) sustain parity of restricted Wright-Fisher models.

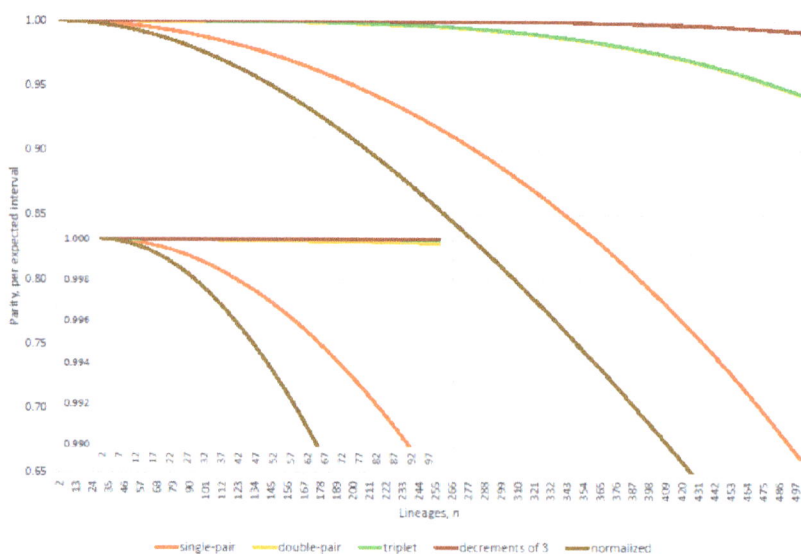

Figure 19. Parity, per expected interval, restricted Wright-Fisher models: inclusively expanded set of ancestral transitions; single-pair, double-pair, triplet, decrement of three (comprises triple-pair, single-pair with triplet, and quadruplet). Normalized curve accords with Equation (17) of the Kingman coalescent. Population size $N = 2 \times 10^5$ and $n = 2, 3, \ldots, 500$ (inset $n = 2, 3, \ldots, 100$).

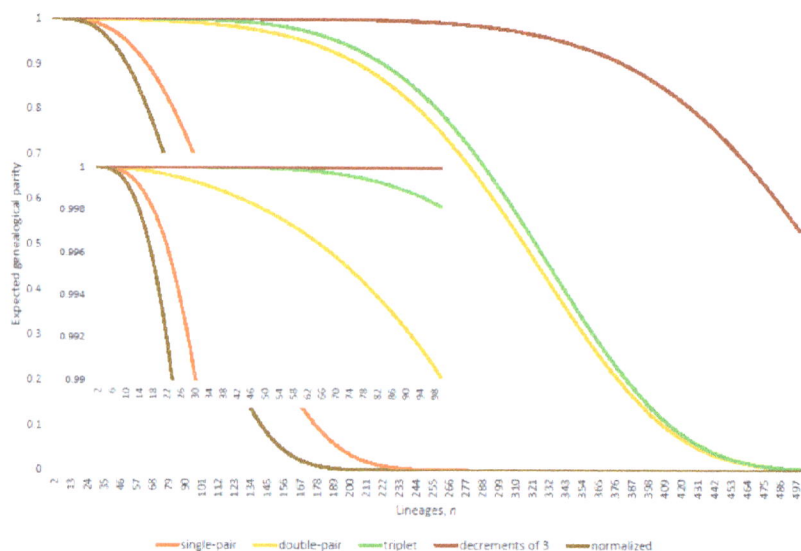

Figure 20. Expected genealogical parity, restricted Wright-Fisher models, inclusively expanded set of ancestral transitions; single-pair, double-pair, triplet, decrement of three (comprises triple-pair, single-pair with triplet, and quadruplet). Normalized curve accords with Equation (17) of the Kingman coalescent. Population size $N = 2 \times 10^5$ and $n = 2, 3, \ldots, 500$ (inset $n = 2, 3, \ldots, 100$).

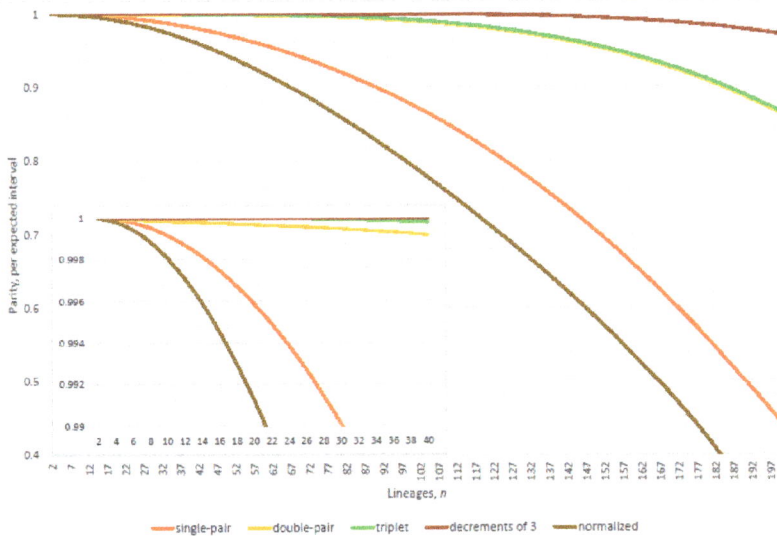

Figure 21. Corresponds to Figure 19 with population size reduced ten-fold. Parity, per expected interval, restricted Wright-Fisher models: inclusively expanded set of ancestral transitions; single-pair, double-pair, triplet, decrement of three (comprises triple-pair, single-pair with triplet, and quadruplet). Normalized curve accords with Equation (17) of the Kingman coalescent. Population size $N = 2 \times 10^4$ and $n = 2, 3, \ldots, 200$ (inset $n = 2, 3, \ldots, 40$).

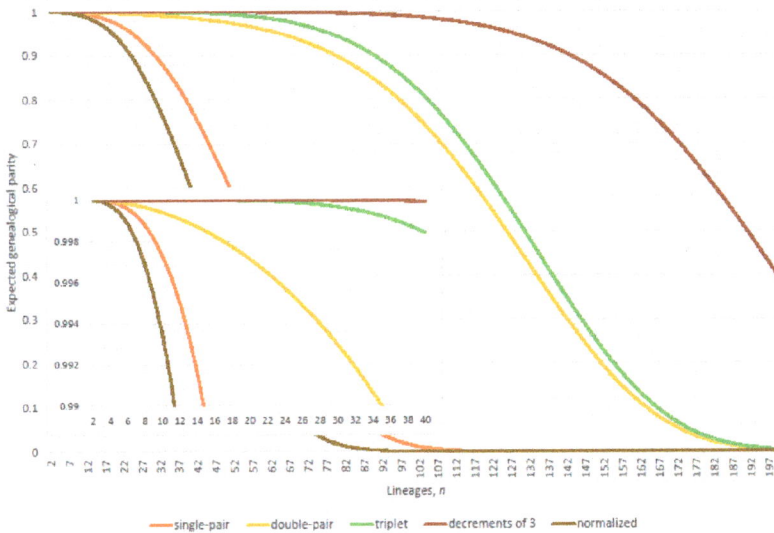

Figure 22. Corresponds to Figure 20 with population size reduced ten-fold. Expected genealogical parity, restricted Wright-Fisher models, inclusively expanded set of ancestral transitions; single-pair, double-pair, triplet, decrements of three (comprises triple-pair, single-pair with triplet, and quadruplet). Normalized curve accords with Equation (17) of the Kingman coalescent. Population size $N = 2 \times 10^4$ and $n = 2, 3, \ldots, 200$ (inset $n = 2, 3, \ldots, 40$).

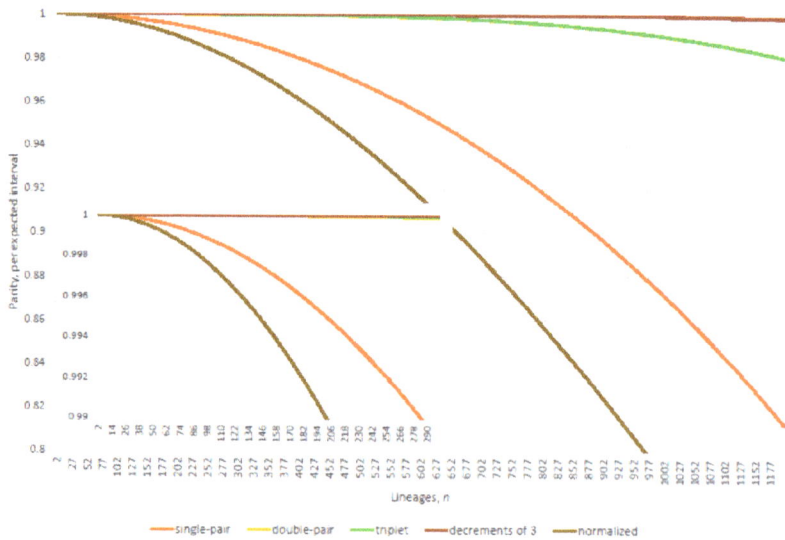

Figure 23. Corresponds to Figure 19 with population size enlarged ten-fold. Parity, per expected interval, restricted Wright-Fisher models: inclusively expanded set of ancestral transitions; single-pair, double-pair, triplet, decrement of three (comprises triple-pair, single-pair with triplet, and quadruplet). Normalized curve accords with Equation (17) of the Kingman coalescent. Population size $N = 2 \times 10^6$ and $n = 2, 3, \ldots, 1200$ (inset $n = 2, 3, \ldots, 300$).

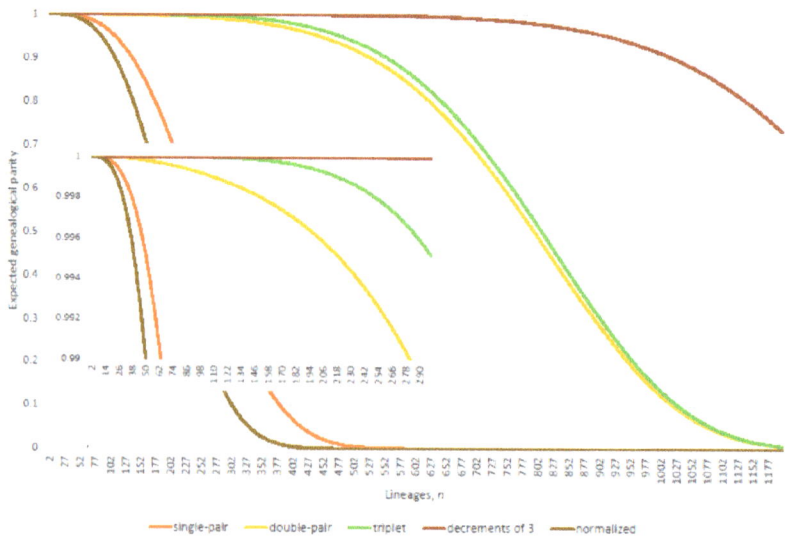

Figure 24. Corresponds to Figure 20 with population size enlarged ten-fold. Expected genealogical parity, restricted Wright-Fisher models, inclusively expanded set of ancestral transitions; single-pair, double-pair, triplet, decrements of three (comprises triple-pair, single-pair with triplet, and quadruplet). Normalized curve accords with Equation (17) of the Kingman coalescent. Population size $N = 2 \times 10^6$ and $n = 2, 3, \ldots, 1200$ (inset $n = 2, 3, \ldots, 300$).

Ancestral process of restricted Wright-Fisher models:

In the single-pair exact ancestral process, parity, per expected interval, exceeds precisely 99% where $n \leq 91$; refer to Table 2. An identical criterion was observed with relative error (type II) of the linearized coalescence probability; refer to Section 2.1. In the single- or double-pair exact ancestral process, parity, per expected interval, exceeds precisely 99% where $n \leq 316$.

Table 2. Restricted Wright-Fisher models; column headings describe inclusively expanded sets of ancestral transitions. Maximum lineages n such that parity, per expected interval, exceeds 99%.

Population Size, N	Single-Pair	Double-Pair	Triplet	Decrements of Three
2000	10	29	34	52
20,000	30	98	102	164
200,000	91	316	319	516
2,000,000	284	1002	1004	1633
20,000,000	895	3174	3178	5159

Values in Table 2 signal a clear loss of parity in the single-pair restricted Wright-Fisher model. Empirical criteria of the single-pair restricted Wright-Fisher model, linearization parity (type I):

- parity, per expected interval, exceeds 99% where (approximately) $n \leq \frac{1}{2}\sqrt{N/6}$

  (Table 2 verified this case where $N = 2000; 20,000; 200,000; 2,000,000$ and $20,000,000$); and
- expected genealogical parity exceeds 99% where $n \leq \frac{1}{2}\sqrt[3]{N}$

  (precise, $N = 20,000$ and $200,000$; plus one, $N = 2,000,000$ and $20,000,000$; minus one, $N = 2000$).

Otherwise, define the *linearization error (type II)*, per generation,

$$\left\{ \left\{ 1 - \frac{\binom{n}{2}}{N} \right\} - \prod_{i=1}^{n-1}\left(1 - \frac{i}{N}\right) \right\} + \left\{ \frac{\binom{n}{2}}{N} - \left[1 - \prod_{i=1}^{n-1}\left(1 - \frac{i}{N}\right) \right] \right\} \equiv 0. \tag{16}$$

The constituent errors in Equation (16) have opposite polarity. Heuristically, reverse the sign of the underestimated non-coalescence probability then add the overestimated coalescence probability to obtain a *normalized error (type II)*. Thus, one minus the normalized error (type II) defines *normalized parity (type II)*, per generation,

$$3 - \frac{n(n-1)}{N} - 2\prod_{i=1}^{n-1}\left(1 - \frac{i}{N}\right). \tag{17}$$

Equation (17) is quantified as n varies, per expected interval and expected cumulative genealogy, according to reduced, mid-range and enlarged constant population sizes in Figures 19–24.

Empirical heuristic criteria of Kingman's coalescent:

- parity, per expected interval, exceeds 99% where $n \leq \frac{1}{4}\sqrt{N/3}$

  (normalized parity criterion, per expected interval: n minus one $N = 2000, 20,000$; plus one $200,000$; plus two $2,000,000$; overestimates maximum lineages by 1.49% when $20,000,000$); and
- expected genealogical parity exceeds 99% where $n \leq \frac{1}{2}\sqrt[3]{N/2}$

  (normalized expected genealogical parity criterion: precise, $N = 2000, 20,000, 200,000$ and $2,000,000$; n plus one when $20,000,000$).

Parity criteria of linearization coalescence error (type II) essentially equals that observed with linearization error (type I); verified with $N = 20,000, 200,000$, and $2,000,000$. Parity based on linearization coalescence error (type II) is realistic since application of the Kingman coalescent usually involves only linearized coalescence probabilities with non-coalescence probabilities implicitly assumed.

Remark 4. *Pr(decrement of 2 | coalescence) rises to 1% at n lineages when parity, per expected interval, falls to 99%; verified as N varies. An alternative interpretation of this coupling is that the fractional cubic root criterion of expected genealogical parity occurs at values of n lineages where multi-coalescence transitions remain probabilistically insignificant.*

4.2. Parity Paradox

The single-pair coalescence probability dominates the expectation of generations between adjacent coalescence events in the genealogy, although inclusion of the double-pair coalescence probability sustains genealogical parity significantly beyond that obtained with single-pair coalescence. The paradox is resolved by two points: (i) relative probability values of single and double-pair coalescence explains the expected inter-arrival generations; and (ii) binomial expansion of the geometric probability for avoidance of omitted multi-coalescence events until the expected inter-arrival generations elapse.

Recall from Section 3.2 that single-pairs dominate expectation of inter-arrival generations. Then, let $G = 1/p_n$, where p_n equals Equation (7). Consider the binomial expansion of parity

$$(x + y + z)^G = (x + y)^G + Gx^{G-1}z + \binom{G}{2}x^{G-2}[2yz + z^2] + \binom{G}{3}[3y^2z + 3yz^2 + z^3] + \cdots + z^G, \quad (18)$$

where x, y, and z denote non-coalescence, single-pair and double-pair coalescence probabilities of Equations (1), (7) and (8), respectively. The left-side of Equation (18) quantifies the long run non-occurrence probability of omitted multi-coalescence events within the expected interval duration, while n lineages remain. Double-pair coalescence yields a non-negligible probability in total, since Equation (18) contains a sum of terms on the order $\frac{1}{2}G^2$ multiplied by Equation (8). Therefore, accumulation of double-pair coalescence probabilities over many generations sustains parity. Hence, parity of the double-pairs restricted Wright-Fisher model is significantly greater than that of the single-pairs restricted Wright-Fisher model. Additional multi-coalescence transition probabilities strengthen parity accordingly.

The conventional standard deviation of the generations expected in between successive coalescence events equals $\frac{\sqrt{q_n}}{p_n}$, where $q_n = 1 - p_n$, and the subscript denotes the dependence of the coalescence probability on n lineages present. Note the conventional variance replaces a pathological mathematical variance of the geometric probability distribution (refer to Appendix B, for derivation of the mathematical variance). The higher moments do not resolve the conundrum that double-pair coalescence sustains genealogical parity, whereas single-pair coalescence determines expected inter-arrival generations. Consider the functional forms of Equations (A6), (A9) and (A10) in two cases: (i) Equation (7); and (ii) the sum of Equations (7) and (8). Therefore, single pair coalescence probability dominates the first, second, (to a lesser extent) third, and fourth moments similarly to the discussion of Section 3.2.

5. Conclusions

Linearization potentially affects the Kingman coalescent in two ways: (i) suppression of multi-coalescence events induces upward size bias; and (ii) inflation of coalescence probabilities due to linearization induces downward size bias. Quantitative analyses demonstrate such affects unlikely. More specifically, genealogical topology is predominantly unaffected from root to tips provided lineage numbers remain small to moderate. This relegates similar conjectured compensatory mechanism [15–17] to regions of many lineages. Many lineages render significant multi-coalescence probabilities and inflated linearized coalescence probabilities, although expected inter-arrival times diminish on external branches, in this region Kingman's coalescent therefore detracts from the exact ancestral process.

Kingman's coalescent is a reasonably robust genealogical model of population genetics, although unsuitable for a wide range of sample sizes dependent on population size. Regions of validity were quantified with restricted versions of the exact ancestral process. Computationally-intensive statistical inference methods usually require many millions of genealogical realizations to converge. Thus, small waiting-time adjustments and slightly inflated coalescence event probabilities could be investigated more fully for significant elaboration of the sample space upon which resultant parametric estimates depend.

Double-pairs and higher combinations of multi-coalescence have proven to be negligible in the region of most significance for timing the genealogy, in both the linearized and exact ancestral processes. In contrast, parity quantifies the long run avoidance of omitted multi-coalescences across many generations as the sample size increases. Multi-coalescence affects the shape towards the tips of large sample genealogies, and then yields only fine-tuning effects of ancestral timing properties. The loss of parity of the Kingman coalescent, under relaxation of its conventional limit of a large population size, was quantified. The resultant empirical criteria, that a valid sample size is less than certain fractional square and cubic roots of population size, were all verified to hold for a wide range of population sizes. Finally, utilizing genomic data for the discovery of ecological evolutionary dynamics represents an important challenge [44] that demands extremely robust statistical models of genealogy applicable to phylogenetics.

Acknowledgments: Research commenced while a postdoctoral scientist with the Computational Biology and Bioinformatics Unit at the Australian National University. Consideration of referee's reports improved the work, especially in that Equations (5) and (6) required correction.

Conflicts of Interest: The author declares no conflicts of interest.

Appendix A

To obtain Equation (5)

$$\sum_{i=i}^{n-1}\left\{i\sum_{j=1,j\neq i}^{n-1}j\right\}=\sum_{i=1}^{n-1}i\left\{\left(\sum_{j=1}^{n-1}j\right)-i\right\}=\binom{n}{2}^2-\frac{n(n-1)(2n-1)}{6}\qquad(A1)$$

it is necessary to multiply Equation (A1) by a factor of $\frac{1}{2}$ to get an equivalent coefficient of the quadratic term N^{-2} in Equation (4), since the expansion above counts permutations. Therefore, the convenient Expression (A1) proves Equation (5).

To obtain Equation (6)

$$
\begin{aligned}
\sum_{i=1}^{n-1}i\sum_{j=1,j\neq i}^{n-1}j\sum_{k=1,k\neq i,j}^{n-1}k \\
=\sum_{i=1}^{n-1}i\left\{\left(\sum_{j=1}^{n-1}j\left[\sum_{k=1}^{n-1}k-j-i\right]\right)-i\sum_{k=1}^{n-1}k-2i\right\} \\
=\left(\sum_{i=1}^{n-1}i\right)^3-\left(\sum_{i=1}^{n-1}i\right)\left(\sum_{j=1}^{n-1}j^2\right)-\left(\sum_{i=1}^{n-1}i^2\right)\left(\sum_{j=1}^{n-1}j\right) \\
-\left(\sum_{i=1}^{n-1}i^2\right)\left(\sum_{k=1}^{n-1}k\right)+2\left(\sum_{i=1}^{n-1}i^3\right)
\end{aligned}
\qquad(A2)
$$

the summations in Equation (A2) yield a general expression of the total

$$\frac{n^3(n-1)^3}{8}-3\frac{n(n-1)}{2}\frac{n(n-1)(2n-1)}{6}+2\frac{n^2(n-1)^2}{4}=\frac{n^2(n-1)^2}{8}(n-2)(n-3).\qquad(A3)$$

It is necessary to multiply Equation (A2) by a factor of $\frac{1}{6}$ to get an equivalent coefficient of the cubic term N^{-3} in Equation (4), since the expansion above counts permutations. Therefore, the convenient Expression (A3) proves Equation (6).

Appendix B

To obtain the correct variance of a geometrically distributed random variable with success probability p, factorize the second moment

$$E\left(X^2\right) = p \sum_{i=1}^{\infty} x^2 q^{x-1} = p\left(1 + 4q + 9q^2 + 16q^3 + 25q^4 + \cdots\right)$$

$$= p(1 + 2q + 3q^2 + 4q^3 + 5q^4 + \cdots$$
$$+q + q^2 + q^3 + q^4 + q^5 + \cdots$$
$$+q + 2q^2 + 3q^3 + 4q^4 + 5q^5 + \cdots$$
$$+q^2 + q^3 + q^4 + q^5 + \cdots$$
$$+2q^2 + 3q^3 + 4q^4 + 5q^5 + \cdots \qquad \text{(A4)}$$
$$+q^3 + q^4 + q^5 + \cdots$$
$$+3q^3 + 4q^4 + 5q^5 + \cdots$$
$$+q^4 + q^5 + \cdots$$
$$+4q^4 + 5q^5 + \cdots$$
$$+q^5 + \cdots$$
$$+\cdots),$$

$$E\left(X^2\right) = p\left(E(X) + qE\left(X^2\right) + q\frac{1}{(1-q)^2}\right), \qquad \text{(A5)}$$

$$E\left(X^2\right) = \frac{1/p}{1 - pq}, \qquad \text{(A6)}$$

since

$$E(X) = p \sum_{i=1}^{\infty} xq^{x-1} = p\left[1 + 2q + 3q^2 + 4q^3 + \cdots\right] = p\left[1 + q + q^2 + q^3 + q^4 + \cdots\right]^2$$
$$= \frac{p}{(1-q)^2} \qquad \text{(A7)}$$

by convergence of the geometric series when $|q| < 1$.
Thus,

$$Var(X) = E\left(X^2\right) - E^2(X) = \frac{1}{p}\left[\frac{1}{1-pq} - \frac{1}{p}\right] = \frac{q}{p^2}\left[\frac{p-1}{1-pq}\right] < 0. \qquad \text{(A8)}$$

The conventional variance accords with that obtained from the adjusted second derivative of the moment generating function, $E\left(e^{tX}\right)$, evaluated at $t = 0$.

Similar factorizations to those of Equation (A4) yield

$$E\left(X^3\right) = p\left[qE\left(X^3\right) + (1 + 2q)E\left(X^2\right) + qE(X)\right] = \frac{1 + 2q}{(1 - pq)^2} + \frac{q}{1 - pq}, \qquad \text{(A9)}$$

$$E\left(X^4\right) = p\left[qE\left(X^4\right) + (1 + 3q)E\left(X^3\right) + 3qE\left(X^2\right) + qE(X)\right]$$
$$= \frac{(1+2q)(1+3q)}{(1-pq)^3} + \frac{3q}{p(1-pq)^2} + \frac{q}{p(1-pq)} \qquad \text{(A10)}$$

References

1. Wakeley, J. *Coalescent Theory: An Introduction*, 1st ed.; Roberts and Company Publishers: Greenwood Village, CO, USA, 2009; ISBN 978-0-9747077-5-4.
2. Hein, J.; Schierup, M.H.; Wiuf, C. *Gene Genealogies, Variation and Evolution: A Primer in Coalescent Theory*, 1st ed.; Oxford University Press: Oxford, UK, 2005; ISBN 0-19-852996-1.

3. Tavaré, S. Ancestral inference in population genetics, Part 1. In *Ecole d'Eté de Probabilités de Saint-Flour XXXI—2001*, 1st ed.; Picard, J., Ed.; Lectures on Probability Theory and Statistics, 1837; Springer: Berlin/Heidelberg, Germany, 2004; pp. 1–188. ISBN 3-540-20832-1.
4. Kingman, J.F.C. On the genealogy of large populations. *J. Appl. Probab.* **1982**, *19*, 27–43. [CrossRef]
5. Kingman, J.F.C. The coalescent. *Stoch. Proc. Appl.* **1982**, *13*, 235–248. [CrossRef]
6. Kingman, J.F.C. *Exchangeability and the evolution of large populations, In Exchangeability in Probability and Statistics*, 1st ed.; Koch, G., Spizzichino, F., Eds.; North-Holland: Amsterdam, The Netherlands, 1982; pp. 97–112, ISBN 04448644032.
7. Kingman, J.F.C. Origins of the coalescent: 1974–1982. *Genetics* **2000**, *156*, 1461–1463. [PubMed]
8. Yang, T.; Deng, H.W.; Niu, T. Critical assessment of coalescent simulators in modelling recombination hotspots in genomic sequences. *BMC Bioinform.* **2014**, *15*, 3. [CrossRef] [PubMed]
9. Allman, E.S.; Degnan, J.H.; Rhodes, J.A. Identifying the rooted species tree from the distribution of unrooted gene trees under the coalescent. *J. Math. Biol.* **2011**, *62*, 833–862. [CrossRef] [PubMed]
10. Steel, M. *Phylogeny: Discrete and Random Processes in Evolution*, 1st ed.; CMBS-NSF Regional Conference Series in Applied Mathematics 89; Society for Industrial and Applied Mathematics (SIAM): Philadelphia, PA, USA, 2016; ISBN 978-1-611974-47-8.
11. Crane, H. The ubiquitous Ewens Sampling Formula. *Stat. Sci.* **2016**, *31*, 1–19. [CrossRef]
12. Crane, H. Rejoinder: The ubiquitous Ewens Sampling Formula. *Stat. Sci.* **2016**, *31*, 37–39. [CrossRef]
13. Kingman, J.F.C. The genealogy of the Wright-Fisher model, appendix II. In *Mathematics of Genetic Diversity*, 1st ed.; CMBS-NSF Regional Conference Series in Applied Mathematics 34; Society for Industrial and Applied Mathematics (SIAM): Philadelphia, PA, USA, 1980; pp. 63–66, ISBN 0-89871-166-5.
14. Felsenstein, J. Trees of genes in populations, chapter 1. In *Reconstructing Evolution: New Mathematical and Computational Advances*, 1st ed.; Steel, M., Gascuel, O., Eds.; Oxford University Press: Oxford, UK, 2007; pp. 3–29, ISBN 978-0-19-920822-7.
15. Wakeley, J.; Takahashi, T. Gene genealogies when the sample size exceeds the effective size of the population. *Mol. Biol. Evol.* **2003**, *20*, 208–213. [CrossRef] [PubMed]
16. Fu, Y.X. Exact coalescent for the Wright-Fisher model. *Theor. Popul. Biol.* **2006**, *69*, 385–394. [CrossRef] [PubMed]
17. Bhaskar, A.; Clark, A.G.; Song, Y.S. Distortion of genealogical properties when the sample is very large. *Proc. Natl. Acad. Sci. USA* **2014**, *111*, 2385–2390. [CrossRef] [PubMed]
18. Wakeley, J. Coalescent theory has many new branches. *Theor. Popul. Biol.* **2013**, *87*, 1–4. [CrossRef] [PubMed]
19. Lessard, S. Recurrence equations for the probability distribution of sample configurations in exact population genetic models. *J. Appl. Probab.* **2010**, *47*, 732–751. [CrossRef]
20. Möhle, M. Robustness results for the coalescent. *J. Appl. Probab.* **1998**, *35*, 438–447. [CrossRef]
21. Möhle, M. Ancestral processes in population genetics—The coalescent. *J. Theor. Biol.* **2000**, *204*, 629–638. [CrossRef] [PubMed]
22. Möhle, M.; Sagitov, S. A classification of coalescent processes for haploid exchangeable population models. *Ann. Probab.* **2001**, *29*, 1547–1562. [CrossRef]
23. Kingman, J.F.C. Random discrete distributions. *J. R. Stat. Soc. B* **1975**, *37*, 1–22.
24. Kingman, J.F.C. Random partitions in population genetics. *Proc. R. Soc. Lond. A* **1978**, *361*, 1–20. [CrossRef]
25. Kingman, J.F.C. The representation of partition structures. *J. Lond. Math. Soc.* **1978**, *18*, 374–380. [CrossRef]
26. Sagitov, S. The general coalescent with asynchronous mergers of ancestral lines. *J. Appl. Probab.* **1999**, *36*, 1116–1125. [CrossRef]
27. Pitman, J. Coalescents with multiple collisions. *Ann. Probab.* **1999**, *27*, 1870–1902. [CrossRef]
28. Sagitov, S. Convergence to the coalescent with simultaneous multiple mergers. *J. Appl. Probab.* **2003**, *40*, 839–854. [CrossRef]
29. Sargsyan, O.; Wakeley, J. A coalescent process with simultaneous multiple mergers for approximating the gene genealogies of many marine organisms. *Theor. Popul. Biol.* **2008**, *74*, 104–114. [CrossRef] [PubMed]
30. Donnelly, P.; Kurtz, T. Particle representations for measure-valued population models. *Ann. Probab.* **1999**, *27*, 166–205. [CrossRef]
31. Birkner, M.; Blath, J.; Capaldo, M.; Etheridge, A.; Möhle, M.; Schweinsberg, J.; Wakolbinger, A. α-stable branching and β-coalescents. *Electron. J. Probab.* **2005**, *10*, 303–325. [CrossRef]
32. Steinrücken, M.; Birkner, M.; Blath, J. Analysis of DNA sequence variation within marine species using β-coalescents. *Theor. Popul. Biol.* **2013**, *87*, 15–24. [CrossRef] [PubMed]

33. Heuer, B.; Sturm, A. On spatial coalescents with multiple mergers in two dimensions. *Theor. Popul. Biol.* **2013**, *87*, 90–104. [CrossRef] [PubMed]

34. Huillet, T.; Möhle, M. On the extended Moran model and its relation to coalescents with multiple collisions. *Theor. Popul. Biol.* **2013**, *87*, 5–14. [CrossRef] [PubMed]

35. Dong, R.; Gnedin, A.; Pitman, J. Exchangeable partitions derived from Markovian coalescents. *Ann. Appl. Probab.* **2007**, *17*, 1172–1201. [CrossRef]

36. Freund, F.; Möhle, M. On the number of allelic types for samples taken from exchangeable coalescents with mutation. *Adv. Appl. Probab.* **2009**, *41*, 1082–1101. [CrossRef]

37. Bertoin, J. The structure of the allelic partition of the total population for Galton-Watson processes with neutral mutations. *Ann. Probab.* **2009**, *37*, 1502–1523. [CrossRef]

38. Burden, C.J.; Simon, H. Genetic drift in populations governed by a Galton-Watson branching process. *Theor. Popul. Biol.* **2016**, *109*, 63–74. [CrossRef] [PubMed]

39. Excoffier, L. fsc26 Manual, online documentation for Fastsimcoal Version 2.6, Swiss Institute of Bioinformatics, Lausanne, Switzerland. 2016. Available online: http://cmpg.unibe.ch/software/fastsimcoal2 (accessed on 23 November 2017).

40. Excoffier, L.; Dupanloup, I.; Huerta-Sánchez, E.; Foll, M. Robust demographic inference from genomic and SNP data. *PLoS Genet.* **2013**, *9*, e1003905. [CrossRef] [PubMed]

41. Excoffier, L.; Foll, M. Fastsimcoal: A continuous-time coalescent simulator of genomic diversity under arbitrarily complex evolutionary scenarios. *Bioinformatics* **2011**, *27*, 1332–1334. [CrossRef] [PubMed]

42. Excoffier, L.; Novembre, J.; Schneider, S. SIMCOAL: A general coalescent program for the simulation of molecular data in interconnected populations with arbitrary demography. *J. Hereditary* **2000**, *91*, 506–509. [CrossRef]

43. Anderson, C.N.K.; Ramakrishnan, U.; Chan, Y.L.; Hadley, E.A. Serial SimCoal: A population genetics model for data from multiple populations and points in time. *Bioinformatics* **2005**, *21*, 1733–1734. [CrossRef] [PubMed]

44. Rudman, S.A.; Barbour, M.A.; Csillérry, K.; Gienapp, P.; Guillaume, F.; Hairston, N.G., Jr.; Hendry, A.P.; Lasky, J.R.; Rafajlović, M.; Räsänen, K.; et al. What genomic data can reveal about eco-evolutionary dynamics. *Nat. Ecol. Evol.* **2018**, *2*, 9–15. [CrossRef] [PubMed]

mathematics

MDPI

Article

The "Lévy or Diffusion" Controversy: How Important Is the Movement Pattern in the Context of Trapping?

Danish A. Ahmed [1,*], Sergei V. Petrovskii [2,*] and Paulo F. C. Tilles [2,3]

[1] Department of Mathematics and Natural Sciences, Prince Mohammad Bin Fahd University, Al-Khobar, Dhahran 34754, Saudi Arabia
[2] Department of Mathematics, University of Leicester, University Road, Leicester LE1 7RH, UK
[3] Departament of Matematica, Universida de Federal de Santa Maria, Santa Maria CEP 97105-900, Brazil; paulo.tilles@ufsm.br
[*] Correspondence: dahmed1@pmu.edu.sa or daa119@outlook.com (D.A.A.); sp237@le.ac.uk (S.V.P.)

Received: 30 March 2018; Accepted: 24 April 2018; Published: 9 May 2018

Abstract: Many empirical and theoretical studies indicate that Brownian motion and diffusion models as its mean field counterpart provide appropriate modeling techniques for individual insect movement. However, this traditional approach has been challenged, and conflicting evidence suggests that an alternative movement pattern such as Lévy walks can provide a better description. Lévy walks differ from Brownian motion since they allow for a higher frequency of large steps, resulting in a faster movement. Identification of the 'correct' movement model that would consistently provide the best fit for movement data is challenging and has become a highly controversial issue. In this paper, we show that this controversy may be superficial rather than real if the issue is considered in the context of trapping or, more generally, survival probabilities. In particular, we show that almost identical trap counts are reproduced for inherently different movement models (such as the Brownian motion and the Lévy walk) under certain conditions of equivalence. This apparently suggests that the whole 'Levy or diffusion' debate is rather senseless unless it is placed into a specific ecological context, e.g., pest monitoring programs.

Keywords: diffusion; random walks; Brownian motion; Lévy walks; stable laws; individual movement; trap counts; pest monitoring

1. Introduction

Pests form a significant threat to agricultural ecosystems worldwide, and therefore, effective and reliable monitoring is required to ease the decision making process for intervention. In agro-ecosystems, monitoring is an essential component of integrated pest management programs (IPM) [1,2], where a control action is implemented if necessary. If the population abundance exceeds a certain predefined threshold level, and given that resource effort and expense is readily available, then intervention becomes imminent. Usually, the control action takes the form of pesticide application, which has many negative implications, such as environmental damage in the form of air, soil and water pollution. Such human-induced pressures on the environment often contribute towards bio-diversity loss and affect the functioning of ecosystems [3–5]. Other major drawbacks, which are not necessarily related, include cancer-related diseases for those handling such chemicals [6,7], increased consumer costs [8], poor efficiency in reaching targeted pests [9], pest resistance to regular use [10] and lethal effects on natural enemies [11], possibly leading to a resurgence of the pest population or a secondary pest to emerge. Therefore, in order to avoid unnecessary pesticide application or the risk of triggering pest outbreaks, accurate evaluation of population abundance is key [12]. Traps are usually installed in the field under controlled experimental conditions as a means to estimate population abundance [13–15]. They are then exposed for the duration of study, insects are caught, traps are normally emptied at regular intervals and the total number that falls into the trap is counted and forms the trap counts. It is precisely these counts that are converted to the pest population density at trap locations and then used to estimate the total pest population size [16,17].

A major ecological challenge is to develop relevant theoretical and mathematical models that can explain patterns and observations obtained from field data [18–20]. This is primarily due to the fact that inherent complexity found in the behavior of animals can be difficult to incorporate. However, insects and some invertebrates are easier to model since they have been thought of as non-cognitive. In the case of either single or multiple traps in the same field, individual insect movement can be modeled successfully using a random walk framework [12]. The earliest attempts were based on Brownian motion, which provided a framework to characterize patterns of movement with broad applications to conservation [21,22], biological invasions [23–25] and, in particular, insect pest monitoring [12,26–28]. Theoretical arguments supported by empirical observations suggest that individuals with limited sensory capabilities tend to follow a Brownian movement pattern, more so at large temporal and spatial scales [29–34]. The corresponding mean field counterpart describes the spatial-temporal population dynamics, which is governed by the diffusion equation [12,27,28,35]. Recently, both Brownian motion (BM) and the diffusion model have been often criticized and deemed to be oversimplified descriptions [36]. Other revised models have attempted to account for possible intermittent stop-start movement [37] or inherent intensive/extensive behavioral changes [38]. Simultaneously, there are also other studies with growing empirical evidence that support an alternative description, which postulates that animal movement can exhibit Lévy walking behavior [39,40]. Lévy walks (LW) are differentiated from Brownian movement, since they allow for arbitrarily large steps, that is the probability of executing a larger step is much higher; which results in a faster movement pattern altogether [26,27,41,42].

The usage of the terms Lévy walks or Lévy flights can vary between disciplines. In the physical sciences, a clear distinction is made, but in the ecological literature, the terminology is often interchanged [43]. Some have stressed the crucial distinction between this [44], and others have taken a more relaxed approach [40]. Lévy flights allow for arbitrarily large steps, which can theoretically result in infinite velocities; which is an unphysical/unrealistic phenomenon. On the other hand, Lévy walks ensure that steps are randomly drawn such that the velocity is constant, or nearly so. In order to avoid confusion, we will use Lévy walks as a reference to a random walk whose step distribution has the property of heavy power-law tails, although technically, this is a Lévy flight (see later Section 4.1 for more details). Note that subsequent results and analysis within this study therefore apply indirectly to Lévy walks.

The Lévy or diffusion controversy has arisen from ongoing debates that provide pro and con arguments for each description [42]. Some cases that provide promising evidence for Lévy-type movement [45] have been later classified as Brownian [41] and then have been reclassified as Lévy afterwards [46]. The confusion arises partly due to different studies providing mixed and often conflicting messages. For example, the movement pattern can switch from Brownian to Lévy in a context-specific scenario where resources are scarce [47]. In another study, Lévy-type characteristics can emerge as a consequence of the fundamental observation that individuals of the same species are non-identical [48]. It is also possible that the underlying movement pattern can be misidentified, since variation in the individual walking behavior of diffusive insects can create the impression of a Lévy flight [49]. Furthermore, composite correlated random walks can produce similar movement patterns as Lévy walks; therefore, current methods fail to reliably differentiate between these two models (although recent attempts have been made to address this issue [50]). On the other hand, diffusive properties can appear for a population of Lévy walking individuals due to strong interactions, e.g., if movement is stopped when individuals encounter each other [51]. Even more recently, the diffusion and Brownian framework has been revisited and shown to be in excellent agreement with field data [28]. In either case, it is still unclear which type of movement is adopted by insects, what conditions can alter the pattern and which mathematical framework is most efficient; hence, the controversy persists.

The idea of using a modified diffusion model as an equivalent framework for Lévy walking individuals was first introduced by Ahmed [35] and later further developed [26,27]. In particular, it was demonstrated that in the context of pest monitoring, trap count patterns could be reproduced when comparing a type of Lévy walk to time-dependent diffusion. Further discussions have highlighted that the ecological basis behind incorporating time-dependent diffusion is not clearly understood and how this is linked to the type of mechanisms [26,52,53]. Our study has been instigated by these fruitful discussions leading to an extension to the previous study by Ahmed and Petrovskii [27]. Our aim is two-fold; that is to propose a diffusion model that consists of parameters that are of ecological significance and can be interpreted with reference to the diffusive properties; also, to investigate if trap counts can be effectively reproduced for a broader class of Lévy walks using diffusion. If so, we question the relative importance of identifying the underlying movement pattern. From a cost perspective, it may make sense to concentrate more on the geometry and design of the experiment rather than the particular movement model.

2. BM vs. LW: Equivalence I

2.1. Brownian Motion

Individual-based models provide a complementary cost-effective methodology to field experiments and can be used to simulate movement and analyze trap counts [12,18,54]. The idea is to replicate these experiments through a virtual environment where supplementary and even alternative information is sought [55]. In the specific case of low-density populations, the magnitude of stochastic fluctuations can be quite large, and an individual-based modeling framework can be particularly useful to describe the movement dynamics, i.e., if dangerous pests are present. Note that movement patterns of the Lévy-type have not been identified in field studies for insects, and therefore, our motivation is from a theoretical viewpoint. Our interest is primarily based on understanding the underlying mechanisms that govern movement; it is sufficient to focus on a 1D conceptual scenario. Despite the fact that this case is hardly realistic in terms of modeling movement in a real field setting, however it does provide a theoretical background for the more realistic 2D case [56]. Furthermore, any unnecessary additional complexity that would arise due to the effects of trap and field geometries is then avoided.

The basic idea in 2D is to model walking or crawling insect movement along a continuous curvilinear path, which can be mapped to a broken line with the position recorded at discrete times [18]. In mathematical terms, the 1D analogue over unbounded space for a population of N individuals records the position $X_i^{(n)}$ of the n-th individual at time t, which is discretized as $t_i = \{t_0 = 0, t_1, t_2, \cdots, t_S = T\}$, $i = 0, 1, \cdots, S, n = 1, 2, \cdots, N$, so that $X_i^{(n)} = X^{(n)}(t_i)$. Here, the script (n) is included to differentiate between movement tracks of different individuals of the same insect. We assume that positions are recorded at regular intervals with constant time increment:

$$\Delta t = t_i - t_{i-1} = \frac{T}{S},$$

where the first observation is recorded at time $t = 0$ with the total number of steps S and total time $t_S = T$. Each individual moves from position $X_{i-1}^{(n)}$ to $X_i^{(n)}$ with step length:

$$l_i^{(n)} = |X_i^{(n)} - X_{i-1}^{(n)}|$$

and velocity:

$$v_i^{(n)} = \frac{|X_i^{(n)} - X_{i-1}^{(n)}|}{\Delta t}$$

defined over each step. The *n*-th individual placed at some position $X_i^{(n)}$ can move either to the right or to the left with resulting position $X_{i+1}^{(n)}$. Assuming that each subsequent step is completely random [1] and generated according to a predefined probability distribution $\phi(\xi)$, then each further position can be determined by:

$$X_{i+1}^{(n)} = X_i^{(n)} + \xi, \; i = 0, 1, \cdots, S, n = 1, 2, \cdots, N, \tag{1}$$

where ξ is a random variable. Trap counts can then be obtained if the individuals that fall within a well-pre-defined region are removed from the system at regular intervals and counted [12,35].

Generally, animal movement is anisotropic due to the mere fact that animals have a front and rear end [57], resulting in a correlated random walk [30]. For simpler movement modes, such as that of insects, we can assume that the walk is uncorrelated and the direction of movement is completely independent of previous directions. Each position is then solely dependent on the previous location, and therefore essentially Markovian [58,59]. Under this assumption, the resulting movement is isotropic, and there is no preferential direction, i.e., no advection or drift of any kind; which can arise in the presence of an attractant. In relation to the step distribution, $\phi(\xi)$ is a symmetric probability density function (pdf) with zero mean, that is $\phi(\xi) = \phi(-\xi)$. In the case of Brownian motion, the corresponding pdf is normal, which reads:

$$\phi_n(\xi) = \frac{1}{\sigma\sqrt{2\pi}} \exp\left(-\frac{\xi^2}{2\sigma^2}\right) \tag{2}$$

with scale parameter σ, which can possibly be dependent on time (the subscript n refers to normal). Alternatively, we write $\xi \sim \mathbf{N}(0,\sigma^2)$, which denotes that ξ is randomly drawn from a normal distribution with mean zero and variance σ^2. In realistic ecological applications, many insects are released into the field instead of a single individual. Using the 1D scenario as a baseline case, the corresponding analogy is to initially distribute N individuals along a finite spatial interval $x \in (0, L)$. Common release methods used in trap count studies are either of two types; that is uniform or a point source. In the uniform case, we prescribe the initial position as $X_0^{(n)} \sim U(0, L)$, where $U(a, b)$ denotes the uniform distribution over the interval $a < x < b$. For a point source, we have $X_0^{(n)} = \tilde{x}$ for all $n = 1, 2, \cdots, N$ at $t = 0$ with $x = \tilde{x}$ as the centralized location of the release point. The resulting movement pattern is then completely determined by the type of initial condition and step distribution $\phi(\xi)$. Note that the population distribution essentially becomes uniform for larger times, and therefore, identical trap counts are obtained irrespective of the type of initial distribution. This effect is realized due to the inherent random movement of individuals in the field. For more detailed information on how the shape of the trap count profile is affected by the type of initial condition, the reader is redirected to [35]. Generally, in most field applications, a uniform initial population distribution can be reasonably assumed. Even in the case when the true release distribution is characterized by multiple point source releases, the effect on trap count variation is somewhat minimal [28]. Henceforth, all simulations in this paper adopt the uniform distribution as an initial condition.

To describe the individual-based model fully, boundary conditions are enforced and defined as follows: an impermeable stop-go 'sticky'-type boundary is installed at the external boundary at $x = L$ [60], such that at any instant in time, if the individual position exceeds this boundary, that is if $X_i^{(n)} > L$, then it remains at location $x = L$. The next position in the process is determined purely by (1), and the individual continues to interplay with the dynamics of the system provided $0 < X_{i+1}^{(n)} < L$ at the next step; otherwise, it either remains at the external boundary, if $X_{i+1}^{(n)} \geq L$ or is deemed to be trapped if $X_{i+1}^{(n)} < 0$. The trap boundary at $x = 0$ introduces a perturbation to the movement and is incorporated

---

[1]   The randomness of animal movement is obviously an idealization, which, however, is well justified under certain conditions, e.g., see [18] for a detailed discussion of this issue.

in the following way: if $X_i^{(n)} < 0$ at any instant in time, then the individual is removed from the system and the total trap count increases by one, functioning as an absorbing boundary. Cumulative trap counts J_t are obtained at each step over discrete times $t = i\Delta t$, resulting in a stochastic trap count trajectory. It may be important to mention that the choice of time step Δt has some significance, since it is known that time scale invariance is lost in the presence of an absorbing and/or stop-go boundary, possibly leading to noticeable differences in trap count recordings [60]. Therefore, Δt is chosen small enough so that trap counts are in line with counts obtained using alternative methods, such as the mean field analytic or numerical solutions (see Section 3 and Appendix A for details). Furthermore, at the same time, Δt is chosen to be sufficiently large, so that the assumption that subsequent steps are uncorrelated is feasible [12,31].

2.2. Condition of Equivalence

In many ecological applications (such as, for instance, conservation and monitoring), it is important to know the probability that a given animal will remain inside a certain domain or area. Since the movement is described by the probability distribution function (pdf) $\phi(\xi)$, one can expect that the probability of remaining inside the domain depends on the properties of ϕ, in particular on its rate of decay at large distances.

We first consider the case where $\phi(\xi)$ is fat-tailed, $\phi(\xi) \sim |\xi|^{-(\alpha+1)}$ with $0 < \alpha < 2$, which is the characteristic exponent for Lévy walks (see Section 4.1 for more details later). We focus on the special case $\alpha = 1$ as it is thought, based both on observations of movement patterns [61,62] and on some evolutionary argument [51], to be ecologically the most relevant. In this case, the pdf for the Lévy walk is described by the Cauchy distribution:

$$\phi_c(x; \gamma) = \frac{\gamma}{\pi} \frac{1}{x^2 + \gamma^2}, \tag{3}$$

where γ is a scale parameter and the subscript c refers to Cauchy.

In the case that at $t = 0$, the animal is at $x = 0$, Function (3) gives the probability density of its position after one step. It is straightforward to see that the pdf of the animal position after i steps, i.e., at time $t_i = i\Delta t$, is given by the same distribution, but with a re-scaled value of the parameter γ, that is:

$$\phi_c(x; i\gamma) = \frac{i\gamma}{\pi} \frac{1}{x^2 + (i\gamma)^2}. \tag{4}$$

From Equation (4), we may define a symmetric interval of interest $x_i(\gamma, \epsilon)$, such that the integral in this interval is always equal to a certain quantity ϵ,

$$\int_{-x_i}^{x_i} \phi_c(x; i\gamma)\, dx = \epsilon, \tag{5}$$

and determine the limits of this interval as a function of the parameters of the process and the arbitrary probability ϵ:

$$x_i(\gamma, \epsilon) = i\gamma \tan\left(\frac{\pi\epsilon}{2}\right). \tag{6}$$

Our goal is to obtain an alternative stochastic process, composed by the sum of random variables from the Gaussian family, which may be comparable to this one in the sense that it replicates the same probability ϵ over the interval of interest (e.g., see [28]). To do that, we will consider the sum $\bar{Y} = Y_1 + Y_2 + \cdots + Y_i$ of normally-distributed random variables, defined by the pdf:

$$\phi_n(y; \Delta_i) = \frac{1}{\Delta_i\sqrt{2\pi}} \exp\left(-\frac{y^2}{2\Delta_i^2}\right), \tag{7}$$

where the variance Δ_i^2 is just the sum of the variances from each random variable Y_k,

$$\Delta_i^2 = \sum_{k=1}^{i} \sigma_k^2. \tag{8}$$

As the relation between the two process is obtained by equating the integrals on the same domain $[-x_i, x_i]$, we may compute the probability ϵ over the Gaussian process using Equation (7),

$$\int_{-x_i}^{x_i} \phi_n \left(y; \Delta_i \right) dy = \mathrm{erf} \left(\frac{x_i}{\Delta_i \sqrt{2}} \right) = \epsilon, \tag{9}$$

and introducing the inverse error function as $\Phi \left(\cdot \right) \equiv \mathrm{erf}^{-1} \left(\cdot \right)$ (see footnote[2]), we relate the two processes by the following relation:

$$\Delta_i^2 \left(\gamma, \epsilon \right) = \frac{x_i \left(\gamma, \epsilon \right)}{2\Phi^2 \left(\epsilon \right)} = \frac{i^2 \gamma^2 \tan^2 \left(\frac{\pi\epsilon}{2} \right)}{2\Phi^2 \left(\epsilon \right)}. \tag{10}$$

In order to determine the behavior of the increments $\sigma_k^2 \left(\gamma, \epsilon \right)$, i.e., the variance of each additional Gaussian variable required to be comparable to the original Cauchy process, we only need to write Equation (8) as:

$$\Delta_i^2 = \sigma_i^2 + \Delta_{i-1}^2, \tag{11}$$

which immediately leads to the expression:

$$\sigma_i^2 \left(\gamma, \epsilon \right) = \frac{\left(2i - 1 \right) \gamma^2 \tan^2 \left(\frac{\pi\epsilon}{2} \right)}{2\Phi^2 \left(\epsilon \right)}. \tag{12}$$

Now, we recall that i is the time of the movement (measured in steps). From Equation (12), we therefore conclude that the probability ϵ to confine the insect performing the Lévy walk over the spatial domain $-L < x < L$ coincides exactly with the probability of the same event in the case where the insect performs the Brownian motion, provided the variance of the Brownian motion (i.e., essentially, the diffusion coefficient) increases linearly with time, $\sigma_i^2 \sim 2t$.

3. Time-Dependent Diffusion

Insect movement is inherently more complex in nature, due to the contribution from both external and internal factors. Typical external factors include environmental effects or stimuli, which can be quite challenging to incorporate from a modeling perspective. Since our interest lies in the actual mechanisms at play, we assume homogeneity in the sense that external factors are absent. In terms of the underlying movement mechanisms, examples of internal factors include individual variation, composite and/or intermittent movement or even time-density-dependent diffusive behavior [63,64]. The main challenge is then to develop a coherent model that can include these different processes and accurately describe the population dynamics. Obviously, the issue becomes more difficult if a combination of these features is present. In the context of insect pest monitoring, diffusion models have been shown to provide a good theoretical framework and the means for trap count interpretation [12,27,35]. In particular, time-dependent diffusion provides an adequate description for more complicated behavior, at least where standard diffusion fails [26,27,35,65]. The notion of insect movement with time-dependent diffusivity is not new and has been observed in field studies [19].

The 1D diffusion equation for the population density $u(x,t)$ with time-dependent diffusion coefficient $D = D(t)$ over the semi-infinite domain $0 < x < \infty$, with initial density $u(x, t = 0) = u_0(x)$ and zero density condition $u(x = 0, t) = 0$ at the trap boundary, reads:

---

[2] This is the inverse of the error function defined by $\mathrm{erf}(z) = \frac{2}{\sqrt{\pi}} \int_0^z \exp \left(-z'^2 \right) dz'$

$$\frac{\partial u}{\partial t} = D(t)\frac{\partial^2 u}{\partial x^2}, \quad u(x, t = 0) = u_0(x), \quad u(x = 0, t) = 0, \quad 0 < x < \infty, \quad t > 0. \tag{13}$$

By introducing a change of the time variable:

$$\tau(t) = \int_0^t D(s)ds, \tag{14}$$

the system of equations (13) transforms to:

$$\frac{\partial u}{\partial \tau} = \frac{\partial^2 u}{\partial x^2}, \quad u(x, \tau = 0) = u_0(x), \quad u(x = 0, \tau) = 0, \quad 0 < x < \infty, \quad \tau > 0. \tag{15}$$

The general solution [26,27,66] is then given by:

$$u(x, \tau) = \int_0^\infty \left(F(x - x', \tau) - F(x + x', \tau)\right) u_0(x')dx' \tag{16}$$

where:

$$F(x, \tau) = \frac{1}{\sqrt{4\pi\tau}} \exp\left(-\frac{x^2}{4\tau}\right) \tag{17}$$

is the fundamental solution of the diffusion equation in (15), which reduces to $F(x, t) = \frac{1}{\sqrt{4\pi Dt}}\exp\left(-\frac{x^2}{4Dt}\right)$ in the specific case of constant diffusivity. The diffusive flux through the boundary at $x = 0$ corresponds to trap counts $j(\tau)$, which can be determined by Fick's law, that is $j(\tau) = -\left.\frac{\partial u(x,\tau)}{\partial x}\right|_{x=0}$ with cumulative trap counts $J(\tau)$ (total flux) given by:

$$J(\tau) = \int_0^\tau j(\tau')d\tau' = \int_0^\infty u_0(x')\text{erfc}\left(\frac{x'}{\sqrt{4\tau}}\right)dx' \tag{18}$$

where $\text{erfc}(z) = 1 - \frac{2}{\sqrt{\pi}}\int_0^z \exp\left(-z'^2\right)dz'$ is the complimentary error function. Therefore, the total number of trap counts $J(t)$ for the system (13) in normal time t is given by,

$$J(t) = \int_0^\infty u_0(x')\text{erfc}\left(\frac{x'}{2\sqrt{\int_0^t D(s)ds}}\right)dx'. \tag{19}$$

In the case of a uniform distribution $u_0(x) = U_0$, (19) reduces to:

$$J(t) = 2U_0\sqrt{\frac{1}{\pi}\int_0^t D(s)ds} \tag{20}$$

with:

$$J(t) = 2U_0\sqrt{\frac{Dt}{\pi}} \tag{21}$$

in the special case with constant diffusivity D, corresponding to standard diffusion.

Equivalence of Trap Counts: Brownian Motion vs. Diffusion in a Semi-Bounded Space

For the diffusion Equation (15), the mean location and mean squared displacement (MSD) are useful statistics that characterize the movement,

$$\langle x(\tau)\rangle = \int_{-\infty}^\infty xF(x, \tau)dx = 0, \quad \left\langle x^2(\tau)\right\rangle = \int_{-\infty}^\infty x^2 F(x, \tau)dx = 2\tau, \tag{22}$$

where $F(x,\tau)$ is defined by (17). It is well known that diffusion is the macroscopic description of Brownian motion [67], where the MSD is equal to the variance of the step distribution $\phi(\xi)$,

$$\left\langle x^2(\tau)\right\rangle = \mathbb{E}\left(\xi^2\right).\tag{23}$$

From this, we obtain a link between the scale parameter in (2) and the diffusion coefficient,

$$\sigma^2(t) = 2\tau = 2\int_0^t D(s)ds.\tag{24}$$

For a discrete time model, one can expect that this remains valid, at least approximately, for a small, but finite value of Δt, that is:

$$\sigma^2(t) \approx 2D(t)\Delta t.\tag{25}$$

In the case of standard diffusion, the MSD grows linearly with time and is related to the scale parameter by:

$$\sigma^2(t) = 2Dt,\tag{26}$$

which is known as the hallmark of Brownian motion [12,18,68]. More generally, for anomalous diffusion, the MSD grows according to some power law relationship:

$$\left\langle x^2(t)\right\rangle \sim t^{2H},\tag{27}$$

where H is the Hurst exponent. Here, $H = \frac{1}{2}$ corresponds to standard diffusion (26), $\frac{1}{2} < H < 1$ corresponds to super-diffusion and $H = 1$ corresponds to ballistic or wavelike motion. A full comprehensive summary of movement properties with reference to H can be found in [59].

To demonstrate equivalence between Brownian motion and an anomalous diffusion model, consider as a baseline case:

$$D(t) = D_0 + D_1 t^{2H-1}, \; H \geq \frac{1}{2}\tag{28}$$

where D_0 is the initial diffusivity and D_1 controls the effect of time dependency for a larger time. This structure is chosen as an example, so that the scale parameter (24) is in accordance with (27), i.e.,

$$\sigma^2(t) = 2\int_0^t \left(D_0 + D_1 s^{2H-1}\right)ds = 2D_0 t + \frac{D_1}{H}t^{2H} \sim t^{2H} \quad \text{provided} \quad H \geq \frac{1}{2}.\tag{29}$$

The analytical solution for the model with an initial uniform distribution can be derived from (20), which reads:

$$J(t) = 2U_0\sqrt{\frac{D_0 t}{\pi}}\left(1 + \frac{D_1}{2D_0 H}t^{2H-1}\right)^{\frac{1}{2}}\tag{30}$$

and approximates the flux for constant diffusion (21) for small time.

In Figure 1, Plot (a), we find that there is almost identical agreement between trap counts obtained from the Brownian individual-based model and the mean field diffusion model, as expected from theory. More specifically, it is shown here that the diffusive flux can be used to reproduce trap count patterns for standard diffusion $H = \frac{1}{2}$, super diffusion $H = \frac{3}{4}$ and ballistic movement $H = 1$, to a high level of accuracy. Intuitively, we expect that this holds for diffusion coefficients that have a more complicated time-dependency. The diffusion coefficient consists of three unknown parameters, namely D_0, D_1 and H. In terms of usage, if initial diffusivity can be measured through experiments, then other parameters can be estimated using the tools outlined in [27], i.e., by approximating the flux rate in the limit $t \to 0$ and relating it to the expected number of individuals trapped after one time step. Note that, since an analytical solution cannot be obtained for the diffusion model (15) over a finite domain, the numerical solution is also shown for consistency. See Appendix A for details on the explicit finite difference scheme used and how the flux is computed at the trap boundary.

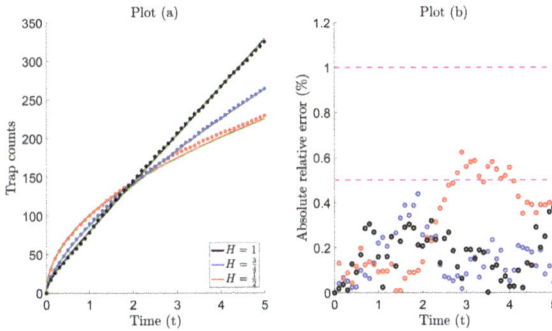

Figure 1. (**a**) Diffusive flux: Solid curves show the total flux $J(t)$ obtained from the diffusion model over time $0 < t < 5$ with the analytic solution given by (30), with fixed diffusion constants $D_0 = 0.05$, $D_1 = 0.15$ and varying Hurst exponents $H = \frac{1}{2}$ (standard diffusion), $H = \frac{3}{4}$ (super-diffusion) and $H = 1$ (ballistic). The solution is defined over the semi-infinite domain $0 < x < \infty$ with initial uniform distribution $u(x, t = 0) = U_0 = 200$ and point trap $u(x = 0, t) = 0$. Trap counts: Bold dots plot cumulative trap counts J_t for Brownian motion with total population $N = 1000$ recorded at times $t = 0, 0.1, 0.2, \cdots, 5$. Discrete time scale parameter is defined by combining (25) and (28), that is $\sigma^2(t) \approx 2(D_0 + D_1 t^{2H-1})\Delta t$ with D_0, D_1, H given above. Each individual executes a total of $S = 5000$ steps with constant time step increment $\Delta t = 0.001$ and total time $T = S\Delta t = 5$. Individuals initially uniformly distributed $X_0^{(n)} \sim U(0, L = 5)$. The trap installed at position $x = 0$ and simulations are conducted with the external boundary condition described in Section 2.1. Trap counts are replicated and averaged over ten realizations to reduce the effect of stochasticity. Numerical solution: The green dashed line represents the mean field numerical solution using the method of explicit finite differences. See Appendix A for further details on the numerical scheme. (**b**) Absolute relative error: $A(t)$ plotted at times $t = hk$, $h = 0.1, k = 0, 1, 2, \cdots, 50$, with average $\bar{A} = 0.306$ (red), 0.157 (blue) and 0.167 (black), to illustrate the magnitude of the discrepancy between the analytic solution and trap counts (for the interpretation of the references to color in this figure legend, and all subsequent figures, the reader is referred to the web version of the article).

We introduce the average absolute error $\bar{A}$ as a means to quantify the discrepancy between the models. Although advanced statistical tools exist to measure differences between stochastic and deterministic processes, for our purposes, this simple statistical metric will suffice and will later prove to be effective. The absolute error (relative to the total population) evaluated at discrete times $t = hk$ is defined by,

$$A(t_k) = \frac{|\text{Diffusive flux} - \text{Trap counts}|}{\text{Total population}} = \frac{|J(hk) - J_{hk}|}{N}, \quad k = 0, 1, 2, \cdots, K \quad (31)$$

with time increment h and total time $T = hK$. The errors are then averaged over $(K+1)$ differences,

$$\bar{A}(t_k) = \frac{1}{N(K+1)} \sum_{k=0}^{K} |J(hk) - J_{hk}|. \quad (32)$$

Plot (b) illustrates the discrepancy using the absolute error, which lies within 0.6 and 0.7% of all total trap counts. Theoretically, the Brownian and corresponding diffusion model are equivalent, and the errors can be dismissed as somewhat negligible, partly due to the accumulation of small computational errors. We expect this error to tend to zero, with the magnitude of stochastic fluctuations decreasing as $N^{-\frac{1}{2}}$ in the limit $N \to \infty$ [69]. Furthermore, longer time simulations (not shown) demonstrate that the discrepancy increases as the effect of external boundary encounters is realized. Therefore, we require that the finite domain is large enough, that is $\frac{L^2}{D}$ is much larger than the typical characteristic trapping time. The quantified errors in Table 1 are probably not useful on their own; however, for now, they function as benchmark values, which indicate an extremely good fit; indicating equivalence. As a general rule of thumb, we classify the level of fit as equivalent $0 < \bar{A} \leq 0.5$, good fit $0.5 < \bar{A} \leq 1$, moderate fit $1 < \bar{A} \leq 1.5$ and poor fit $\bar{A} > 1.5$. These will be useful later as a point of reference, when comparisons are made between various diffusion models, in an attempt to reproduce Lévy trap count data.

Table 1. Tabulated values of the average absolute relative error $\bar{A}$ as defined by (32), to compare the fit between the anomalous diffusion model (30) and trap counts obtained from Brownian motion (see Figure 1).

	Standard Diffusion $H = \frac{1}{2}$	Super Diffusion $H = \frac{3}{4}$	Ballistic Diffusion $H = 1$
Brownian trap counts	0.306	0.157	0.167

4. BM vs. LW: Equivalence II

4.1. Stable Laws

In the case of Brownian motion, the step distribution is normal (2), and the end tails decay exponentially fast (thin tail). A large number of studies has shown that animal movement can follow a more complicated movement pattern where the step distribution decays much more slowly according to some type of inverse power law (heavy or fat tail), known as Lévy walks [32,45,70]. As a result, individuals have a greater chance of executing 'rare' large steps, and therefore, the properties of the random walk are altered. The biological consequence is such that the overall movement pattern is faster in comparison to what is typically observed in Brownian motion. Lévy walks can be characterized by Lévy α-stable distributions, simply known as stable laws. A distribution is said to be stable if the sum (or, more generally, a linear combination with positive weights) of two independent random variables has the same distribution up to a scaling factor and shift [44,71]. The mechanisms behind the resulting movement are governed by the step distribution, which is completely described using four parameters, namely a tail index $\alpha \in (0,2]$, skewness parameter $\beta \in [-1,1]$, scale parameter $\gamma \in (0,\infty)$ and location parameter $\delta \in \mathbb{R}$. The asymptotic behavior of the end tails is,

$$\phi(\xi) \sim |\xi|^{-(\alpha+1)}, \; \xi \to \pm\infty, \tag{33}$$

where $\alpha \in (0,2]$ determines the rate at which the tails of the distribution taper off. For $\alpha \leq 0$, the distribution cannot be normalized, and therefore, the pdf cannot be defined. For $\alpha \geq 2$, the end tails decay sufficiently fast at large $|\xi|$, ensuring that all moments exist and the central limit theorem (CLT) applies, that is the probability density of the walker after S steps converges to the normal distribution as $S \to \infty$. The generalized central limit theorem (gCLT) states that the sum of identically-distributed random variables with distributions having inverse power law tails converges to one of the stable laws, of which the normal distribution is a special case. For the range $0 < \alpha < 2$, the gCLT applies, and the condition on the second moment is relaxed; that is, second moments diverge, and the tails are asymptotically equivalent to a Pareto law. Since their first introduction, usage of stable laws has been overlooked and somewhat neglected, mainly due to difficulties arising from an infinite variance. However, there are now well-developed and readily-available algorithms that can be exploited for simulation runs [72,73], and thus, stable laws are increasingly being considered, particularly in movement ecology [40].

Stable laws can be parametrized in $\mathcal{Z}$ different, but equivalent ways, and currently, there exists at least eleven different variations, which has led to much confusion [74,75]. Each type has an advantage over the others, and the parameter $\mathcal{Z}$ is often chosen based on the purpose of use, i.e., simulation-based studies, data fitting or the study of algebraic/analytic properties. Since our focus is primarily based on obtaining trap counts from simulations, we choose the $\mathcal{Z}=0$ parametrization and henceforth use this setting. We adopt the notation introduced by [71]; that is, the random variable ξ is drawn from the stable distribution:

$$\mathbf{S}(\alpha,\beta,\gamma,\delta;\mathcal{Z}).$$

Since explicit pdfs are not available for all values of $\alpha\in(0,2]$, the distribution is often described in terms of the characteristic function: the inverse Fourier transform of the pdf,

$$\ln\mathbb{E}e^{i\omega\xi}=\begin{cases}-\gamma^\alpha|\omega|^\alpha\left[1+i\beta\tan\frac{\alpha\pi}{2}\cdot\mathrm{sign}(\omega)\cdot(|\gamma\omega|^{1-\alpha}-1)\right]+i\delta\omega,&\alpha\neq1\\-\gamma|\omega|\left[1+\frac{2i\beta}{\pi}\cdot\mathrm{sign}(\omega)\cdot\log|\gamma\omega|\right]+i\delta\omega,&\alpha=1\end{cases} \tag{34}$$

For an isotropic random walk, the pdf of the step distribution is symmetrical, and therefore, both the skewness and location parameters are fixed with $\beta=\delta=0$. The resulting distribution is known as an α-stable symmetric Lévy distribution, which is then completely characterized solely by the index α and scale parameter γ. For brevity, we adopt the notation:

$$\mathbf{S}(\alpha,\beta=0,\gamma,\delta=0;\mathcal{Z}=0)=\mathbf{S}(\alpha,\gamma),$$

where it is understood that all parameters are zero except α and γ. The resulting characteristic function in (34) reads:

$$\mathbb{E}e^{i\omega\xi}=\exp\left(-\gamma^\alpha|\omega|^\alpha\right), \tag{35}$$

which is a useful way to mathematically describe all (symmetric) stable distributions, since a closed form or analytical expression does not exist for all indices α, with the exception of the normal $\alpha=2$ and Cauchy $\alpha=1$ cases;

Normal $\quad\xi\sim\mathbf{S}(2,\gamma_n)\qquad\qquad\phi_n(\xi)=\dfrac{1}{2\gamma_n\sqrt{\pi}}\exp\left(-\dfrac{\xi^2}{4\gamma_n^2}\right),\;\gamma_n=\dfrac{\sigma}{\sqrt2}\qquad(36)$

Holtsmark $\quad\xi\sim\mathbf{S}\left(\dfrac32,\gamma_h\right)\qquad\phi_h(\xi)\quad$ Cannot be expressed in closed form $\qquad(37)$

Cauchy $\quad\xi\sim\mathbf{S}(1,\gamma_c)\qquad\qquad\phi_c(\xi)=\dfrac{\gamma_c}{\pi(\gamma_c^2+\xi^2)}\qquad(38)$

Symmetric-Lévy $\quad\xi\sim\mathbf{S}\left(\dfrac12,\gamma_l\right)\qquad\phi_l(\xi)\quad$ Cannot be expressed in closed form $\qquad(39)$

where the subscripts n,h,c,l have been included to distinguish between the different distributions. For the normal distribution, σ is the standard deviation shown in the step distribution (2), with γ_n defined in this way due to the choice of parametrization. This relation can be easily derived through the characteristic function (35). Note that, some authors use the term 'Lévy distribution' to refer to stable laws; however, more commonly, it refers to $\alpha=\frac12,\beta=1$, which is a skewed distribution, defined for $\xi\geq0$ [71]. Since we consider symmetric distributions, i.e., $\beta=0$, we will refer to (39) as the 'symmetric'-Lévy distribution. In some cases, the pdf can be expressed analytically, even if it cannot be written in closed form, e.g., the Holtsmark distribution (37) can be written using hyper-geometric functions (if symmetric) or the Whittaker function (if skewed). Furthermore, the symmetric-Lévy distribution can be expressed in terms of special functions, such as Fresnel integrals [74]. However, these representations are bulky and not useful in the context of this study.

Figure 2 illustrates pdfs for symmetric stable laws with the decay of end tails characterized by different indices $\alpha=\frac12,1,\frac32,2$, with faster decay rates as α increases and exponential fast decay in the case of the normal distribution.

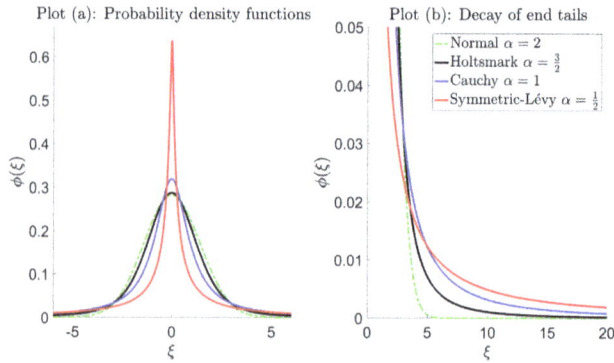

Figure 2. (**a**) Probability density functions $\phi(\zeta)$ for stable laws; normal $\alpha = 2$, Holtsmark $\alpha = \frac{3}{2}$, Cauchy $\alpha = 1$ and symmetric-Lévy $\alpha = \frac{1}{2}$ with fixed scaling parameter $\gamma = 1$ chosen for illustrative purposes (see (36)–(39)). (**b**) Comparison of end tails for different α, defined by (33).

4.2. Equivalence of Trap Counts: Cauchy Walk vs. Diffusion

Standard diffusion provides an oversimplified description [76], partly due to an underlying assumption that all individuals are identical in terms of their movement capabilities. In reality, this is not entirely true, and it is known that individuals in the field do not possess an equal ability for self-motion. Even in the case of a population of an identical insect species type, distinct traits can significantly affect movement abilities, such as body mass, length of wings or more generally shape and size [77]. To overcome this assumption, modified diffusion models have been successfully introduced. For example, [48] take into account that the diffusion coefficient can vary according to some type of diffusivity distribution function, rather than being constant. As a result, it is shown that by introducing the concept of a statistically-structured population, the fat tails that are inherent in Lévy walks can appear due to the fact that individuals of the same species are non-identical. Therefore, the mechanism of fat tails' formation in a real population is always present, even if it can sometimes be induced by a mixture of other processes. This is not the only approach that attempts to explain the phenomena of fat tails appearing. We propose an alternative model (see later Section 4.3), where diffusivity varies continuously with time. On an individual level, the interpretation is such that distinct diffusive rates are adopted, and therefore, the model takes into account individual variation. However, at the population level, when rates are aggregated, the movement is governed by time-dependent diffusion.

In Section 3, it was demonstrated that equivalent trap counts can be obtained for Brownian movement using an anomalous diffusion model (30). Here, we test whether this same model can explain trap counts from non-Brownian movement, with particular interest in the level of discrepancy.

Figure 3, Plot (a), compares trap counts J_t from the Cauchy walk [3] against the diffusive flux $J(t)$ for exponents; standard diffusion $H = \frac{1}{2}$, super-diffusion $H = \frac{3}{4}$ and ballistic diffusion $H = 1$. A non-linear curve fitting tool is used to determine the best-fit parameters D_0, D_1 by fitting the analytic solution (30) against trap count recordings in the least squares sense (see Table B1 for the complete list of trap count recordings). Plot (b) illustrates the discrepancy between the trap counts and diffusion model. The relative error is shown (instead of the absolute relative error used previously in Figure 1) to distinguish between the time intervals when trap counts are either over- or under-estimated; here, a positive relative error corresponds to the diffusive flux forming an overestimation, and vice versa.

---

[3]   This specific type of random walk is of significant interest in foraging theory since an inverse square power-law distribution of flight lengths provides an optimal strategy to detect target sites provided that the sites are sparse and can be revisited [61]. Furthermore, see Section 2.2.

Table 2 quantifies the fit and compares whether the diffusion model can reproduce trap counts as effectively as what was previously seen in the Brownian case (see Figure 1).

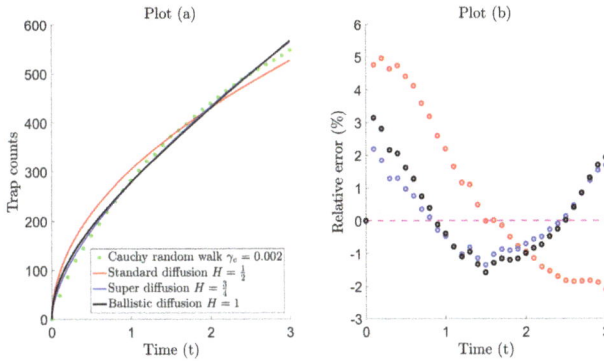

Figure 3. (a) Comparison of trap counts J_t for the Cauchy walk $S(\alpha = 1, \gamma_c = 0.002)$ (see (38)) against the diffusive flux $J(t)$ for anomalous diffusion for the three cases; standard diffusion $H = 0.5, \{D_0, D_1\} = \{0.1431, 1.6730\}$, super-diffusion $H = 0.75, \{D_0, D_1\} = \{0.7263, 1.1776\}$ and ballistic diffusion $H = 1, \{D_0, D_1\} = \{1.228, 0.5844\}$ (see (30)). Trap counts were averaged over five realizations to reduce the effect of stochasticity. Total number of individuals $N = 1000$ uniformly distributed over a finite domain $L = 5$ with population density $U_0 = 200$. Each individual executes a total of $S = 3000$ steps with constant time step increment $\Delta t = 0.001$ and total time $T = S\Delta t = 3$. (b) Relative error (%) measures the discrepancy between trap counts for the random walk and diffusion model defined by $\frac{J(t)-J_t}{N}$ plotted at times $t = 0, 0.1, 0.2, \cdots, 3$.

Table 2. Tabulated values of the average absolute relative error $\bar{A}$ as defined by (32), to compare the fit between the anomalous diffusion model (30) and trap counts obtained from Brownian motion (see Figure 1) and the Cauchy walk (see Figure 3).

	Standard Diffusion $H = \frac{1}{2}$	Super Diffusion $H = \frac{3}{4}$	Ballistic Diffusion $H = 1$
Brownian trap counts	0.306	0.157	0.167
Cauchy trap counts	2.035	0.849	1.120

From Figure 3, Plot (b), it is clear that standard diffusion fails to predict trap counts adequately, with a maximum relative error of about 5%. This is expected since standard diffusion corresponds to the random walk model with a normal step distribution whose end tails decay exponentially, unlike the Cauchy distribution. On comparison, Table 2 shows that both super diffusion and ballistic diffusion provide a good/moderate fitting, respectively. Petrovskii et al. [26] demonstrated that in the case of some simple diffusion models, such as linear dependency on time $D(t) = at + b$ and its variation $D(t) = at + bt^{\frac{1}{3}}$ (a, b constant), both of which constitute ballistic motion, provide a reasonable fit to trap count data. Figure 3 also confirms a reasonable fit for the ballistic case where the relative error lies within approximately 3%; however, still, an evident discrepancy is noticed. Trap counts are much better reproduced by super-diffusion, in this case $H = \frac{3}{4}$ with relative error within approximately 2%. This is somewhat expected, since it is well known that generally, super diffusion has long been acquainted with Lévy walks [43,78]. Despite this promising accuracy, note that the diffusion flux tends to alternate; in the sense that trap counts are overestimated for small time, underestimated for intermediate time and overestimated again on a larger time scale; for both the super diffusive and ballistic case. This phenomenon is also realized for other Hurst exponents in the super-diffusive and ballistic regime (simulations not shown here), even when a variety of scale parameters is considered.

Although the accuracy of the matching is somewhat ecologically acceptable in either case, a diffusion coefficient that yields trap counts at a higher level of precision is sought.

In a recent study, Ahmed and Petrovskii [27] demonstrated that the time-dependency of the diffusion coefficient could be inherently more complex than that proposed by anomalous diffusion. Subsequently, a time-dependent diffusion model was developed to provide an alternative framework for a Cauchy walk with the step distribution (38). In particular, passive trap counts were reproduced effectively, and the study indicated that in the case of a Cauchy walk, the problem of trap count interpretation can be addressed with a high precision based on the diffusion equation. However, some drawbacks with this model [4] include, firstly, that the complicated structure of the diffusion coefficient leads to practicality issues, since it is not expressed in closed form. Secondly, the model consists of multiple unknown parameters with little room for interpretation, i.e., the ecological significance of parameters or how they relate to the movement pattern is unclear. Finally, the study is constrained to Cauchy walks, and it is not known whether the diffusion model is effective at predicting trap counts for a broader range of tail indices α. With this background, we attempt to address the following: Can trap counts obtained from a system of genuine Lévy walkers be accurately reproduced using the diffusion equation, in particular, with a greater accuracy than what is observed for anomalous diffusion in Figure 3? If so, what is the structure of the diffusion coefficient, and how can the behavior of the resulting diffusion profiles be explained from an ecological viewpoint in relation to any parameters?

4.3. Proposed Diffusion Coefficient

Observations of trap count patterns (such as those typically observed in Figure 3) suggest that the coefficient proposed should consist of some type of growth function $G(t)$, which should behave as a controlling mechanism for diffusivity on a short and/or intermediate time scale. In addition to this, a suitable decay function should also be introduced to induce a dampening effect to ensure that trap counts are not overestimated for larger times, typically observed when the movement process grows faster than standard diffusion. Intuitively, we propose the following structure:

$$D(t) = \underbrace{D_0}_{\substack{\text{Initial}\\\text{diffusivity}}} + \underbrace{G(t)}_{\substack{\text{Growth}\\\text{function}}} \cdot \underbrace{e^{-\nu t}}_{\substack{\text{Exponential}\\\text{decay}}} \tag{40}$$

with $G(0) = 0$, i.e., growth is zero at $t = 0$ so that initial diffusivity is defined as $D(0) = D_0$, with obvious meaning. Here, the growth function is subject to exponential decay causing the diffusivity to be damped over larger times with damping coefficient ν. Subsequently, the diffusivity returns to an initial state D_0 in the large time limit as $t \to \infty$, provided the growth function is not faster than exponential growth, that is $\lim_{t\to\infty} G(t)e^{-\nu t} = 0$. The corresponding diffusive flux for an initial uniform population density U_0 across a semi-infinite domain $x > 0$ with zero density condition at $x = 0$ (described in Section 3) can be derived using (20), which reads:

$$J(t) = 2U_0 \sqrt{\frac{D_0 t}{\pi} + \frac{1}{\pi} \int_0^t G(s)e^{-\nu s}ds}. \tag{41}$$

A number of possible candidates for the growth function exist in the literature, but are often applied to model population dynamics. Examples of such include logistic, Gompertz, von Bertalanffy and generalized or hyper-logistic growth [79]. The simplest of these is the logistic type, and an example of an application is the Rosenzweig and MacArthur [80] model for predator-prey interactions with

---

[4]   See Ahmed and Petrovskii [27] for a detailed description of the model previously proposed.

logistically growing prey. We propose that the growth function $G(t)$ in (40) grows logistically, as a means to model diffusivity, rather than the typical use of modeling populations, so that:

$$G(t) = D_1 t \left(1 - \frac{t}{k}\right) \tag{42}$$

with corresponding diffusion coefficient:

$$D(t) = D_0 + D_1 t \left(1 - \frac{t}{k}\right) e^{-vt}. \tag{43}$$

The movement dynamics are then completely governed by set parameters $\{D_0, D_1, k, v\}$, all positive. This growth function is a parabolic function of time, which increases from $G = 0$ at time $t = 0$, until a maximum $G_{max} = \frac{1}{4} D_1 k$ is attained at time $t = \frac{k}{2}$, and then decreases until the growth diminishes, i.e., $G = 0$ at time $t = k$. The corresponding diffusion coefficient is still valid for negative growth over the interval $k < t < t^*$ provided $D(t) > 0$, where t^* is a solution of $\ln D_1(t^* - k) - vt^* = \ln D_0 k$. For fixed D_1, the value of k controls the maximum growth capacity and also determines the instant in time when growth alternates from positive to negative. In the special case, for sufficiently small time with large k, the term $\frac{t}{k}$ in (43) is negligible, and the growth function is approximately linear $G(t) \approx D_1 t$, and in the limiting case as $k \to \infty$,

$$G(t) = D_1 t \tag{44}$$

with corresponding diffusion coefficient,

$$D(t) = D_0 + D_1 t e^{-vt} \tag{45}$$

which now depends on three parameters $\{D_0, D_1, v\}$. The corresponding diffusive flux can be derived for the logistic model (42) from (41), which results in:

Model 1: $\quad J(t) = \dfrac{2U_0}{\sqrt{\pi}} \left(D_0 t + \dfrac{D_1}{v^2}\left[1 - (1 + vt)e^{-vt}\right] + \dfrac{D_1}{kv^3}\left[(v^2 t^2 + 2vt + 2)e^{-vt} - 2\right]\right)^{\frac{1}{2}} \tag{46}$

and simplifies to:

Model 2: $\quad J(t) = \dfrac{2U_0}{\sqrt{\pi}} \left(D_0 t + \dfrac{D_1}{v^2}\left[1 - (1 + vt)e^{-vt}\right]\right)^{\frac{1}{2}} \tag{47}$

in the reduced linear case (44). Henceforth, we will refer to this as Models 1 and 2, respectively.

Figure 4, Plot (a), illustrates the logistic growth function as a parabolic profile for different values of k, with linear growth as $k \to \infty$. Figure 4, Plot (b), shows how the diffusion coefficient behaves for particular parameter values. In ecological terms, the mechanistic process is such that insect diffusivity increases from the initial value D_0 until maximum diffusivity D_{max} is attained at time $t = \frac{k}{2} + \frac{1}{v} - \frac{k}{2}\sqrt{1 + \frac{4}{(kv)^2}}$. We presume that this increase in diffusion rate can induce faster movement, which can be comparable (at some level) to the pattern inherent in Lévy walks. Following this, the diffusivity begins to decrease until it reaches a minimum level D_{min} at time $t = \frac{k}{2} + \frac{1}{v} + \frac{k}{2}\sqrt{1 + \frac{4}{(kv)^2}}$ and then asymptotically approaches the initial state D_0 in the large time limit. In the special case of the linear growth function, insect diffusivity reaches $D_{max} = D_0 + \frac{D_1}{ev}$ at time $t = \frac{1}{v}$ with no subsequent local minimum and the same asymptotic behavior as $t \to \infty$. Figure 4, Plot (c), illustrates the flux for each corresponding diffusion coefficient in Plot (b), and it is precisely these diffusion Models 1 and 2 ((46)–(47)) that will be tested in Section 4.4 against Lévy trap count data.

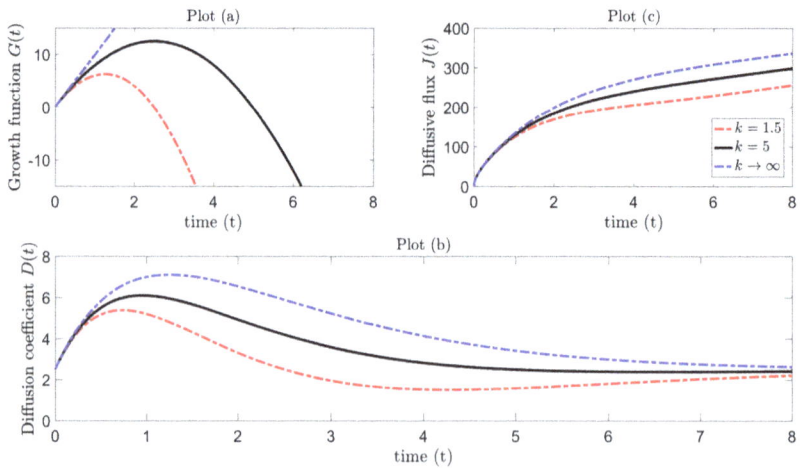

Figure 4. (a) Growth function: Logistic $G(t) = D_1 t \left(1 - \frac{t}{k}\right)$, which reduces to linear growth $G(t) = D_1 t$ as $k \to \infty$. (b) Diffusion coefficient: $D(t) = D_0 + D_1 t \left(1 - \frac{t}{k}\right) e^{-vt}$, which reduces to $D(t) = D_0 + D_1 t e^{-vt}$ as $k \to \infty$. Parameter values (chosen for illustrative purposes): $D_0 = 2.5$, $D_1 = 10$, $v = 0.8$ with different values $k = 1.5, 5$, including the limiting case $k \to \infty$. (c) Diffusive flux given by ((46)–(47)).

4.4. Reproducing Lévy Trap Counts Using Diffusion

In this section, we test Models 1 (46) and 2 (47) against Lévy trap count data (see Figure 5 and Table B1). The simulation setting, alongside the initial and boundary conditions, is precisely that which is outlined in Section 2.1, with the difference that the steps ζ are now randomly drawn from those stable laws, defined in ((37)–(39)). Furthermore, the assumptions that the walk is uncorrelated and unbiased in a homogeneous environment still apply. In a system of N individuals executing a Lévy walk, the position of the n-th individual at the $(i + 1)$-th step can be described by:

$$X_{i+1}^{(n)} = X_i^{(n)} + \zeta, \quad i = 0, 1, \cdots, S, \ n = 1, 2, \cdots, N, \quad \zeta \sim S(\alpha, \gamma), \quad \alpha \in (0, 2). \tag{48}$$

The trap count is expected to grow faster with time, compared to what is usually recorded for Brownian movement. This is due to the frequency of long jumps increasing, and therefore, the contribution from remote parts of the population to the trap count also increases. For our purposes, we simulate trap counts for the tail indices $\alpha = \frac{3}{2}, 1, \frac{1}{2}$, referring to the Holtsmark (37), Cauchy (38) and symmetric-Lévy (39) distributions, previously introduced in Section 4.1. The movement dynamics are completely governed by the scale parameters $\gamma_h, \gamma_c, \gamma_l$. Although, comparing random walks prior to simulation runs can reveal information on parameter selection [81] (also see Section 2.2), for our purposes, it suffices to arbitrarily select three distinct scale parameters for each case, i.e., $\gamma_h = 0.01, 0.02, 0.04$, $\gamma_c = 0.0005, 0.002, 0.003$ and $\gamma_l = 1 \times 10^{-6}, 4 \times 10^{-6}, 2 \times 10^{-5}$. Here, parameters are chosen so that trap count data are obtained with a reasonable level of variation (see Table B1). The diffusive flux curves $J(t)$ given by Model 1 (46) and Model 2 (47) are then fitted (in the least squares sense) against these trap counts using a non-linear curve fitting tool, and the best-fit parameters are estimated, listed in Table 3.

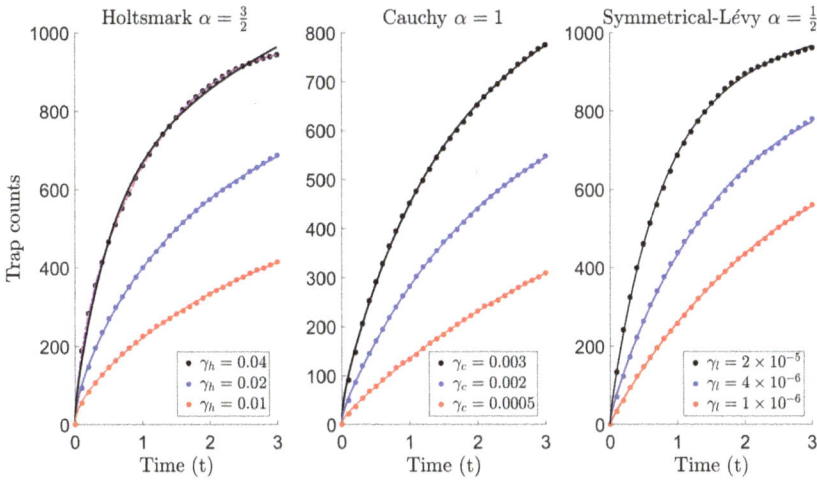

Figure 5. Simulation details: In accordance with the simulation setting in Section 2.1, $N = 1000$ individuals are initially uniformly distributed along a 1D spatial domain $0 < x < L = 5$. After one time step $\Delta t = 0.001$, each individual executes a single step, with the subsequent position defined by the recurrence relation (48). A total number of $S = 3000$ steps is executed, with the total time of exposure $T = S\Delta t = 3$. Prior to the simulation run, an impermeable external boundary is installed at $x = 5$ ensuring that no individual can escape or enter the system at this end (no-migration/immigration), by forcing the condition: if $X_i^{(n)} > 5$ at any instant in time, then $X_i^{(n)} = 5$. The point trap at $x = 0$ functions in the following way: if $X_i^{(n)} < 0$ at any instant in time, then the individual is removed from the system, and the accumulated trap count increases by one. Consequently, the number of individuals in the population decrease as time flows, and an increasing stochastic trap count trajectory is formed. Trap counts: Bold dots depict cumulative trap counts J_t recorded for the cases (a) Holtsmark, (b) Cauchy and (c) symmetric-Lévy at times $t = 0, 0.1, 0.2, \cdots, 3$. Different scale parameters are considered for each respective case. Furthermore, trap counts are averaged over five realizations to reduce the effect of stochasticity (for the full list of recordings, see Table B1). Diffusive flux: Curves $J(t)$ shown for Model 2 in all three cases (red, blue and black curves). Model 1 shown only for the case corresponding to Holtsmark $S(\alpha = 1.5, \gamma_h = 0.04)$ (magenta curve). All best-fit parameters are listed in Table 3.

Table 3. Best-fit parameters using a non-linear curve fitting tool (in the least squares sense) by fitting Model 1 (46) and Model 2 (47) against cumulative trap counts (see Table B1 for the complete list of recordings). The diffusion coefficients in Figure 6 are those plotted with highlighted parameters in the table below.

	D_0	D_1	ν	k	D_0	D_1	ν
$\gamma_h = 0.01$	1.974	3.186	0.615	1286.032	1.9745	3.1834	0.6158
0.02	5.798	16.488	1.102	2846.675	5.7984	16.4881	1.1025
0.04	22.552	26.262	0.631	**1.235**	11.1403	133.0602	2.3539
$\gamma_c = 0.0005$	0.286	2.083	0.327	2598.906	0.2861	2.0828	0.3276
0.002	1.374	10.971	0.687	1918.029	1.3743	10.9709	0.6874
0.003	5.218	27.968	1.036	2589.831	5.2184	27.9676	1.0365
$\gamma_l = 1 \times 10^{-6}$	0.271	11.137	0.541	2608.857	0.2715	11.1369	0.5405
4×10^{-6}	2.099	34.539	0.935	2613.735	2.0992	34.5388	0.9345
2×10^{-5}	4.671	152.605	1.847	5202.465	4.6706	152.6045	1.8471

Figure 5 illustrates the fitting between the diffusive flux $J(t)$ and Lévy trap count data J_t, for the Holtsmark $\alpha = \frac{3}{2}$, Cauchy $\alpha = 1$ and symmetric-Lévy $\alpha = \frac{1}{2}$ distributions, respectively. Model 2 is shown and found to form an almost identical fit, with the exception of the Holtsmark case with $\gamma_h = 0.04$. In this special case, we find that Model 2 eventually overestimates trap counts, which is more apparent for larger γ_h. Here, we find that Model 1 forms a better fit, and this can also be realized upon inspection of the best-fit parameters in Table 3. The parameter k is significant (compare the order of magnitude of boxed value ($k = 1.235$) with other k), and therefore, the term $\frac{t}{k}$ behaves as some type of correction term, which slows the diffusivity rate. The corresponding growth function is of the logistic type, with diffusion coefficient $D(t) = D_0 + D_1 t \left(1 - \frac{t}{k}\right) e^{-\nu t}$. In all other cases, k is relatively large, and therefore, on a short time scale, the term $\frac{t}{k}$ in this diffusion coefficient is negligible. As a result, the growth function is then approximately linear, and Model 1 reduces to Model 2 with diffusion coefficient $D(t) = D_0 + D_1 t e^{-\nu t}$.

Figure 6 shows a plot of the diffusion coefficients, corresponding to each case in Figure 5. The diffusive profiles tend to follow a particular pattern. Typically, the diffusivity begins at some initial value D_0, begins to increase until it peaks at D_{max} and then subsequently decays to re-approach the initial value for a larger time (where the latter is not typically essential as the interest is in short time dynamics). In the special case, seen in Plot (a), the rate of decay for Model 2 (black curve) is slower than required, resulting in larger diffusivity and explains the overestimation previously observed in Figure 5, Plot (a), for the Holtsmark case with $\gamma_h = 0.04$. On comparing the diffusion coefficients for both models, we find that trap counts are better estimated using a profile with larger initial diffusivity, a smaller peak and a faster decay (see the dashed curve).

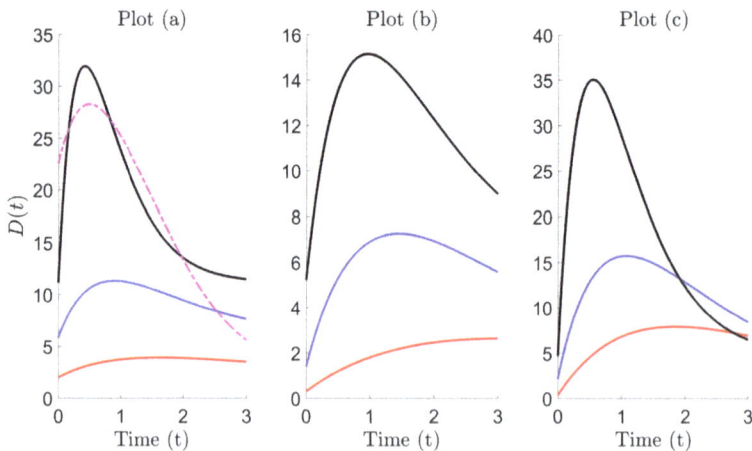

Figure 6. Plots (a), (b), (c) Solid curves (red, blue black) show the diffusion coefficients given by (45) for Model 2. Dashed curve in (a) shows the diffusion coefficient given by (43) for Model 1. Best-fit parameters used are those highlighted in Table 3.

Figure 7 illustrates that a high level of accuracy is maintained, where the error lies roughly within 1% using (i) Model 1 for the Holtsmark case with $\gamma_h = 0.04$ (magenta circles in Plot (a)) and (ii) Model 2 for all other cases. On comparing the absolute relative error in Table 4, we find that the proposed diffusion models significantly improve trap count prediction more than what is obtained from super diffusion. Moreover, the numerical values in Table 4 are indicative of equivalence (compare the boxed values to the others), since these values lie within the interval $0 < \bar{A} \le 0.5$, also previously seen when comparing standard diffusion to Brownian motion, which are theoretically equivalent movement processes (see Table 1). Evidently, these proposed diffusion models can be used effectively to reproduce trap counts for a system of Lévy walking individuals to a remarkable level of accuracy, yielding almost identical counts.

Table 4. Tabulated values of the average absolute relative error $\bar{A}$ as defined by (32), to compare the fit between Models 1 (46) and 2 (47) and trap counts. $\bar{A}$ is also included for the anomalous diffusion model; see Figure 3 and Table 2. Boxed values signify 'equivalence' between the diffusion and Lévy movement models.

Diffusion	$\gamma_h = 0.01$	0.02	0.04	$\gamma_c = 0.0005$	0.002	0.003	$\gamma_l = 10^{-6}$	4×10^{-6}	2×10^{-5}
Standard $H = \frac{1}{2}$					2.035				
Super $H = \frac{3}{4}$					0.849				
Ballistic $H = 1$					1.120				
Model 1 (46)			0.139						
Model 2 (47)	0.162	0.198	0.839	0.123	0.135	0.299	0.194	0.370	0.394

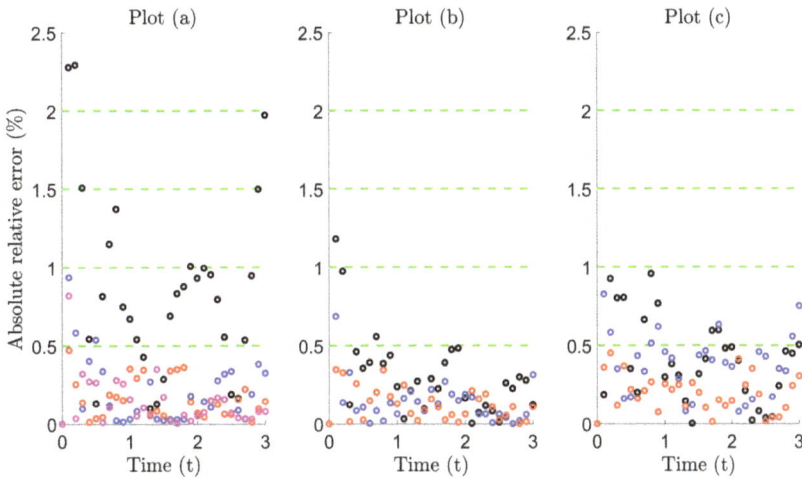

Figure 7. Absolute relative error between trap counts and diffusive flux for the cases (**a**) Holtsmark, (**b**) Cauchy and (**c**) symmetric-Lévy. Each color corresponds to those cases with scale parameters shown in Figure 5.

5. Discussion

The concept of Lévy walks emerging from time-dependent diffusion in the physical or biological sciences in not uncommon. For example, Ott et al. [82] argued that anomalous diffusion of tracer particles in systems of polymer-like breakable micelles ('living polymers') provides an experimental realization of a Lévy walk. A more recent example is that of Chen et al. [83], who showed from experiments that active transport within living cells described by time-dependent Brownian walks can self-organize into (truncated) Lévy walks. Other examples demonstrating this concept can be found elsewhere in the literature, e.g., swarming bacteria [84], pollen dispersal [85], etc. Despite this, from an ecological viewpoint, such as insect trapping, the motivation behind time-dependent diffusion, and how this is linked to the type of mechanisms involved, to date, has not been clearly understood [52].

In this study, the diffusion coefficient introduced (43) consists of parameters of ecological significance with obvious meaning, in the sense that they are not all arbitrary and are related to the underlying movement dynamics. One possible explanation of the type of diffusive pattern seen in Figure 6 can be arrived at through the concept of differential energetics [86–88]. If an individual executes a large step, then this will incur higher energy costs than a small step, so energy expenditure for different step lengths is a given. In relation to basic individual traits, such as body mass, a heavier body requires a larger force and a larger energy expense to change direction or execute a larger step. Therefore, it may be expected that the frequency of moving long distances is lower for heavier individuals. Furthermore, for those individuals that do prefer to take large steps, this could possibly induce larger rest pauses contributing towards intermittent behavior [89], from which the decrease in the rate of diffusivity can be explained. On the other hand, the mechanisms behind the movement behavior could involve a degree of spatial synchronization of individual variation. In the case of movement with a physiological origin, this could result from a diffusivity distribution as indicated by Petrovskii and Morozov [48] or, alternatively, even time-dependent diffusion. Furthermore, with reference to behavioral effects, a sudden event may trigger a change in behavior in one individual that results in swarming behavior, leading to a change at the population level [90]. It must also be realized that the build up of the insect population and following changes in diffusivity can also be a result of transient environmental factors, e.g., temperature [91], as sudden temperature changes can excite movement or even lead to erratic behavior. If one or more of these factors can bring about diffusive movement varying with time, then as a consequence, the pattern produced by an insect population performing Brownian motion may be indistinguishable from the pattern produced by insects performing Lévy walks. Altogether, it seems that identifying a genuine Lévy walk may be more challenging than previously thought. In light of this, it is not surprising that the Lévy or diffusion controversy has been persistent, with strong evidence arguing for either side; in this study, we have demonstrated that both sides are somewhat equivalent, at least in the context of trapping.

On a final note, we would like to mention some limitations and suggest possible further research directions. Although this study is purely theoretical and attempts to answer some important issues in movement ecology, it is limited to the 1D spatial scale. In a more practical scenario, it would be interesting to see a similar study in a more realistic 2D domain, which would be more relevant to field studies with the trapping of walking/crawling insects [12,28]. The problem would then have enhanced complexity due to the introduction of a single trap of different possible shapes and sizes. For example, Reynolds [92] suggests that trap size will become a relevant quantity when the analysis is extended from the 1D case to higher dimensions. The movement pattern would then be altered, as the effects of trap and field boundaries are realized by enforcing a confined or restricted space [60]. In particular, interest would lie in how the time-dependent diffusion model could be further refined in order to produce trap counts at a high level of accuracy within these different geometries. The diffusion model can then be checked and tested for good agreement against insect trap counts in the real field, given that there is evidence of Lévy-type movement beforehand. Developing a corresponding model in 2D would be the next obvious step to take.

Attempting to model more realistic scenarios would further increase the level of model complexity. For example, in the real field, very rarely, a single trap is installed; rather, a multiple trapping system is implemented [93]. In terms of modeling, the geometry of the domain becomes intrinsically complicated, and the effects of perturbations from each trap may become difficult to unravel. In addition, removing the assumption that the random walk is unbiased would result in directed movement. This can be related to baited traps, which are widely used in practice to increase the frequency of captures [94,95]. Application of a certain agent can induce behavioral responses and attract insects to the trap, such as light, odors or pheromone. Since installation of baited traps considerably alters the insect behavior, they are much more difficult to model [96], and the corresponding theory is largely absent. In this case, the convenient mathematical framework would consist of advection-diffusion equations, where possibly, a time-dependent diffusivity may be used.

6. Concluding Remarks

In this paper, we show that the diffusion coefficients (43) and the reduced version (45) can be incorporated into a time-dependent diffusion model, which can then be used to predict, almost identically, trap counts from a system of individuals who undergo a Lévy walk. Moreover, we show that this can be achieved for a broad range of Lévy tail indices. Furthermore, we find that these proposed diffusion models are much more accurate and effective when compared to super diffusion. Alongside the development of the models, we explore the biological basis for time-dependent diffusion in more detail and interpret parameters in relation to diffusive patterns. We argue that, if these inherently different movement models yield almost identical trap counts, then how important is the movement pattern in the context of trapping? This study suggests that the movement pattern is not that important after all, and rather, emphasis should be on the ecological context, at least in integrated pest monitoring programs.

Author Contributions: D.A.A. provided the theoretical results, conducted the simulations and wrote majority of the paper. S.V.P. reviewed the paper and contributed by providing a detailed revision, which substantially improved the manuscript. P.F.C.T. wrote Section 2.2 and provided the details therein.

Acknowledgments: The authors are thankful to Rod Blackshaw and Carly Benefer (Plymouth, U.K.) for fruitful discussions related to the biological interpretation in the discussion, end of Section 5. We also thank Katriona Shea (Penn State University, USA) for her brief comments on the main message of the paper. D.A.A. gratefully acknowledges the support given by Prince Mohammad Bin Fahd University (KSA) through the Phase II grant, which was essential for the completion of this work. P.F.C.T. and S.V.P. gratefully acknowledge support from The Royal Society (U.K.) through Grant No. NF161377. P.F.C.T. was also supported by Sao Paulo Research Foundation (FAPESP–Brazil), Grant No. 2013/07476-0, and partially supported by CAPES, Brazil.

Conflicts of Interest: The authors declare no conflict of interest.

Appendix A. Mean Field Numerical Solution

Consider the 1D diffusion equation for the population density $u(x,t)$ with time-dependent diffusion coefficient $D = D(t)$ over the finite domain $0 < x < L$, with initial uniform density $u(x, t = 0) = U_0$. The boundary conditions include the zero density condition $u(x = 0, t) = 0$ at the trap boundary and no-flux condition $\frac{\partial u(L,t)}{\partial x} = 0$ at the external boundary. To summarize,

$$\frac{\partial u}{\partial t} = D(t)\frac{\partial^2 u}{\partial x^2}, \quad u(x, t = 0) = U_0, \quad u(x = 0, t) = 0, \quad \frac{\partial u(L, t)}{\partial x} = 0 \tag{A1}$$

An analytical solution can only be derived for the system (A1) over the semi-infinite domain $L = \infty$, previously demonstrated in Section 3. In the case of finite L, a numerical solution can be sought using the method of explicit finite differences [97,98] by introducing a uniform computational grid. We discretize the spatial and temporal scales,

$$t_{k+1} = k\Delta t, \quad k = 0, 1, 2, \cdots \tag{A2}$$

$$x_1 = 0, \quad x_{n+1} = x_n + \Delta x, \quad n = 1, 2, \cdots, N \tag{A3}$$

with constant time Δt and spatial Δx increments. For brevity, using the notation:

$$u(x_n, t_k) = u_n^k, \; u(x_n, t_k + \Delta t) = u_n^{k+1} \quad \text{and} \quad u(x_n + \Delta x, t_k) = u_{n+1}^k$$

we can approximate the partial derivatives as:

$$\frac{\partial u}{\partial t} \approx \frac{u_n^{k+1} - u_n^k}{\Delta t}, \quad \frac{\partial^2 u}{\partial x^2} \approx \frac{u_{n+1}^k - 2u_n^k + u_{n-1}^k}{(\Delta x)^2}. \tag{A4}$$

Here, we have used a forward difference for $\frac{\partial u}{\partial t}$ and a central difference for $\frac{\partial^2 u}{\partial x^2}$. The numerical scheme is said to be explicit since the solution at the $(k+1)^{st}$ time step, namely u_n^{k+1}, is given explicitly in terms of the values u_n^k from the previous time layer t_k. Discretizing the diffusion equation in (A1) and rearranging, we obtain the recurrence relation:

$$u_n^{k+1} = u_n^k + \frac{D(k\Delta t)\Delta t}{(\Delta x)^2}(u_{n+1}^k - 2u_n^k + u_{n-1}^k). \tag{A5}$$

The theoretical approximation related to this numerical scheme is $O(\Delta x^2 + \Delta t)$, and in the case of short time dynamics of trap counts, we are interested in the solution for small time t, where we can assume that the approximation error $O(\Delta t)$ is negligible. The reader is redirected to [99] for a discussion on local/global truncation and round-off errors. The spatial-temporal increments must also satisfy the Courant–Friedrichs–Lewy condition:

$$\frac{D(k\Delta t)\Delta t}{(\Delta x)^2} < \frac{1}{2}. \tag{A6}$$

to ensure stability. Other numerical schemes such as the method of implicit finite differences relax this condition; however, since our interest lies in calculating the flux through the trap boundary for small times, the explicit scheme suffices with simpler computation. The initial condition can be written as $u_n^0 = u(x_n, t_0) = U_0$, and the discretization of the trap boundary reads $u_1^k = 0$ with the no-flux condition $u_{N+1}^k = u_N^k$ at the external boundary. The flux $j(t)$ through the trap boundary at time t is given by $j(t) = -D\frac{\partial u(x=0,t)}{\partial x}$. To compute this, we approximate the derivative $\frac{\partial u}{\partial x} \approx \frac{u_{n+1}^k - u_n^k}{\Delta x}$, and at the trap location $x = 0$ corresponding to grid node $n = 1$, this reduces to $\frac{\partial u(x=0,t)}{\partial x} \approx \frac{u_2^k - u_1^k}{\Delta x}$. Therefore, the flux through the boundary at time t_k is given by:

$$j(t_k) = D(k\Delta t)\frac{|u_2^k - u_1^k|}{\Delta x} \tag{A7}$$

Here, we take the absolute value instead of omitting the '−' sign, which would be required since the flux is in the opposite direction of the positive x-axis. A linear approximation is used in (A7); alternatively, a more accurate way to compute the flux uses a quadratic polynomial [56], which yields an error in line with the numerical scheme, i.e., $O(\Delta x^2)$. For our simulations, the linear approximation suffices, and we choose the spatial step Δx to be sufficiently small to ensure that any accumulated errors are negligible. The cumulative flux passing through the trap boundary between times t_k and t_{k+1} can be computed as:

$$J^{k,k+1} = j(t_k)\Delta t = \frac{\Delta t D(k\Delta t)|u_2^k - u_1^k|}{\Delta x} \tag{A8}$$

The cumulative flux J^{k+1} from time $t > 0$ is then computed by summing:

$$J^{k+1} = J^k + \frac{\Delta t D(k\Delta t)|u_2^k - u_1^k|}{\Delta x} \tag{A9}$$

It is precisely this flux through the trap boundary that is used to model the cumulative trap counts in the diffusion model.

Appendix B. Trap Count Recordings

Table B1. Trap count recordings for the cases (a) Holtsmark $S(\alpha = \frac{3}{2}, \gamma_h)$, (b) Cauchy $S(\alpha = 1, \gamma_c)$ and (c) symmetric-Lévy $S(\alpha = \frac{1}{2}, \gamma_l)$, with tail exponent α and scale parameter γ. Simulation details with all parameter values are given in the caption of Figure 5.

Time (t)	$\gamma_h = 0.01$	0.02	0.04	$\gamma_c = 0.0005$	0.002	0.003	$\gamma_l = 1 \times 10^{-6}$	4×10^{-6}	2×10^{-5}
0.1	54	93.8	188	21.6	48.6	91	32.6	69.4	134.2
0.2	83.2	147.2	283.4	36.4	86.4	147.8	60.6	123.4	241.6
0.3	106.6	195.6	355.2	53.2	120.2	206.2	94	173	324.8
0.4	127.6	235.6	413.8	68.4	145	252.6	123.2	222.2	399.4
0.5	146.2	270.6	466	78.4	171	291.2	147.4	263.6	460.8
0.6	163.2	299.6	510.4	91.6	194.4	328.4	171.6	305	513
0.7	180.6	326.2	551.8	103.8	218.6	364.4	196	340.4	560.
0.8	195.6	351.6	588.8	116.6	240.2	394.8	219.6	376.4	603.
0.9	209.8	376.8	629.6	126	262.6	425.4	238.4	409.6	645.8
1.0	225.6	400.8	661	133.8	282.2	451.6	258	438.2	686.4
1.1	238.2	421.8	689.6	143.2	302.6	476	279	466.2	717.4
1.2	251.4	442.4	715.2	155.4	321.6	498.4	298.6	491.6	746.4
1.3	260.8	460.4	740.6	166.4	335.8	521.2	321.4	513	773.2
1.4	271	482	760.4	175.2	355	545	341.4	536.2	797
1.5	282.6	499	783	184.2	372.6	563.8	358.6	555.2	820.2
1.6	290.6	516.2	804	194.2	384.6	584	374.8	575.8	839
1.7	301	531.4	821.2	204.8	398	600.8	390.2	596	856.8
1.8	311	546.4	836.4	214.2	413	617.4	407.8	612.2	871.2
1.9	323	562.8	851.6	222.6	427	633.8	422.8	632	883
2.0	333.2	574	864	231.8	439.6	652.6	435	648.6	894.8
2.1	342.6	586.2	877.2	333.2	574	864	446.6	668.4	904.6
2.2	350.4	599	888.8	241.8	452.2	669	462	681	912.4
2.3	359.2	609.2	898.8	246.4	465.2	683.8	474.2	694.2	919.4
2.4	369.4	620.2	907.6	254.2	476.6	695.2	488.6	704	928.2
2.5	376.2	631	914.8	263	488.2	708.2	502.6	716.4	934.6
2.6	386.2	645.8	921.8	271.6	499.2	721.2	515	730.4	941.6
2.7	394.4	655.4	928.4	280	509	735.2	525.6	747.4	946.2
2.8	399.4	666.8	934.4	288	518.2	745.8	538	756.4	950.2
2.9	407.6	678.2	938.8	294.2	527.2	757	550	768.2	956.2
3.0	415	686.8	943.8	301.2	536.8	766.8	560.8	779.2	961.2

References

1. Burn, A. *Integrated Pest Management*; Academic Press: New York, NY, USA, 1987.
2. Kogan, M. Integrated pest management: Historical perspectives and contemporary developments. *Annu. Rev. Entomol.* **1998**, *43*, 243–270. [CrossRef] [PubMed]
3. Millennium Ecosystem Assessment (MEA). *Ecosystems and Human Well-Being: Biodiversity Synthesis of the Millennium Ecosystem Assessment*; Millennium Ecosystem Assessment World Resources Institute: Washington, DC, USA, 2005.
4. Cardinale, B.; Duffy, J.; Gonzalez, A.; Hooper, D.; Perrings, C.; Venail, P.; Narwani, A.; Mace, G.; Tilman, D.; Wardle, D.; et al. Biodiversity loss and its impact on humanity. *Nature* **2012**, *486*, 59–67. [CrossRef] [PubMed]
5. Ahmed, D.; van Bodegom, P.; Tukker, A. Evaluation and selection of functional diversity metrics with recommendations for their use in life cycle assessments. *Int. J. Life Cycle Assess.* **2018**, doi:10.1007/s11367-018-1470-8. [CrossRef]
6. Pimentel, D.; Greiner, A. Environmental and socio-economic costs of pesticide use. In *Techniques for Reducing Pesticide Use: Economic and Environmental Benefits*; Pimentel, D., Ed.; John Wiley and Sons: New York, NY, USA, 1997.
7. Alavanja, M.; Ross, M.; Bonner, M. Increased cancer burden among pesticide applicators and others due to pesticide exposure. *CA Cancer J. Clin.* **2013**, *62*, 120–142. [CrossRef] [PubMed]
8. Bourguet, D.; Guillemaud, T. *Sustainable Agriculture Reviews*; The Hidden and External Costs of Pesticide Use; Springer: Berlin, Germany, 2016; Volume 19, pp. 35–120.

9. Pimentel, D. Amounts of pesticides reaching target pests: Environmental impacts and ethics. *J. Agric. Environ. Ethics* **1995**, *8*, 17–29. [CrossRef]

10. Alyokhin, A.; Baker, M.; Mota-Sanchez, D.; Dively, G.; Grafius, E. Colorado potato beetle resistance to insecticides. *Am. J. Potato Res.* **2008**, *85*, 395–413. [CrossRef]

11. Sohrabi, F.; Shishehbor, P.; Saber, M.; Mosaddegh, M. Lethal and sub-lethal effects of imidacloprid and buprofezin on the sweet potato whitefly parasitoid Eretmocerus mundus (*Hymenoptera*: *Aphelinidae*). *Crop. Prot.* **2013**, *45*, 98–103. [CrossRef]

12. Petrovskii, S.; Ahmed, D.; Blackshaw, R. Estimating Insect Population Density. *Ecol. Complex* **2012**, *10*, 69–82. [CrossRef]

13. Holland, J.M.; Perry, J.N.; Winder, L. The within-field spatial and temporal distribution of arthropods in winter wheat. *Bull. Entomol. Res.* **1999**, *89*, 499–513. [CrossRef]

14. Ferguson, A.W.; Klukowski, Z.; Walczak, B.; Perry, J.N.; Mugglestone, M.A.; Clark, S.J.; Williams, I. The spatio-temporal distribution of adult Ceutorhynchus assimilis in a crop of winter oilseed rape in relation to the distribution of their larvae and that of the parasitoid Trichomalus perfectus. *Entomol. Exp. Appl.* **2000**, *95*, 161–171. [CrossRef]

15. Alexander, C.J.; Holland, J.M.; Winder, L.; Woolley, C.; Perry, J.N. Performance of sampling strategies in the presence of known spatial patterns. *Ann. Appl. Biol.* **2005**, *146*, 361–370. [CrossRef]

16. Byers, J.A.; Anderbrant, O.; Lofqvist, J. Effective attraction radius: A method for comparing species attractants and determining densities of flying insects. *J. Chem. Ecol.* **1989**, *15*, 749–765. [CrossRef] [PubMed]

17. Raworth, D.A.; Choi, M. Determining numbers of active carabid beetles per unit area from pitfall-trap data. *Entomol. Exp. Appl.* **2001**, *98*, 95–108. [CrossRef]

18. Turchin, P. *Quantitative Analysis of Movement: Measuring and Modelling Population Redistribution in Animals and Plants*; Sinauer Associates: Sunderland, MA, USA, 1998.

19. Okubo, A.; Levin, S. *Diffusion and Ecological Problems: Modern Perspectives*; Springer: New York, NY, USA, 2001.

20. Lewis, M.; Maini, P.; Petrovskii, S. *Dispersal, Individual Movement and Spatial Ecology*; Springer: Berlin, Germany, 2013.

21. Levin, S.; Cohen, D.; Hastings, A. Dispersal strategies in patchy environments. *Theor. Popul. Biol.* **1984**, *26*, 165–180. [CrossRef]

22. Reichenbach, T.; Mobilia, M.; Frey, E. Mobility promotes and jeopardizes biodiversity in rock-paper-scissors games. *Nature* **2007**, *448*, 1046–1049. [CrossRef] [PubMed]

23. Hengeveld, R. *Dynamics of Biological Invasions*; Chapman and Hall: London, UK, 1989.

24. Shigesada, N.; Kawasaki, K. *Biological Invasions: Theory and Practice*; Oxford University Press: Oxford, UK, 1997.

25. Petrovskii, S.; Brian, L. *Exactly Solvable Models of Biological Invasion*; Chapman and Hall/CRC: Boca Raton, FL, USA, 2006.

26. Petrovskii, S.; Petrovskaya, N.; Bearup, D. Multiscale approach to pest insect monitoring: Random walks, pattern formation, synchronization and networks. *Phys. Life Rev.* **2014**, *11*, 467–525. [CrossRef] [PubMed]

27. Ahmed, D.; Petrovskii, S. Time Dependent Diffusion as a Mean Field Counterpart of Lévy Type Random Walk. *Math. Model. Nat. Phenom.* **2015**, *10*, 5–26. [CrossRef]

28. Bearup, D.; Benefer, C.; Petrovskii, S.; Blackshaw, R. Revisiting Brownian motion as a description of animal movement: A comparison to experimental movement data. *Methods Ecol. Evol.* **2016**, *7*, 1525–1537. [CrossRef]

29. Skellam, J. Random dispersal in theoretical populations. *Biometrika* **1951**, *38*, 196–218. [CrossRef] [PubMed]

30. Kareiva, P.; Shigesada, N. Analyzing insect movement as a correlated random walk. *Oecologia* **1983**, *56*, 234–238. [CrossRef] [PubMed]

31. Kareiva, P. Local movement in herbivorous insecta: Applying a passive diffusion model to mark-recapture field experiments. *Oecologia* **1983**, *57*, 322–327. [CrossRef] [PubMed]

32. Reynolds, A.; Smith, A.; Menzel, R.; Greggers, U.; Reynolds, D.; Riley, J. Displaced honey bees perform optimal scale-free search flights. *Ecology* **2007**, *88*, 1955–1961. [CrossRef] [PubMed]

33. Hapca, S.; Crawford, J.; Young, I. Anomalous diffusion of heterogeneous populations characterized by normal diffusion at the individual level. *J. R. Soc. Interface* **2009**, *6*, 111–122. [CrossRef] [PubMed]

34. Jansen, V.; Mashanova, A.; Petrovskii, S. Model selection and animal movement: "Comment on Lévy walks evolve through interaction between movement and environmental complexity". *Science* **2012**, *335*, 918. [CrossRef] [PubMed]
35. Ahmed, D. Stochastic and Mean Field Approaches for Trap Count Modelling and Interpretation. Ph.D. Thesis, Leicester University, Leicester, UK, 2015.
36. Petrovskii, S.; Morozov, A.; Li, B. On a possible origin of the fat-tailed dispersal in population dynamics. *Ecol. Complex.* **2008**, *5*, 146–150. [CrossRef]
37. Mashanova, A.; Olive, T.; Jansen, V. Evidence for intermittency and a truncated power law from highly resolved aphid movement data. *J. R. Soc. Interface* **2010**, *7*, 199–208. [CrossRef] [PubMed]
38. Knell, A.; Codling, E. Classifying area-restricted search (ARS) using a partial sum approach. *Theor. Ecol.* **2012**, *5*, 325–329. [CrossRef]
39. Sims, D.; Southall, E.; Humphries, N.; Hays, G.; Bradshaw, C.; Pitchford, J.; James, A.; Ahmed, M.; Brierley, A.; Hindell, M.; et al. Scaling laws of marine predator search behavior. *Nature* **2008**, *451*, 1098–1102. [CrossRef] [PubMed]
40. Viswanathan, G.; Afanasyev, V.; Buldryrev, S.; Havlin, S.; da Luz, M.R.; Stanley, H. *The Physics of Foraging*; Cambridge University Press: Cambridge, UK, 2011.
41. Edwards, A.; Phillips, R.; Watkins, N.; Freeman, M.; Murphy, E.; Afanasyev, V.; Buldyrev, S.; da Luz M.G.; Raposo, E.; Stanley, H.; et al. Revisiting Lévy flight search patterns of wandering albatrosses, bumblebees and deer. *Nature* **2007**, *449*, 1044–1048. [CrossRef] [PubMed]
42. Benhamou, S. How many animals really do the Lévy walk? *Ecology* **2007**, *88*, 1962–1969. [CrossRef] [PubMed]
43. Shlesinger, M.; Zaslavsky, G.; Klafter, J. Strange kinetics. *Nature* **1993**, *363*, 31–37. [CrossRef]
44. Zaburdaev, Z.; Denisov, S.; Klafter, J. Lévy walks. *Rev. Mod. Phys.* **2015**, *87*, 483. [CrossRef]
45. Viswanathan, G.; Afanasyev, V.; Buldryrev, S.E.A. Lévy flight search patterns of wandering albatrosses. *Nature* **1996**, *381*, 413–415. [CrossRef]
46. Reynolds, A. Olfactory search behavior in the wandering albatross is predicted to give rise to Lévy flight movement patterns. *Anim. Behav.* **2012**, *83*, 1225–1229. [CrossRef]
47. Bartumeus, F.; Catalan, J. Optimal search behavior and classic foraging theory. *J. Phys. A* **2009**, *132*, 569–580. [CrossRef]
48. Petrovskii, S.; Morozov, A. Dispersal in a Statistically Structured Population. *Am. Nat.* **2008**, *173*, 278–289. [CrossRef] [PubMed]
49. Petrovskii, S.; Mashanova, A.; Jansen, V. Variation in individual walking behavior creates the impression of a Lévy flight. *Proc. Natl. Acad. Sci. USA* **2011**, *108*, 8704–8707. [CrossRef] [PubMed]
50. Auger-Méthé, M.; Derocher, A.; Plank, M.; Codling, E.; Lewis, M. Differentiating the Lévy walk from a composite correlated random walk. *Methods Ecol. Evol.* **2015**, *6*, 1179–1189. [CrossRef]
51. De Jager, M.; Weissing, F.; Herman, P.; Nolet, B.; vande Koppel, J. Response to Comment on "Lévy walks evolve through interaction between movement and environmental complexity". *Science* **2012**, *335*, 918. [CrossRef]
52. Codling, E. Pest insect movement and dispersal as an example of applied movement ecology. Comment on "Multiscale approach to pest insect monitoring: Random walks, pattern formation, synchronization, and networks" by Petrovskii, Petrovskaya and Bearup. *Phys. Life Rev.* **2014**, *11*, 533–535. [CrossRef] [PubMed]
53. Petrovskii, S.; Petrovskaya, N.; Bearup, D. Multiscale ecology of agroecosystems is an emerging research field that can provide a stronger theoretical background for the integrated pest management. Reply to comments on "Multiscale approach to pest insect monitoring: Random walks, pattern formation, synchronization, and networks". *Phys. Life Rev.* **2014**, *11*, 536–539. [PubMed]
54. Grimm, V.; Railsback, S. *Individual Based Modelling and Ecology*; Princeton University Press: Princeton, NJ, USA, 2005.
55. Petrovskii, S.; Petrovskaya, N. Computational ecology as an emerging science. *Interface Focus* **2012**, *2*, 241–254. [CrossRef] [PubMed]
56. Bearup, D.; Petrovskaya, N.; Petrovskii, S. Some analytical and numerical approaches to understanding trap counts resulting from pest insect immigration. *Math. Biosci.* **2015**, *263*, 143–160. [CrossRef] [PubMed]

57. Pyke, G. Understanding movements of organisms: It's time to abandon the Lévy foraging hypothesis. *Methods Ecol. Evol.* **2015**, *6*, 1–16. [CrossRef]
58. Weiss, G. *Aspects and Applications of the Random Walk*; North Holland Press: Amsterdam, The Netherlands, 1994.
59. Codling, E.; Plank, M.; Benhamou, S. Random walk models in biology. *J. R. Soc. Interface* **2008**, *5*, 813–834. [CrossRef] [PubMed]
60. Bearup, D.; Petrovskii, S. On time scale invariance of random walks in confined space. *J. Theor. Biol.* **2015**, *367*, 230–245. [CrossRef] [PubMed]
61. Viswanathan, G.; Buldyrev, S.; Havlin, S. Optimizing the success of random searches. *Nature* **1999**, *401*, 911–914. [CrossRef] [PubMed]
62. Knighton, J.; Dapkey, T.; Cruz, J. Random walk modeling of adult Leuctra ferruginea (stonefly) dispersal. *Ecol. Inf.* **2014**, *19*, 1–9. [CrossRef]
63. Nathan, R.; Getz, W.; Revilla, E.; Holyoak, M.; Kadmon, R.; Saltz, D.; Smouse, P. A movement ecology paradigm for unifying organismal movement research. *Proc. Natl. Acad. Sci. USA* **2008**, *105*, 19052–19059. [CrossRef] [PubMed]
64. Benhamou, S. Of scales and stationarity in animal movements. *Ecol. Lett.* **2014**, *17*, 261–272. [CrossRef] [PubMed]
65. Murray, J. *Mathematical Biology: 1. An Introduction*, 3rd ed.; Springer: Berlin, Germany, 2002.
66. Crank, J. *The Mathematics of Diffusion*, 2nd ed.; Oxford University Press: Oxford, UK, 1975.
67. Einstein, A. Über die von der molekularkinetischen Theorie der Wärme geforderte Bewegung von in ruhenden Flüssigkeiten suspendierten Teilchen. *Ann. Phys.* **1905**, *17*, 549–560. [CrossRef]
68. Sornette, D. *Critical Phenomena in Natural Sciences*, 2nd ed.; Springer: Berlin, Germay, 2004.
69. Balescu, R. *Equilibrium and Non-Equilibrium Statistical Mechanics*; John Wiley: New York, NY, USA, 1975.
70. Kölzsch, A.; Alzate, A.; Bartumeus, F.; de Jager, M.; Weerman, E.; Hengeveld, G.; Naguib, M.; Nolet, B.; van de Koppel, J. Experimental evidence for inherent Lévy search behavior in foraging animals. *Proc. R. Soc. B* **2015**, *282*. [CrossRef] [PubMed]
71. Nolan, J. *Stable Distributions—Models for Heavy Tailed Data*; Birkhauser: Boston, MA, USA, 2015; In Progress: Chapter 1. Available online: academic2.american.edu/~jpnolan (accessed on 1 December 2017).
72. Chambers, J.; Mallows, C.; Stuck, B. A method for simulating stable random variables. *JASA* **1976**, *71*, 340–344. [CrossRef]
73. Weron, R. On the Chambers-Mallows-Stuck method for simulating skewed stable random variables. *Stat. Probabil. Lett.* **1996**, *28*, 165–171. See also Weron, R. Correction to: On the Chambers-Mallows-Stuck Method for Simulating Skewed Stable Random Variables, Research Report HSC/96/1, Wroc Law University of Technology, 1996. Available online: http://www.im.pwr.wroc.pl/~hugo/Publications.html (accessed on 1 December 2017). [CrossRef]
74. Zolotarev, V. *One-Dimensional Stable Distributions*; American Mathematical Society: Providence, RI, USA, 1986.
75. Samorodnitsky, G.; Taqqu, M. *Stable Non-Gaussian Random Processes*; Chapman and Hall: Boca Raton, FL, USA, 1994.
76. Holmes, E.E. Are diffusion models too simple? A comparison with telegraph models of invasion. *Am. Nat.* **1993**, *142*, 779–795. [CrossRef] [PubMed]
77. Brose, U. Body-mass constraints on foraging behavior determine population and food-web dynamics. *Funct. Ecol.* **2010**, *24*, 28–34. [CrossRef]
78. Klafter, J.; Sokolov, I. Anomalous diffusion spreads its wings. *Phys. World* **2005**, *18*, 29–32. [CrossRef]
79. Malchow, H.; Petrovskii, S.; Venturino, E. *Spatiotemporal Patterns in Ecology and Epidemiology: Theory, Models, and Simulations*; Chapman and Hall/CRC: Boca Raton, FL, USA, 2008.
80. Rosenzweig, M.; MacArthur, R. Graphical representation and stability conditions of predator-prey interaction. *Am. Nat.* **1963**, *97*, 209–223. [CrossRef]
81. Choules, J.; Petrovskii, S. Which Random Walk is Faster? Methods to Compare Different Step Length Distributions in Individual Animal Movement. *Math. Model. Nat. Phenom.* **2017**, *12*, 22–45. [CrossRef]
82. Ott, A.; Bouchaud, J.; Langevin, D.; Urbach, W. Anomalous diffusion in "living polymers": A genuine Lévy flight? *Phys. Rev. Lett.* **1990**, *65*, 2201–2204. [CrossRef] [PubMed]
83. Chen, K.; Wang, B.; Granick, S. Memoryless self-reinforcing directionality in endosomal active transport within living cells. *Nat. Mater.* **2015**, *14*, 589–593. [CrossRef] [PubMed]

84. Ariel, G.; Rabani, A.; Benisty, S.; Partridge, J.; Harshey, R.; Be'er, A. Swarming bacteria migrate by Lévy Walk. *Nat. Commun.* **2015**, *6*, 8396. [CrossRef] [PubMed]

85. Vallaeys, V.; Tyson, R.; Lane, W.; Deleersnijder, E.; Hanert, E. A Lévy-flight diffusion model to predict transgenic pollen dispersal. *J. R. Soc. Interface* **2017**, *14*. [CrossRef] [PubMed]

86. Harold, H. *Energetics of Desert Invertebrates*; Springer: Berlin, Germany, 1996.

87. El Baidouri, F.; Venditti, C.; Humphries, S. Independent evolution of shape and motility allows evolutionary flexibility in Firmicutes bacteria. *Nat. Ecol. Evol.* **2016**, *9*, 0009. [CrossRef] [PubMed]

88. Halsey, L. Terrestrial movement energetics: Current knowledge and its application to the optimising animal. *J. Exp. Biol.* **2016**, *219*, 1424–1431. [CrossRef] [PubMed]

89. Kramer, D.; McLaughlin, R. The Behavioral Ecology of Intermittent Locomotion. *Am. Zool.* **2015**, *41*, 137–153. [CrossRef]

90. Jervis, M. *Insects as Natural Enemies: A Practical Perspective*; Springer: Berlin, Germany, 2005; p. 281.

91. Tyson, R. Pest control: A modeling approach. Comment on "Multiscale approach to pest insect monitoring: Random walks, pattern formation, synchronization, and networks" by S. Petrovskii, N. Petrovskaya, D. Bearup. *Phys. Life Rev.* **2014**, *11*, 526–528. [CrossRef] [PubMed]

92. Reynolds, A. Extending Lévy search theory from one to higher dimensions: Lévy walking favours the blind. *Proc. Math. Phys. Eng. Sci.* **2015**, *471*. [CrossRef] [PubMed]

93. Brennan, K.; Majer, J.; Reygaert, N. Determination of an Optimal Pitfall Trap Size for Sampling Spiders in a Western Australian Jarrah Forest. *J. Insect Conserv.* **1999**, *3*, 297–307. [CrossRef]

94. Elkinton, J.; Lance, D.; Boettner, G.; Khrimian, A.; Leva, N. Evaluation of pheromone-baited traps for winter moth and Bruce spanworm (*Lepidoptera: Geometridae*). *J. Econ. Entomol.* **2011**, *104*, 494–500. [CrossRef] [PubMed]

95. Tuf, I.; Chmelík, V.; Dobroruka, I.; Hábová, L.; Hudcová, P.; Šipoš, J.; Stašiov, S. Hay-bait traps are a useful tool for sampling of soil dwelling millipedes and centipedes. *Zookeys* **2015**, *510*, 197–207. [CrossRef] [PubMed]

96. Yamanaka, T.; Tatsuki, S.; Shimada, M. An individual-based model for sex-pheromone-oriented flight patterns of male moths in a local area. *Ecol. Model.* **2003**, *161*, 35–51. [CrossRef]

97. Morton, K.; Mayers, D. *Numerical Solution of Partial Differential Equations: An Introduction*; Cambridge University Press: Cambridge, UK, 1994.

98. Holmes, M. *Introduction to Numerical Methods in Differential Equations*; Springer: Berlin, Germany, 2006.

99. Strauss, W. *Partial Differential Equations: An Introduction*; John Wiley and Sons: Hoboken, NJ, USA, 2008.

Σ **mathematics**

MDPI

Article

Effects of Viral and Cytokine Delays on Dynamics of Autoimmunity

Farzad Fatehi, Yuliya N. Kyrychko and Konstantin B. Blyuss *

Department of Mathematics, University of Sussex, Brighton BN1 9QH, UK; F.Fatehi@sussex.ac.uk (F.F.);
Y.Kyrychko@sussex.ac.uk (Y.N.K.)
*   Correspondence: k.blyuss@sussex.ac.uk; Tel.: +44-1273-872878

Received: 30 March 2018; Accepted: 24 April 2018; Published: 28 April 2018

Abstract: A major contribution to the onset and development of autoimmune disease is known to come from infections. An important practical problem is identifying the precise mechanism by which the breakdown of immune tolerance as a result of immune response to infection leads to autoimmunity. In this paper, we develop a mathematical model of immune response to a viral infection, which includes T cells with different activation thresholds, regulatory T cells (Tregs), and a cytokine mediating immune dynamics. Particular emphasis is made on the role of time delays associated with the processes of infection and mounting the immune response. Stability analysis of various steady states of the model allows us to identify parameter regions associated with different types of immune behaviour, such as, normal clearance of infection, chronic infection, and autoimmune dynamics. Numerical simulations are used to illustrate different dynamical regimes, and to identify basins of attraction of different dynamical states. An important result of the analysis is that not only the parameters of the system, but also the initial level of infection and the initial state of the immune system determine the progress and outcome of the dynamics.

Keywords: mathematical model; immune response; autoimmunity; time delays

1. Introduction

An immune system can only be viewed as effective when it can robustly identify and destroy pathogen-infected cells, while also distinguishing such cells from healthy cells. In the case of breakdown of immune tolerance, the immune system fails to discriminate between self-antigens and foreign antigens, which results in autoimmune disease, i.e., undesired destruction of healthy cells. Under normal conditions, once foreign epitopes are presented on antigen presenting cells (APCs) to T cells, this results in the proliferation of T cells and eliciting their effector function. While this mechanism is responsible for a successful clearance of infection, cross-reactivity between epitopes associated with foreign and self-antigens can lead to a T cell response against healthy host cells [1,2].

For many autoimmune diseases, the disease occurs in a specific organ or part of the body, such as retina in uveitis, central nervous system in multiple sclerosis, or pancreatic β-cells in type-1 diabetes [3–5]. It is extremely difficult to identify the specific causes of autoimmunity in individual patients, as it usually has contributions from a number of internal and external factors, including a genetic predisposition, age, previous immune challenges, exposure to pathogens etc., [6–9]. Even though genetic predisposition is known to play a very significant role, it is believed that some additional environmental triggers are required for the onset of autoimmunity, and these are usually represented by infections [10,11]. A very recent work has experimentally identified a gut bacterium that, when present in mice and humans, can migrate to other parts of the body, facilitating subsequent triggering of autoimmune disease in those organs [12]. Various mechanisms of onset of pathogen-induced autoimmune disease have been identified, including bystander activation [13] and molecular mimicry [14,15], which is particularly important in the context of autoimmunity caused by viral infections.

A number of mathematical models have looked into dynamics of onset and development of autoimmune disease. Segel et al. [16] analysed interactions between effector and regulatory T cells in the context of T cell vaccination, without explicitly specifying possible causes of autoimmunity. Similar models were later studied by Borghans et al. [17,18] who demonstrated possible onset of autoimmune state, defined as stable above-threshold oscillations in the number of autoreactive cells, as a result of interactions between regulatory and autoreactive T cells. León et al. [19–21] and Carneiro et al. [22] have studied interactions between different T cells, with an emphasis on the suppressing role of regulatory T cells in the dynamics of immune response and control of autoimmunity. Alexander and Wahl [23] have also looked into the role of regulatory T cells, in particular focusing on their interactions with professional APCs and effector cells for the purpose of controlling immune response. Iwami et al. [24,25] explicitly included a separate compartment representing the viral population in their models of immune response, and showed that the functional form of the growth function for susceptible host cells can have a significant effect on the resulting immune dynamics. Despite being able to explain the emergence of autoimmunity as a by-product of immune response to infection, these models were not able to exhibit another practically important dynamical regime of normal viral clearance. For the case of pathogen-induced autoimmunity arising through bystander activation, Burroughs et al. [26–28] have developed a model that investigates interactions between T cells and interleukin-2 (IL-2), an important cytokine, in mediating the onset of autoimmunity.

Among various parts of the immune system involved in coordinating an effective immune response, a particularly significant role is known to be played by the T cells, with experimental evidence suggesting that regulatory T cells are vitally important for controlling autoimmunity [29–32]. To account for this fact in mathematical models, Alexander and Wahl [23] and Burroughs et al. [26,27] have explicitly included a separate compartment for regulatory T cells that are activated by autoantigens and suppress the activity of autoreactive T cells. Another framework for modelling the effects of T cells on autoimmune dynamics is by using the idea that T cells have different or *tunable activation thresholds* (TAT), which result in different immune functionality of the same T cells, and also allow T cells to adjust their response to simulation by autoantigens. This approach was proposed for the analysis of the peripheral and central T cell dynamics [33–35], it has also been used to study differences in activation/response thresholds that are dependent on the activation state of the T cell [36]. Murine and human experiments have confirmed that activation of T cells can indeed change dynamically during their circulation [37–40]. To model this feature, van den Berg and Rand [41], and Scherer et al. [42] developed stochastic models for the tuning of activation thresholds.

Blyuss and Nicholson [43,44] have proposed a mathematical model of autoimmunity resulting from immune response to a viral infection through a mechanism of molecular mimicry. This model explicitly includes the virus population and two types of T cells with different activation thresholds, and it also accounts for a biologically realistic scenario where infection and autoimmune response can occur in different organs of the host. Besides the normal viral clearance and chronic infection, in some parameter regime the model also exhibits an autoimmune state characterised by stable oscillations in the amounts of cell populations. From a clinical perspective, such behaviour is to be expected, as it is associated with relapses and remissions that have been observed in a number of autoimmune diseases, such as autoimmune thyroid disease, MS, and uveitis [45–47]. One deficiency of this model is the fact that the oscillations associated with autoimmune regime can only occur if the amount of free virus and the number of infected cells are also exhibiting oscillations, while in clinical and laboratory settings, autoimmunity usually occurs after the initial infection has been fully cleared. To overcome this limitation, Fatehi et al. [48] have recently developed a more advanced model that also includes regulatory T cells and cytokines, which has allowed the authors to obtain a more realistic representation of immune response and various dynamical regimes. A particularly important practical insight provided by this model is the observation that it is not only the system parameters, but also the initial level of infection and the initial state of the immune system, that determine whether the host will just successfully clear the infection, or will proceed to develop autoimmunity. Approaching the same problem from another perspective, Fatehi et al. [49] have investigated the role of stochasticity in driving the dynamics of immune response and determining which of the immune states is more likely to be attained. The authors have also determined an experimentally important characterisation of autoimmune state, as provided by the dependence of variance in cell populations on various system parameters.

In this paper, we develop and analyse a model of autoimmune dynamics, with particular focus on the role of time delays associated with different aspects of immune response, as well as an inhibiting effect of regulatory T cells on secretion of IL-2. In the next section, we introduce the model and discuss its basic properties. Section 3 contains systematic analysis of all steady states, including conditions for their feasibility and stability. Section 4 contains a bifurcation analysis of the model and demonstrates various types of behaviour that the system exhibits depending on parameters and initial conditions, which includes identification of attraction basins of various states. The paper concludes in Section 5 with the discussion of results.

2. Model Derivation

To understand how interactions between different parts of the immune system and the cytokine can lead to autoimmunity, we consider a model illustrated in a diagram shown in Figure 1. In this model, unlike earlier work of Blyuss and Nicholson [43,44], we consider a situation where a viral infection and possible autoimmunity occur in the same organ of the host. The healthy host cells, whose number is denoted by $A(t)$, in the absence of infection are assumed to grow logistically with linear growth rate r and carrying capacity N, and they acquire infection at rate β from the infected cells $F(t)$. Since experimental evidence suggests that antibodies play a secondary role compared to T cells [50], and autoimmunity can develop even in the absence of B cells [51], we do not include antibody response in the model, but focus solely on the dynamics of T cell populations. Naïve (inactivated) T cells $T_{in}(t)$ are assumed to be in homeostasis [24,25,43], and once they are activated through interaction with infected cells, which occurs at rate α, a proportion p_1 of them will go on to differentiate into additional regulatory T cells, a fraction p_2 will become normal activated T cells $T_{nor}(t)$ able to destroy infected cells at rate μ_F upon recognition of foreign antigen present on their surface. The remaining proportion of $(1 - p_1 - p_2)$ of T cells will become autoreactive T cells $T_{aut}(t)$ that, in light of their lower activation threshold will be eliminating both infected cells and healthy host cells at rate μ_a due to the above-mentioned cross-reactivity between some epitopes in self- and foreign antigens. Regulatory T cells $T_{reg}(t)$ are assumed to be in their own homeostasis [52], and their main contribution to immune dynamics lies in suppressing autoreactive T cells at rate δ_1. To reduce the dimensionality of the model, it is assumed that the process of viral production is occurring very fast compared to other characteristic timescales of the model, thus the viral population can be represented by its quasi-steady-state approximation, i.e., it is taken to be proportional to the number of infected cells, and this eliminates the need for a separate compartment for free virus.

A number of different cytokines mediate immune response to infection, and in the context of T cell dynamics, a particular important role is played by interleukin 2 (IL-2), represented by the variable $I(t)$ in the model, which acts to enhance the proliferation of T cells, which, in turn, secrete further IL-2. One of the actions of regulatory T cells is to suppress the expression of IL-2 [53], which is only produced by the activated T cells, but not by the regulatory T cells [54,55]. To represent this mathematically, we will assume that T_{nor} and T_{aut} produce IL-2 at rates σ_1 and σ_2, and conversely, IL-2 enhances proliferation of T_{reg}, T_{nor} and T_{aut} at rates ρ_1, ρ_2, and ρ_3. We include in the model suppression of IL-2 by regulatory T cells at rate δ_2, in a manner similar to Burroughs et al. [28].

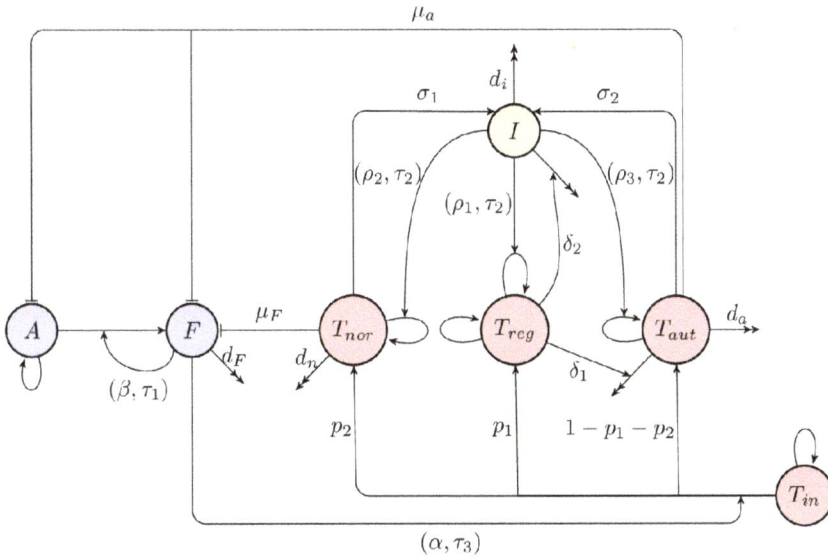

Figure 1. A schematic diagram of immune response to infection. Blue indicates host cells (susceptible and infected), red denotes different T cells (naïve, regulatory, normal activated, and autoreactive T cells), yellow shows cytokines (interleukin 2). τ_i inside each of the subnetworks shows the time delay associated with that process.

While the production of new virus particles by infected cells is assumed to be fast, we explicitly include in the model time delay τ_1 associated with the actual process of infection, which includes multiple stages of the *eclipse phase* of viral life cycle, such as virus attachment, cell penetration and uncoating [56,57]. We also include the time delay τ_2 associated with simulation and proliferation of T cells by IL-2, and the time delay τ_3 between antigen encounter and resulting T cell expansion [58]. With the above assumptions, the complete model takes the form

$$\frac{dA}{dt} = rA\left(1 - \frac{A}{N}\right) - \beta AF - \mu_a T_{aut} A,$$

$$\frac{dF}{dt} = \beta A(t - \tau_1) F(t - \tau_1) - d_F F - \mu_F T_{nor} F - \mu_a T_{aut} F,$$

$$\frac{dT_{in}}{dt} = \lambda_{in} - d_{in} T_{in} - \alpha T_{in} F,$$

$$\frac{dT_{reg}}{dt} = \lambda_r - d_r T_{reg} + p_1 \alpha T_{in}(t - \tau_3) F(t - \tau_3) + \rho_1 I(t - \tau_2) T_{reg}(t - \tau_2),$$

$$\frac{dT_{nor}}{dt} = p_2 \alpha T_{in}(t - \tau_3) F(t - \tau_3) - d_n T_{nor} + \rho_2 I(t - \tau_2) T_{nor}(t - \tau_2),$$

$$\frac{dT_{aut}}{dt} = (1 - p_1 - p_2)\alpha T_{in}(t - \tau_3) F(t - \tau_3) - d_a T_{aut} - \delta_1 T_{reg} T_{aut} + \rho_3 I(t - \tau_2) T_{aut}(t - \tau_2),$$

$$\frac{dI}{dt} = \sigma_1 T_{nor} + \sigma_2 T_{aut} - \delta_2 T_{reg} I - d_i I.$$

Introducing non-dimensional variables

$$\hat{t} = rt, \quad A = N\hat{A}, \quad F = N\hat{F}, \quad T_{in} = \frac{\lambda_{in}}{d_{in}}\hat{T}_{in}, \quad T_{reg} = \frac{\lambda_{in}}{d_{in}}\hat{T}_{reg},$$

$$T_{nor} = \frac{\lambda_{in}}{d_{in}}\hat{T}_{nor}, \quad T_{aut} = \frac{\lambda_{in}}{d_{in}}\hat{T}_{aut}, \quad I = \frac{\lambda_{in}}{d_{in}}\hat{I},$$

where

$$\hat{\beta} = \frac{\beta N}{r}, \quad \hat{\mu}_a = \frac{\mu_a \lambda_{in}}{r d_{in}}, \quad \hat{d}_F = \frac{d_F}{r}, \quad \hat{\mu}_F = \frac{\mu_F \lambda_{in}}{r d_{in}}, \quad \hat{d}_{in} = \frac{d_{in}}{r}, \quad \hat{\alpha} = \frac{\alpha N}{r},$$

$$\hat{\lambda}_r = \frac{\lambda_r d_{in}}{\lambda_{in} r}, \quad \hat{d}_r = \frac{d_r}{r}, \quad \hat{d}_n = \frac{d_n}{r}, \quad \hat{d}_a = \frac{d_a}{r}, \quad \hat{\rho}_i = \frac{\rho_i \lambda_{in}}{r d_{in}}, \quad i = 1, 2, 3,$$

$$\hat{\delta}_1 = \frac{\delta_1 \lambda_{in}}{r d_{in}}, \quad \hat{\delta}_2 = \frac{\delta_2 \lambda_{in}}{r d_{in}}, \quad \hat{\sigma}_1 = \frac{\sigma_1}{r}, \quad \hat{\sigma}_2 = \frac{\sigma_2}{r}, \quad \hat{d}_i = \frac{d_i}{r},$$

yields a rescaled model

$$
\begin{aligned}
\frac{dA}{dt} &= A\,(1 - A) - \beta A F - \mu_a T_{aut} A, \\
\frac{dF}{dT} &= \beta A\,(T - \tau_1)\,F\,(T - \tau_1) - d_F F - \mu_F T_{nor} F - \mu_a T_{aut} F, \\
\frac{dT_{in}}{dT} &= d_{in}\,(1 - T_{in}) - \alpha T_{in} F, \\
\frac{dT_{reg}}{dT} &= \lambda_r - d_r T_{reg} + p_1 \alpha T_{in}\,(T - \tau_3)\,F\,(T - \tau_3) + \rho_1 I\,(T - \tau_2)\,T_{reg}\,(T - \tau_2), \\
\frac{dT_{nor}}{dT} &= p_2 \alpha T_{in}\,(T - \tau_3)\,F\,(T - \tau_3) - d_n T_{nor} + \rho_2 I\,(T - \tau_2)\,T_{nor}\,(T - \tau_2), \\
\frac{dT_{aut}}{dT} &= (1 - p_1 - p_2)\alpha T_{in}\,(T - \tau_3)\,F\,(T - \tau_3) - d_a T_{aut} - \delta_1 T_{reg} T_{aut} + \rho_3 I\,(T - \tau_2)\,T_{aut}\,(T - \tau_2), \\
\frac{dI}{dT} &= \sigma_1 T_{nor} + \sigma_2 T_{aut} - \delta_2 T_{reg} I - d_i I,
\end{aligned}
\tag{1}
$$

where all hats in variables and parameters have been dropped for simplicity of notation, and all parameters are assumed to be positive. It is easy to show that this system is well-posed, i.e., solutions with non-negative initial conditions remain non-negative for all $t \geq 0$.

As a first step in the analysis of model (1), we look at its steady states

$$S^* = \left(A^*, F^*, T_{in}^*, T_{reg}^*, T_{nor}^*, T_{aut}^*, I^* \right),$$

that can be found by equating to zero the right-hand sides of Equation (1) and solving the resulting system of algebraic equations, deferring the discussion of conditionally stable steady states to Section 3. First, we consider a situation where there are no infected cells at a steady state, i.e., $F^* = 0$, which immediately implies $T_{in}^* = 1$. In this case, there are four possible combinations of steady states depending on whether T_{nor}^* and T_{aut}^* are each equal to zero or positive. If $T_{nor}^* = T_{aut}^* = 0$, there are two steady states

$$S_1^* = \left(0, 0, 1, \frac{\lambda_r}{d_r}, 0, 0, 0 \right), \quad S_2^* = \left(1, 0, 1, \frac{\lambda_r}{d_r}, 0, 0, 0 \right),$$

of which S_1^* is always unstable, and S_2^* is a disease-free conditionally stable steady state.

For $T_{nor}^* \neq 0$ and $T_{aut}^* = 0$, we again have two steady states

$$S_3^* = \left(0, 0, 1, \frac{\lambda_r \rho_2}{\rho_2 d_r - \rho_1 d_n}, T_{nor}^*, 0, \frac{d_n}{\rho_2} \right), \quad S_4^* = \left(1, 0, 1, \frac{\lambda_r \rho_2}{\rho_2 d_r - \rho_1 d_n}, T_{nor}^*, 0, \frac{d_n}{\rho_2} \right),$$

where $T_{nor}^* = \dfrac{d_n\,(\lambda_r \delta_2 \rho_2 + d_i d_r \rho_2 - d_i d_n \rho_1)}{\rho_2 \sigma_1 (\rho_2 d_r - \rho_1 d_n)}$, but they are both unstable for any values of parameters.

In the case when $T^*_{nor} = 0$ and $T^*_{aut} \neq 0$, we have two further steady states S^*_5 and S^*_6,

$$S^*_5 = \left(0, 0, 1, T^*_{reg}, 0, \frac{\left(d_i + \delta_2 T^*_{reg} \right) \left(d_a + \delta_1 T^*_{reg} \right)}{\rho_3 \sigma_2}, \frac{d_a + \delta_1 T^*_{reg}}{\rho_3} \right),$$

$$S^*_6 = \left(A^*, 0, 1, T^*_{reg}, 0, \frac{\left(d_i + \delta_2 T^*_{reg} \right) \left(d_a + \delta_1 T^*_{reg} \right)}{\rho_3 \sigma_2}, \frac{d_a + \delta_1 T^*_{reg}}{\rho_3} \right),$$

where $A^* = 1 - \dfrac{\mu_a \left(d_i + \delta_2 T^*_{reg} \right) \left(d_a + \delta_1 T^*_{reg} \right)}{\rho_3 \sigma_2}$, and

$$T^*_{reg} = \frac{d_r \rho_3 - \rho_1 d_a \pm \sqrt{(d_r \rho_3 - \rho_1 d_a)^2 - 4\rho_1 \delta_1 \lambda_r \rho_3}}{2\rho_1 \delta}.$$

The steady state S^*_5 has $A^* = 0$, which implies the death of host cells, whereas the steady state S^*_6 corresponds to an autoimmune regime. The steady state S^*_7 with $T^*_{nor} \neq 0$ and $T^*_{aut} \neq 0$ exists only for a particular combination of parameters, namely, when

$$\delta_1 \rho_2^2 \lambda_r = (\rho_3 d_n - \rho_2 d_a)(\rho_2 d_r - \rho_1 d_n),$$

and is always unstable. Finally, when $F^* \neq 0$, the system (1) can have a steady state S^*_8 with all of its components being positive, but it does not appear possible to find a closed form expression for this state.

In summary, besides the unconditionally unstable steady states, the model (1) has at most for conditionally stable steady states: the disease-free steady state S^*_2, the steady state with the death of host cells S^*_5, the autoimmune steady state S^*_6, and the persistent or chronic steady state S^*_8.

3. Stability Analysis of the Steady States

3.1. Stability Analysis of the Disease-Free Steady State

Linearising the system (1) near the disease-free steady state S^*_2 yields the following equation for characteristic roots λ

$$\lambda + d_F - \beta e^{-\lambda \tau_1} = 0. \tag{2}$$

If $d_F < \beta$, the above equation always has a real positive root for any value $\tau_1 \geq 0$, implying that the disease-free steady state is always unstable for any value of the time delays. If, however, the condition $d_F > \beta$ holds, the disease-free steady state is stable for $\tau_1 = 0$. To find out whether it can lose stability for $\tau_1 > 0$, we look for solutions of Equation (2) in the form $\lambda = i\omega$. Separating real and imaginary parts yields

$$d_F = \beta \cos(\omega \tau_1),$$

$$\omega = -\beta \sin(\omega \tau_1).$$

Squaring and adding these two equations gives the following equation for potential Hopf frequency ω

$$\omega^2 + d_F^2 - \beta^2 = 0.$$

since $d_F > \beta$, this equation does not have real roots for ω, suggesting that there can be no roots of the form $\lambda = i\omega$ of the characteristic Equation (2). This implies that in the case $d_F > \beta$ the disease-free steady state S^*_2 is stable for all values of the time delay $\tau_1 \geq 0$.

3.2. Stability Analysis of the Death, Autoimmune and Chronic Steady States

The steady state S_5^* (respectively, S_6^*) is stable if

$$P < \frac{d_a + \delta_1 T_{reg}^*}{\rho_3} < \frac{d_n}{\rho_2}, \tag{3}$$

and all roots of the following equation have negative real part

$$\Delta(\tau_2, \lambda) = p_2(\lambda)e^{-2\lambda\tau_2} + p_1(\lambda)e^{-\lambda\tau_2} + p_0(\lambda) = 0, \tag{4}$$

where

$$p_2(\lambda) = \frac{\rho_1 \left(d_a + \delta_1 T_{reg}^*\right)^2}{\rho_3} \left(\lambda + 2d_i + \delta_2 T_{reg}^*\right),$$

$$p_1(\lambda) = -\frac{\left(d_a + \delta_1 T_{reg}^*\right)}{\rho_3} \left\{ (\rho_1 + \rho_3)\lambda^2 + \left[\rho_1 \left(d_i + d_a + \delta_1 T_{reg}^*\right) + \rho_3 \left(d_r + 2d_i + 2\delta_2 T_{reg}^*\right)\right] \lambda \right.$$
$$\left. + d_i(\rho_1 d_a + 2d_r\rho_3) + \delta_2 T_{reg}^* \left(-\rho_1\delta_1 T_{reg}^* + 2d_r\rho_3\right) \right\},$$

$$p_0(\lambda) = (\lambda + d_r)\left(\lambda + d_i + \delta_2 T_{reg}^*\right)\left(\lambda + d_a + \delta_1 T_{reg}^*\right),$$

and

$$P = \begin{cases} \dfrac{\sigma_2}{\mu_a \left(d_i + \delta_2 T_{reg}^*\right)}, & \text{for } S_5^*, \\[4mm] \dfrac{\sigma_2 \left(\beta - d_F\right)}{\mu_a \left(1 + \beta\right)\left(d_i + \delta_2 T_{reg}^*\right)}, & \text{for } S_6^*. \end{cases}$$

This steady state undergoes a steady-state bifurcation if

$$\frac{d_a + \delta_1 T_{reg}^*}{\rho_3} = P, \quad \text{or} \quad \frac{d_a + \delta_1 T_{reg}^*}{\rho_3} = \frac{d_n}{\rho_2}, \quad \text{or} \quad \delta_1\rho_1 \left(T_{reg}^*\right)^2 = \lambda_r\rho_3. \tag{5}$$

For $\tau_2 = 0$ these steady states are stable if T_{reg}^* satisfies (3) and

$$\delta_1\rho_1 \left(T_{reg}^*\right)^2 > \lambda_r\rho_3,$$
$$a_5 \left(T_{reg}^*\right)^5 + a_4 \left(T_{reg}^*\right)^4 + a_3 \left(T_{reg}^*\right)^3 + a_2 \left(T_{reg}^*\right)^2 + a_1 T_{reg}^* + a_0 > 0, \tag{6}$$

where

$$a_5 = -\delta_1\delta_2(\delta_1\rho_1 - \delta_2\rho_1 + \delta_2\rho_3), \quad a_4 = d_a\delta_2(\delta_2\rho_2 - \delta_1\rho_1 - \delta_2\rho_3) - d_i\delta_1(\delta_1\rho_1 - \delta_2\rho_1 + 2\delta_2\rho_3),$$
$$a_3 = -d_i\delta_1(d_a\rho_1 + d_i\rho_3) + d_a d_i\delta_2(\rho_1 - 2\rho_3) + \lambda_r\delta_2(\delta_1\rho_1 + \delta_2\rho_3),$$
$$a_2 = -d_a d_i^2\rho_3 + \lambda_r\delta_2(d_a\rho_1 + 2d_i\rho_3), \quad a_1 = \lambda_r\rho_3(d_i^2 + \delta_2\lambda_r), \quad a_0 = d_i\rho_3\lambda_r^2.$$

To investigate whether stability can be lost for $\tau_2 > 0$, we use an iterative procedure described in [59,60] to determine a function $F(\omega)$, whose roots give the Hopf frequency associated with purely imaginary roots of Equation (4). Substituting $\lambda = i\omega$ into Equation (4), we define $\Delta^{(1)}(\tau_2, \lambda)$ as

$$\Delta^{(1)}(\tau_2, \lambda) = \overline{p_0(i\omega)}\Delta(\tau_2, i\omega) - p_2(i\omega)e^{-2i\omega\tau_2}\overline{\Delta(\tau_2, i\omega)} = p_0^{(1)}(i\omega) + p_1^{(1)}(i\omega)e^{-i\omega\tau_2},$$

where

$$p_0^{(1)}(i\omega) = |p_0(i\omega)|^2 - |p_2(i\omega)|^2,$$
$$p_1^{(1)}(i\omega) = \overline{p_0(i\omega)}p_1(i\omega) - \overline{p_1(i\omega)}p_2(i\omega),$$

and the bar denotes the complex conjugate. If we define

$$F(\omega) = \left| p_0^{(1)}(i\omega) \right|^2 - \left| p_1^{(1)}(i\omega) \right|^2,$$

then $\Delta(\tau_2, i\omega) = 0$ whenever ω is a root of $F(\omega) = 0$. The function $F(\omega)$ has the explicit form

$$F(\omega) = \omega^{12} + b_{10}\omega^{10} + b_8\omega^8 + b_6\omega^6 + b_4\omega^4 + b_2\omega^2 + b_0,$$

with

$$b_0 = \frac{\left(\delta_1 d_a + T_{reg}^*\right)^4}{\rho_3{}^4}\left(d_i + \delta_2 T_{reg}^*\right)\left(2T_{reg}^*\delta_1\rho_1 + d_a\rho_1 - d_r\rho_3\right)$$
$$\left[\left(d_i + \delta_2 T_{reg}^*\right)(d_a\rho_1 + 3d_r\rho_3) + 2d_i\rho_1\left(d_a + \delta_1 T_{reg}^*\right)\right]$$
$$\left[\rho_1\left(d_a + \delta_1 T_{reg}^*\right)\left(2d_i + \delta_2 T_{reg}^*\right) - d_r\rho_3\left(d_i + \delta_2 T_{reg}^*\right)\right]^2.$$

The explicit formulae for other coefficients of $F(\omega)$ can be found in Appendix A. Introducing $s = \omega^2$, the equation $F(\omega) = 0$ can be equivalently rewritten as follows,

$$h(s) = s^6 + b_{10}s^5 + b_8s^4 + b_6s^3 + b_4s^2 + b_2s + b_0 = 0. \tag{7}$$

Without loss of generality, suppose that Equation (7) has six distinct positive roots denoted by s_1, s_2, ... , s_6, which means that the equation $F(\omega) = 0$ has six positive roots

$$\omega_i = \sqrt{s_i}, \quad i = 1,2,...,6.$$

Substituting $\lambda_k = i\omega_k$ into Equation (4) gives

$$\tau_{k,j} = \frac{1}{\omega_k}\left[\arctan\left(\frac{\omega_k\left((\rho_1 + \rho_3)\omega_k^4 + f_2\omega_k^2 + f_0\right)}{\left(\rho_3 Z - d_r\rho_1 - \rho_3^2 I^* - \rho_1\delta_2 T_{reg}^*\right)\omega_k^4 + g_2\omega_k^2 + g_0}\right) + j\pi\right],$$

for $k = 1,2,...,6$, $j = 0,1,2,...$, where

$$f_0 = -\rho_1{}^2\rho_3{}^2 I^{*3} Z - \rho_1\rho_3(2\rho_1 + 3\rho_3) I^{*2} Z^2 + \rho_1\rho_3 T_{reg}^*(-\delta_1\rho_1 + 3\delta_2\rho_1 + \delta_2\rho_3) I^{*2} Z$$
$$- T_{reg}^*{}^2\delta_2{}^2\rho_1{}^2\rho_3 I^{*2} - T_{reg}^*\delta_1\rho_1\rho_3 I^* Z^2 + d_r\rho_3\left(-\delta_1\rho_1 T_{reg}^* + d_r\rho_3\right) I^* Z$$
$$+ d_r\left(-T_{reg}^*\delta_1\rho_1 + 2d_r\rho_3\right) Z^2,$$
$$f_2 = -\rho_1{}^2\rho_3 I^{*2} + \rho_3{}^2 I^* Z + (\rho_1 + 2\rho_3) Z^2 + \rho_1 T_{reg}^*(\delta_1 - \delta_2) Z - d_r\left(T_{reg}^*\delta_2\rho_1 - d_r\rho_3\right),$$
$$g_0 = \rho_1{}^2\rho_3{}^2 I^{*3}\left(2Z^2 - 3T_{reg}^*\delta_2 Z + T_{reg}^*{}^2\delta_2{}^2\right) + \rho_1\rho_3\left(-2T_{reg}^*\delta_1\rho_1 + 3d_r\rho_3\right) I^{*2} Z^2$$
$$+ \rho_1\rho_3\delta_2 T_{reg}^*\left(T_{reg}^*\delta_1\rho_1 - d_r\rho_3\right) I^{*2} Z + d_r\rho_3\left(T_{reg}^*\delta_1\rho_1 - 2d_r\rho_3\right) I^* Z^2,$$
$$g_2 = \rho_3 I^*\left(\rho_1{}^2\rho_3 I^{*2} - \rho_1{}^2 I^* Z - 2\rho_3 Z^2 - T_{reg}^*\delta_1\rho_1 Z - d_r{}^2\rho_3\right) - \rho_1\left(d_r + \delta_1 T_{reg}^*\right) Z^2$$
$$+ d_r\left(-T_{reg}^*\delta_1\rho_1 + T_{reg}^*\delta_2\rho_1 + d_r\rho_3\right) Z,$$

and

$$I^* = \frac{d_a + \delta_1 T_{reg}^*}{p_3}, \quad Z = d_i + \delta_2 T_{reg}^*.$$

This allows us to find

$$\tau^* = \tau_{k_0,0} = \min_{1 \le k \le 6} \{\tau_{k,0}\}, \quad \omega_0 = \omega_{k_0},$$

as the first time delay for which the roots of the characteristic Equation (4) cross the imaginary axis. To determine whether these steady states actually undergo a Hopf bifurcation at $\tau_2 = \tau^*$, we have to compute the sign of $d\mathrm{Re}[\lambda(\tau^*)]/d\tau_2$. For $\tau = \tau^*$, $\lambda(\tau^*) = i\omega_0$, and we also define $s_0 = \omega_0^2$.

Lemma 1. *Suppose* $h'(s_0) \ne 0$ *and* $p_0^{(1)}(i\omega_0) \ne 0$. *Then the following transversality condition holds*

$$\mathrm{sgn}\left\{ \left. \frac{d\,\mathrm{Re}(\lambda)}{d\,\tau_2} \right|_{\tau_2=\tau^*} \right\} = \mathrm{sgn}[p_0^{(1)}(i\omega_0)h'(s_0)].$$

Proof. Considering $p_j(i\omega_0) = x_j(\omega_0) + iy_j(\omega_0)$ for $j = 0, 1, 2$, we have

$$p_0^{(1)}(i\omega_0) = x_0^2 + y_0^2 - x_2^2 - y_2^2,$$
$$p_1^{(1)}(i\omega_0) = (x_0 x_1 + y_0 y_1 - x_1 x_2 - y_1 y_2) + (x_0 y_1 + x_2 y_1 - x_1 y_0 - x_1 y_2)i,$$

where all x_j and y_j are expressed in terms of system parameters and steady state values of the variables. Substituting these expressions into $\Delta(\tau_2, i\omega_0) = 0$ and $\Delta^{(1)}(\tau_2, i\omega_0) = 0$, and then separating real and imaginary parts gives

$$\begin{cases} x_2 \cos(2\omega_0\tau^*) + y_2 \sin(2\omega_0\tau^*) + x_1 \cos(\omega_0\tau^*) + y_1 \sin(\omega_0\tau^*) = -x_0, \\ y_2 \cos(2\omega_0\tau^*) - x_2 \sin(2\omega_0\tau^*) + y_1 \cos(\omega_0\tau^*) - x_1 \sin(\omega_0\tau^*) = -y_0, \\ (x_0 x_1 + y_0 y_1 - x_1 x_2 - y_1 y_2)\cos(\omega_0\tau^*) + (x_0 y_1 + x_2 y_1 - x_1 y_0 - x_1 y_2)\sin(\omega_0\tau^*) \\ \qquad = -x_0^2 - y_0^2 + x_2^2 + y_2^2, \\ (x_0 y_1 + x_2 y_1 - x_1 y_0 - x_1 y_2)\cos(\omega_0\tau^*) - (x_0 x_1 + y_0 y_1 - x_1 x_2 - y_1 y_2)\sin(\omega_0\tau^*) = 0. \end{cases}$$

Solving this system of equations provides the values of $\sin(\omega_0\tau^*)$, $\cos(\omega_0\tau^*)$, $\sin(2\omega_0\tau^*)$, and $\cos(2\omega_0\tau^*)$. Taking the derivative of Equation (4) with respect to τ_2, one finds

$$\left(\frac{d\lambda}{d\tau_2}\right)^{-1} = \frac{p_2'(\lambda)e^{-2\lambda\tau_2} + p_1'(\lambda)e^{-\lambda\tau_2} + p_0'(\lambda)}{\lambda\left(2p_2(\lambda)e^{-2\lambda\tau_2} + p_1(\lambda)e^{-\lambda\tau_2}\right)} - \frac{\tau_2}{\lambda}.$$

Hence,

$$\left(\left.\frac{d\,\mathrm{Re}(\lambda)}{d\,\tau_2}\right|_{\tau_2=\tau^*}\right)^{-1} = \mathrm{Re}\left\{\frac{p_2'(\lambda)e^{-2\lambda\tau_2} + p_1'(\lambda)e^{-\lambda\tau_2} + p_0'(\lambda)}{\lambda\left(2p_2(\lambda)e^{-2\lambda\tau_2} + p_1(\lambda)e^{-\lambda\tau_2}\right)}\right\}_{\tau_2=\tau^*} - \mathrm{Re}\left\{\frac{\tau_2}{\lambda}\right\}_{\tau_2=\tau^*}$$

$$= \mathrm{Re}\left\{\frac{p_2'(i\omega_0)e^{-2i\omega_0\tau_2} + p_1'(i\omega_0)e^{-i\omega_0\tau_2} + p_0'(i\omega_0)}{i\omega_0\left(2p_2(i\omega_0)e^{-2i\omega_0\tau_2} + p_1(i\omega_0)e^{-i\omega_0\tau_2}\right)}\right\}$$

$$= \frac{1}{\omega_0}\mathrm{Im}\left\{\frac{p_2'(i\omega_0)e^{-2i\omega_0\tau_2} + p_1'(i\omega_0)e^{-i\omega_0\tau_2} + p_0'(i\omega_0)}{2p_2(i\omega_0)e^{-2i\omega_0\tau_2} + p_1(i\omega_0)e^{-i\omega_0\tau_2}}\right\}$$

$$= \frac{1}{\Lambda\omega_0}\Big[-x_2x_2' - y_2y_2' + x_0x_0' + y_0y_0' + (x_2y_1' - y_2x_1' + x_0y_1' - x_1'y_0)\sin(\omega_0\tau^*)$$

$$\quad + (x_0x_1' + y_0y_1' - x_1'x_2 - y_1'y_2)\cos(\omega_0\tau^*) + (x_2y_0' - x_0'y_2 + x_0y_2' - x_2'y_0)\sin(2\omega_0\tau^*)$$

$$\quad + (x_0x_2' + y_0y_2' - x_0'x_2 - y_0'y_2)\cos(2\omega_0\tau^*)\Big],$$

where

$$\Lambda = \left| 2p_2(i\omega_0)e^{-2i\omega_0\tau_2} + p_1(i\omega_0)e^{-i\omega_0\tau_2} \right|^2.$$

Substituting the values of $\sin(\omega_0\tau^*)$, $\cos(\omega_0\tau^*)$, $\sin(2\omega_0\tau^*)$, and $\cos(2\omega_0\tau^*)$ found earlier gives

$$\left(\frac{d\operatorname{Re}(\lambda)}{d\tau_2} \bigg|_{\tau_2=\tau^*} \right)^{-1} = \frac{1}{\Lambda\omega_0} \frac{F'(\omega_0)}{2\,p_0^{(1)}(i\omega_0)} = \frac{h'(s_0)}{\Lambda\,p_0^{(1)}(i\omega_0)}.$$

Therefore

$$\operatorname{sgn}\left\{ \frac{d\operatorname{Re}(\lambda)}{d\tau_2} \bigg|_{\tau_2=\tau^*} \right\} = \operatorname{sgn}\left\{ \left(\frac{d\operatorname{Re}(\lambda)}{d\tau_2} \bigg|_{\tau_2=\tau^*} \right)^{-1} \right\} = \operatorname{sgn}\left\{ \frac{h'(s_0)}{\Lambda\,p_0^{(1)}(i\omega_0)} \right\}$$

$$= \operatorname{sgn}[p_0^{(1)}(i\omega_0)h'(s_0)],$$

which completes the proof. □

We can now formulate the main result concerning stability of the steady states S_5^* and S_6^*.

Theorem 1. *Suppose the value of T_{reg}^* satisfies conditions (3) and (6). If Equation (7) has at least one positive root s_0, and $p_0^{(1)}(i\omega_0)h'(s_0) > 0$ with $\omega_0 = \sqrt{s_0}$, then the steady state S_5^* (respectively, S_6^*) is stable for $0 \leq \tau_2 < \tau^*$, unstable for $\tau_2 > \tau^*$, and undergoes a Hopf bifurcation at $\tau_2 = \tau^*$.*

Since T_{reg}^* satisfies conditions (3) and (6), the steady state S_5^*/S_6^* is stable for $\tau_2 = 0$. Lemma 1 then ensures that τ^* is the first positive value of the time delay τ_2, for which the roots of the characteristic Equation (4) cross the imaginary axis with positive speed. Hence, the steady state S_5^*/S_6^* is stable for $0 \leq \tau_2 < \tau^*$, unstable for $\tau_2 > \tau^*$, and undergoes a Hopf bifurcation at $\tau_2 = \tau^*$.

Remark 1. *A similar result can be formulated for a subcritical Hopf bifurcation of the steady state S_5^*/S_6^* at some higher value of τ_2.*

The only remaining steady state is the persistent (chronic) equilibrium S_8^* with all of its components being positive. Since it did not prove possible to find a closed form expression for this steady state, its stability also has to be studied numerically.

4. Numerical Stability Analysis and Simulations

To investigate the role of different parameters in the dynamics of model (1), in this section we perform a detailed numerical bifurcation analysis and simulations of this model. Stability of different steady states is determined numerically by computing the largest real part of the characteristic eigenvalues, which is achieved by using a pseudospectral method implemented in a traceDDE suite in MATLAB [61].

Analytical results from the previous section suggest that at $\beta = d_F$, the disease-free steady state S_2^* undergoes a transcritical bifurcation. For $\beta < d_F$, the disease-free steady state S_2^* is stable, and the chronic steady state is infeasible. On the contrary, when $\beta > d_F$, the disease-free steady state S_2^* is unstable, and in this case it makes sense to investigate stability of the chronic steady state. Therefore, these two cases are considered separately, and as a first step we fix the baseline values as given in Table 1. For this choice of parameters, we have $d_F - \beta > 0$, implying that S_2^* is always stable, and Figure 2 illustrates how the stability of S_5^* and S_6^* is affected by parameters. This figure indicates that the steady states S_5^* and S_6^* are only biologically feasible if the regulatory T cells do not grow too rapidly and do not clear autoreactive T cells too quickly. Importantly, Figure 2 shows that the value of the rate δ_2 of clearance of IL-2 by regulatory T cells does not have any effect on the thresholds of λ_r and δ_1, where the steady states S_5^* and S_6^* lose their feasibility. Moreover, if λ_r and δ_1

are small, then increasing the rate δ_2 at which Tregs inhibit the production of IL-2 makes S_6^* become unfeasible, resulting in a stable steady state S_5^*, which has the zero population of host cells A. On the other hand, the steady state S_6^* associated with autoimmune responses is favoured for higher values of δ_1 and λ_r. In the case stable periodic solutions around these steady states, increasing δ_2 results in the disappearance of oscillations and stabilisation of the associated steady state. At the intersection of the lines of Hopf bifurcation and the steady-state bifurcation, as determined by Theorem 1 and conditions (5), one has the co-dimension two fold-Hopf (also known as zero-Hopf or saddle-node Hopf) bifurcation [62].

Table 1. Table of parameter values.

Parameter	Value	Parameter	Value
β	1	ρ_3	2
μ_a	20	d_n	1
d_F	1.1	d_a	0.001
μ_F	6	δ_1	0.0025
d_{in}	1	δ_2	0.001
α	0.4	σ_1	0.15
λ_r	3	σ_2	0.33
d_r	0.4	d_i	0.6
p_1	0.4	τ_1	1.4
p_2	0.4	τ_2	0.6
ρ_1	10	τ_3	0.6
ρ_2	0.8		

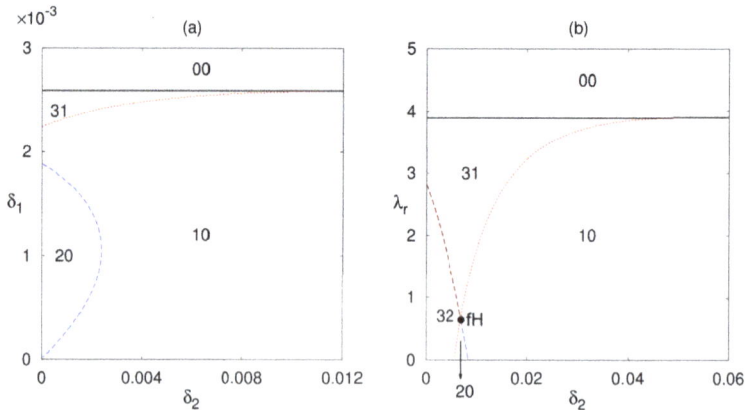

Figure 2. (a) Regions of feasibility and stability of the steady states S_5^* and S_6^* with parameter values from Table 1; and (b) with $\mu_a = 10$. Black and red curves indicate the boundaries of feasibility and the steady-state bifurcation, whereas dashed lines (blue/brown) show the boundaries of Hopf bifurcation of the steady states S_5^* and S_6^*, respectively, with 'fH' indicating the fold-Hopf bifurcation. The first digit of the index refers to S_5^*, while the second corresponds to S_6^*, and they indicate that in that parameter region the respective steady state is unfeasible (index is '0'), stable (index is '1'), unstable via Hopf bifurcation with a periodic solution around this steady state (index is '2'), or unstable via a steady-state bifurcation (index is '3'). In all plots, the condition $\beta < d_F$ holds, so the disease-free steady state S_2^* is also stable.

Since our earlier analysis showed that stability of the steady states S_5^*/S_6^* is affected by the time delay τ_2, in Figure 3 we consider stability of these equilibria depending on τ_2 and the rate δ_2. For the steady state S_5^*, if the effect of IL-2 on promoting proliferation of T cells is fast (i.e., τ_2 is small), there is a large range of δ_2, starting with some very low values, for which S_5^* is stable. Increasing the time delay τ_2 results in the Hopf bifurcation of this steady state as described in Theorem 1. One should note that for intermediate values of δ_2, the steady state S_5^* undergoes stability switches, whereby increasing the delay τ_2 further results in a subcritical Hopf bifurcation, which stabilises S_5^*, but after some number of such stability switches eventually the steady state S_5^* is unstable. For higher still values of δ_2, the steady state S_5^* remains stable for an entire range of τ_2 values, and the only way to lose its stability is via a steady state bifurcation as given by (5). In the case of autoimmune steady state S_6^*, the situation is somewhat different in that increasing δ_2 beyond some critical values makes this steady state biologically infeasible. At the same time, for an entire range of δ_2 values where it is feasible, this steady state exhibits a single loss of stability through a Hopf bifurcation for some critical value of the time delay τ_2, in agreement with Theorem 1.

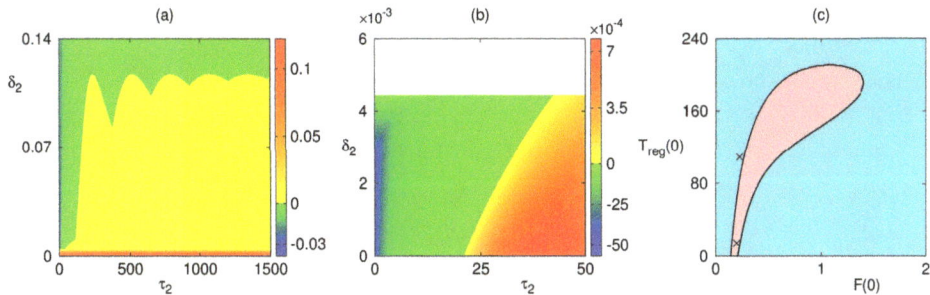

Figure 3. Stability of the steady states S_5^* (**a**), and S_6^* (**b**) with parameter values from Table 1. White area shows the region where the steady state S_6^* is infeasible. Colour code denotes max[Re(λ)] for the steady states when they are feasible. In all plots the condition $d_F > \beta$ holds, so the disease-free steady state S_2^* is stable. Basins of attraction of different steady states depending on the initial conditions (**c**), with other initial conditions specified in (8), and parameter values from Table 1, except for $\tau_2 = 18$. Cyan and pink areas are the basins of attraction of S_2^* and S_6^*, respectively.

As mentioned earlier, for parameter values used in Figure 3, the disease-free steady state S_2^* is stable. Hence, the system exhibits a bi-stability between a disease-free state and either stable steady states S_5^*/S_6^*, or periodic solutions around these steady states. To investigate this bi-stability, we choose parameter values as in Table 1 except for $\tau_2 = 18$, which corresponds to a stable steady state S_6^*, and we fix initial conditions for state variables as follows,

$$(A(0), T_{in}(0), T_{nor}(0), T_{aut}(0), I(0)) = (0.9, 0.8, 0, 0, 0), \tag{8}$$

except for initial amounts of infected cells and regulatory T cells that are allowed to vary. Figure 3c illustrates the bi-stability between S_2^* and S_6^* in terms of their basins of attraction. It is worth noting that recently significant research in approximation theory and meshless interpolation has focused on developing techniques for detection and analysis of attraction basins [63–68]. Figure 3c suggests that for very large initial amounts of regulatory T cells, the system converges to the disease-free steady state. It also indicates that if the initial amount of infected cells is very small or is bigger than some specific value, then the infection will be cleared. Interestingly, increasing the initial amount of the regulatory T cells results in a larger range of initial amounts of infection, for which the system tends to a stable autoimmune state S_6^*. In Figure 3b we discovered that increasing τ_2 makes the autoimmune steady state S_6^* undergo a Hopf bifurcation, in which case the system will exhibit a bi-stability between stable S_2^* and a periodic solution around S_6^*. Our numerical investigation suggests that the shape of basins of attraction in this case is qualitatively similar to that shown in Figure 3c, with the basin of attraction of the stable steady state S_6^* being replaced by the basin of attraction of the periodic solution around this steady state.

Figure 4 shows temporary evolution of the system (1) in the regime of bi-stability between a stable disease-free steady state and a stable autoimmune steady state S_6^* (similar pattern of behaviour is exhibited in the case of bi-stability between S_2^* and S_5^*). It also illustrates how the system develops a periodic solution around the steady state S_6^* for a higher value of τ_2. Periodic oscillations around the steady state S_6^* biologically correspond to a genuine autoimmune state: after the initial infection is cleared, the system exhibits sustained endogenous oscillations, characterised by periods of significant reduction in the number of organ cells through a negative action of autoreactive T cells, separated by periods of quiescence. This type of behaviour is often observed in clinical manifestations of autoimmune disease [44–47]. This result has substantial biological significance as effectively it suggests that even for the same kinetic parameters of immune response, the ultimate state of the system, which can be either a successful clearance of infection without lasting consequences, or progression to autoimmunity, also depends on the strength of the initial infection and of the initial state of the immune system, as represented by the initial number of regulatory T cells.

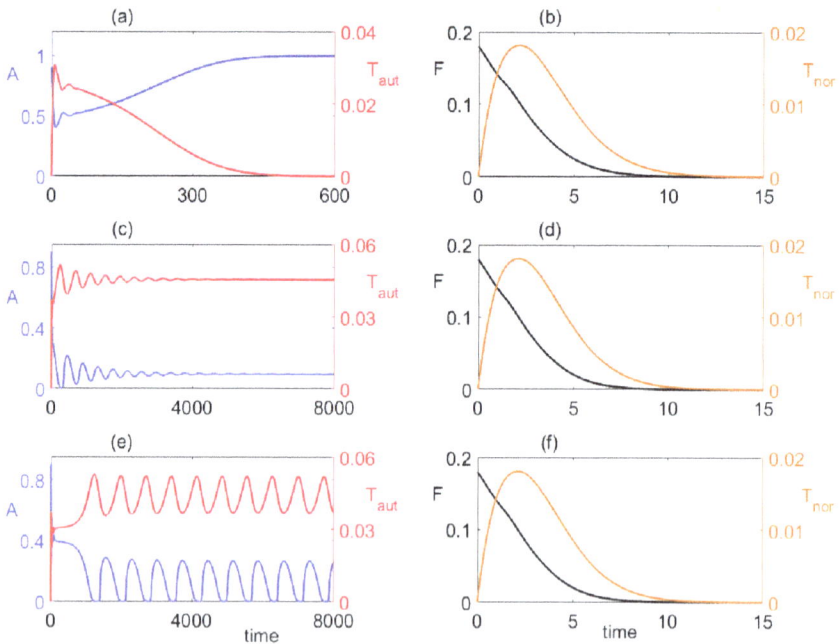

Figure 4. Numerical solutions of the model with parameters values from Table 1, except for $\tau_2 = 18$. (**a,b**) Stable disease-free steady state S_2^* for $F(0) = 0.18$, and $T_{reg}(0) = 100$. (**c,d**) Transient oscillations settling on a stable steady state S_6^* for $F(0) = 0.18$, and $T_{reg}(0) = 10$. (**e,f**) Autoimmune dynamics represented by periodic oscillations around the steady state S_6^* for $\tau_2 = 32$, $F(0) = 0.18$, and $T_{reg}(0) = 10$.

Next we consider a situation where $\beta > d_F$, so the disease-free steady state is unstable, and the system can have three steady states S_5^*, S_6^* and S_8^*. Our earlier results [48] suggest that in the case where regulatory T cells do not inhibit the production of IL-2, i.e., for $\delta_2 = 0$, the steady state S_6^* is stable. Figure 5 shows regions of feasibility and stability of these steady states depending on δ_2 and τ_2 in this case. One observes that S_5^* and S_6^*, whose stability boundaries are determined by Theorem 1, exhibit the same behaviour as in Figure 3, namely, for S_5^* increasing τ_2 causes multiple stability switches for smaller values of δ_2, and the steady state is unstable for very small δ_2 and stable for large δ_2; in contrast, S_5^* exhibits a single loss of stability via Hopf bifurcation at some critical value of the time delay τ_2, which itself increases with δ_2. Behaviour of S_8^* is similar to that of S_5^* in that there are multiple stability switches for increasing value of τ_2 and small to intermediate values of δ_2, while for high values of δ_2, the chronic steady state S_8^* is stable for all values of τ_2. Figure 5d divides the δ_2-τ_2 plane into different regions based on feasibility and stability of these steady states and shows that increasing δ_2 makes the autoimmune steady state S_6^* infeasible. In other regions, the system can exhibit a bi-stability between a stable steady state S_8^* and either a stable steady state S_5^*, or a periodic solution around S_5^*.

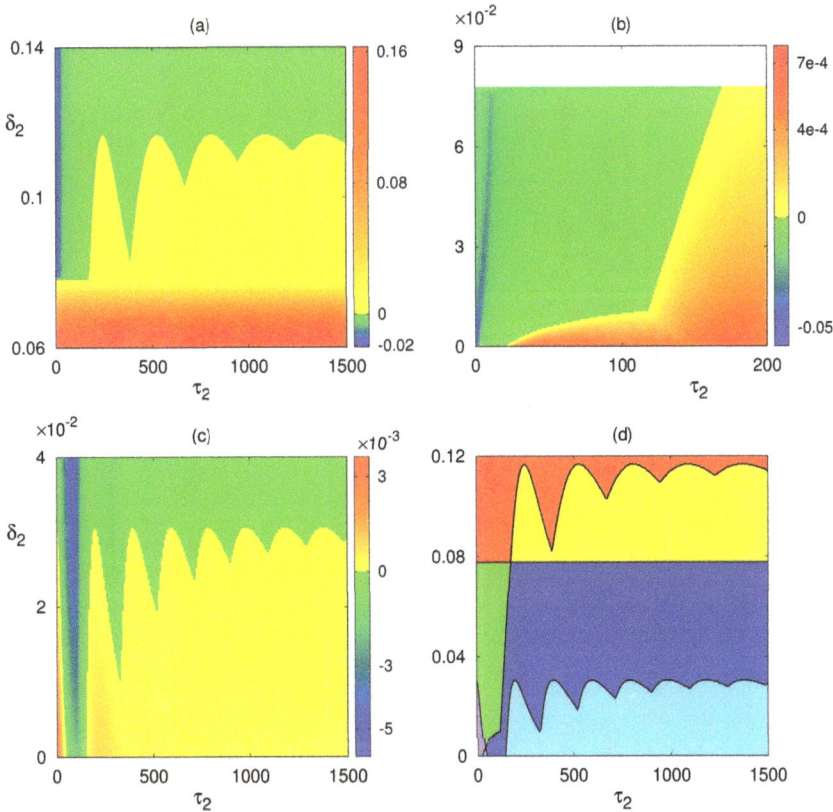

Figure 5. Stability of S_5^* (**a**), S_6^* (**b**), and S_8^* (**c**), with parameter values from Table 1, except for $\beta = 1.4$ and $\sigma_2 = 1$, so that $\beta > d_F$. White area shows the region where the steady state is infeasible. Colour code denotes max[Re(λ)] for each steady states when it is feasible. (**d**) Summary of stability results. Green indicates the region where S_6^* and S_8^* are stable, and S_5^* is unstable, whereas red is the area where S_5^* and S_8^* are stable, and S_6^* is infeasible. Yellow is where S_8^* is stable, S_5^* is unstable, and S_6^* is infeasible. Purple shows the region where S_6^* is stable, but S_5^* and S_8^* are unstable. Blue and cyan indicate the regions where S_5^* and S_6^* are unstable, but S_8^* is stable or unstable, respectively.

Figure 6 illustrates the basins of attraction of the steady states S_5^*, S_6^* and S_8^*, as well as periodic solutions around S_8^*. Figure 6a shows the basins of attraction of the steady states S_5^* and S_8^* and demonstrates that if the initial number of regulatory T cells or infected cells is sufficiently high, or the initial amount of infected cells is very low, the immune response neither eliminates infection nor clears autoreactive T cells, and the system approaches the stable steady state S_5^*. Figure 6b illustrates bi-stability between the stable steady state S_6^* and a periodic solution around S_8^*, and has a different behaviour to than shown in Figure 6a. This figure suggests that for a specific range of $F(0)$ the system converges to a stable autoimmune state S_6^* for all values of $T_{reg}(0)$. However, if the initial number of infected cells is very high or very low, the system instead develops a periodic solution around the steady state S_8^* associated with chronic infection.

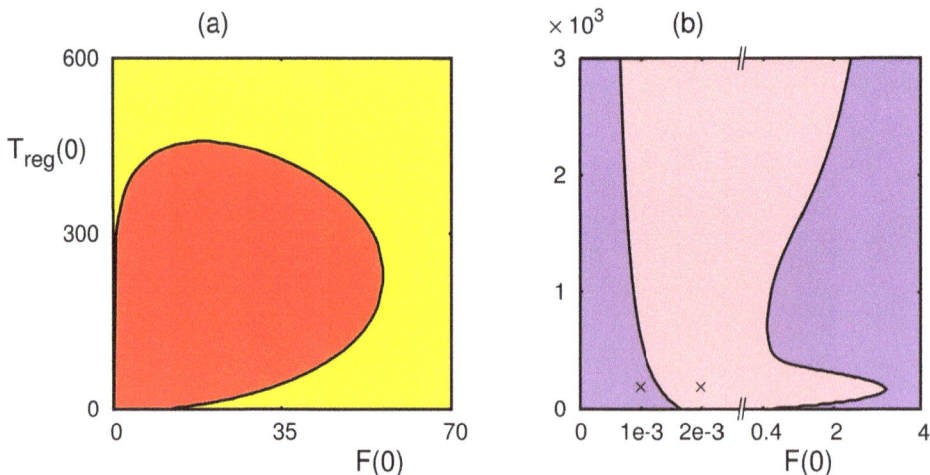

Figure 6. Bi-stability analysis of the steady states S_5^*, S_6^*, and S_8^* with the same parameter values as in Figure 5, except for (**a**) $\delta_2 = 0.1$, (**b**) $\delta_2 = 0.02$. Yellow indicates the basin of attraction of the chronic steady state S_8^*, purple is the basin of attraction of periodic solutions around S_8^*. Red and pink are the basins of attraction of the steady states S_5^* and S_6^*, respectively.

Figure 7 illustrates a regime of bi-stability between a stable steady state S_6^* and a periodic solution around S_8^* for combinations of initial conditions indicated by crossed in Figure 6b. It also illustrates how the system develops a stable solution around the steady state S_8^* for a higher value of τ_2. This figure shows that by increasing the initial number of infected cells the behaviour of the system changes, as it then approaches the autoimmune steady state S_6^*. Interestingly, one can observe that for high values of $F(0)$ the system can eliminate the infection, but it cannot clear the autoreactive T cells, in which case the system converges to S_6^*. On the other hand, for a smaller number of infected cells the system develops a periodic solution around the endemic steady state.

Figure 8 shows how the stability of the chronic infection steady state S_8^* changes with respect to time delays. Figure 8a indicates that for small values of τ_2 (i.e., when the influence of IL-2 on proliferation of T cells is occurring quite rapidly), the steady state S_8^* is stable, and increasing the time delay τ_1 associated with viral eclipse phase does not have an effect on its stability. At the same time, if τ_2 exceeds some specific value, by increasing τ_1 the chronic steady state switches between being stable or unstable. Figure 8b demonstrates a different behaviour, suggesting that for each value of τ_1, there is small range of τ_3 values where S_8^* is stable, but for smaller and larger values of τ_3 it is unstable. For intermediate values of the eclipse phase delay τ_1, there is an additional narrow range of τ_3 values where S_8^* is stable. Figure 8c illustrates that for very small, respectively very large, values of τ_3, the chronic infection steady state is stable, respectively unstable for any value of τ_2; for intermediate values of τ_3, this steady state undergoes a finite number of stability switches for increasing values of τ_2 and eventually becomes unstable.

It should be noted Figure 8 shows that unlike τ_1 and τ_2, once the steady state S_8^* loses stability via Hopf bifurcation due to increasing τ_3, it cannot regain stability for higher values of τ_3.

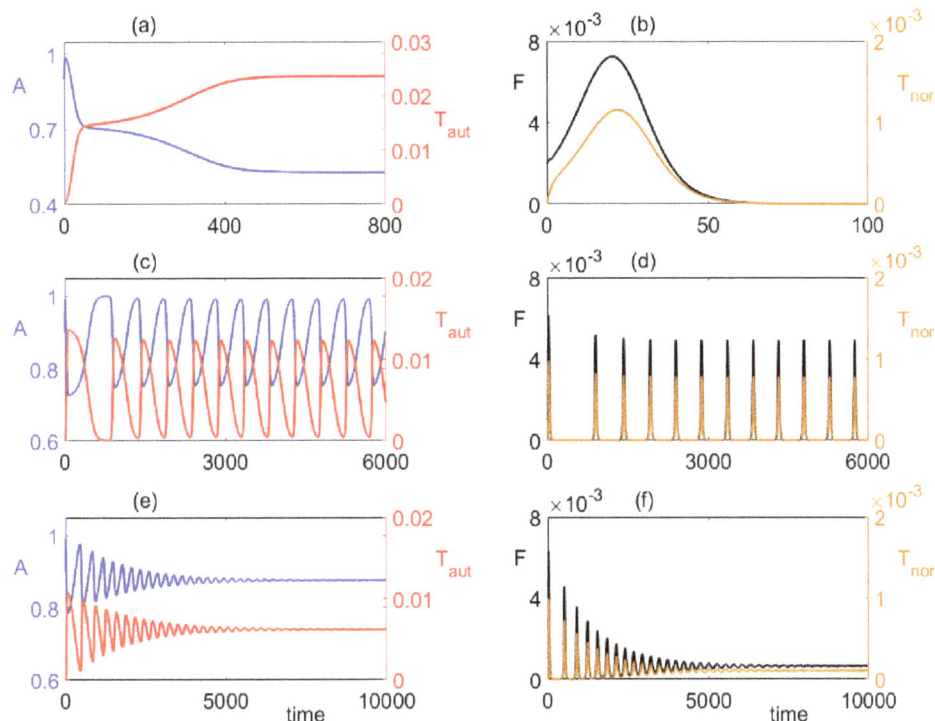

Figure 7. Numerical solutions of the model with same parameters values as Figure 6b. (**a,b**) Stable steady state S_6^* for $F(0) = 0.002$ and $T_{reg}(0) = 200$. (**c,d**) Periodic oscillations around the steady state S_8^* for $F(0) = 0.001$ and $T_{reg}(0) = 200$. (**e,f**) Transient oscillations settling on a stable steady state S_8^* for $\tau_2 = 25$, $F(0) = 0.001$ and $T_{reg}(0) = 200$.

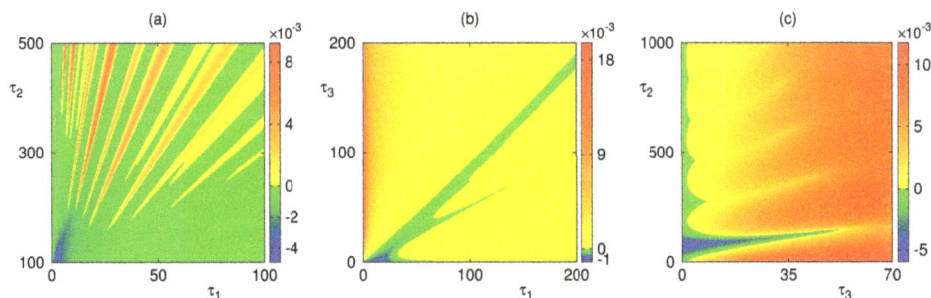

Figure 8. Colour code denotes $\max[\mathrm{Re}(\lambda)]$ for the endemic steady state S_8^* depending on (**a**) $\tau_1 - \tau_2$; (**b**) $\tau_1 - \tau_3$; and (**c**) $\tau_3 - \tau_2$, with the parameter values taken from Table 1, except for $\beta = 1.4$, $\sigma_2 = 1$, and $\delta_2 = 0.04$.

5. Conclusions

In this paper, we have developed and analysed a time-delayed model of immune response to a viral infection, which accounts for T cells with different activation thresholds, a cytokine mediating T cell proliferation, as well as regulatory T cells. Particular attention is payed to the dual suppressive role of regulatory T cells in terms of reducing the amount of autoreactive T cells, and also inhibiting IL-2. To achieve better biological realism of the model, we have explicitly included time delays associated with the eclipse phase of the virus life cycle, stimulation/proliferation of T cells by IL-2, and suppression of IL-2 by regulatory T cells. Depending on the values of parameters, the system can have four steady states: the disease-free state, the state characterised by the death of host cells, the autoimmune state, and a state of chronic infection. We have established conditions for stability and steady-state or Hopf bifurcations of these steady states in terms of system parameters.

In the case where the natural death rate of infected cells exceeds the infection rate, the immune system is able to clear the infection, and the disease-fee steady state is stable. In this regime, the system can also support the autoimmune steady state or the steady state with the death of host cells, either of which can be stable, or give rise to a periodic solution emerging via Hopf bifurcation. In the opposite case, when the natural death rate of infected cells is smaller than the infection rate, the disease-fee steady state is unstable, but it is possible to have a bi-stability between the other three steady states or periodic solutions around them. To better understand bi-stability between different dynamical regimes, we have used numerical simulations to identify basins of attraction of different steady states and periodic solutions depending on the initial level of infection and the initial number of regulatory T cells. The fact that for the same parameter values the system can exhibit bi-stability between a disease-free steady state and an autoimmune state, represented by sustained periodic oscillations following the clearance of infection, is very important from a clinical point of view, as effectively it suggests that the progress and eventual outcome of viral infection is also determined by the strength of infection and the initial state of the immune system. One counter-intuitive observation is that in the case of bi-stability with a disease-free steady state, for higher initial numbers of regulatory T cells, the autoimmune steady state is actually stable for a wider range of initial levels of infection. In this regime of bi-stability, increasing the time delay associated with the positive impact of IL-2 on proliferation of T cells results in the loss of stability of autoimmune steady state and emergence of autoimmune dynamics, characterised by stable periodic oscillations. On the contrary, in the case where the disease-free steady state is unstable, increasing this time delay results in stabilisation of the chronic infection.

There are several directions in which the work presented in this paper can be extended. One direction is exploration of the contributions from other components of immune response, more specifically, antibodies and memory T cells, to the onset and progress of autoimmunity [69,70]. This is particularly important from the perspective that clinically the onset of autoimmune disease is often taking place on a much longer scale than the timescale of a regular immune response to a viral infection, so memory T cells can be expected to play a more substantial role. While our model has focused on one specific growth cytokine IL-2, a number of other cytokines, such as IL-7 [71], TNFβ and IL-10 [29], are known to significantly affect homeostasis and proliferation of different types of T cells, as well as mediate their efficiency in eliminating the infection. Including these immune mediators explicitly in the model can provide further significant insights into the dynamics of immune response, as has been recently demonstrated on the example of a detailed model of immune response to hepatitis B [72]. Another biologically relevant and mathematically challenging problem is the investigation of the interplay between stochasticity, which is known to be an intrinsic feature of immune response [49,73], and effects of time delays associated with various aspects of immune dynamics.

Author Contributions: The authors made equal contributions in the article. All authors read and approved the final manuscript.

Acknowledgments: F.F. acknowledges the support from Chancellor's Studentship from the University of Sussex.

Conflicts of Interest: The authors declare no conflict of interest.

Appendix A

Coefficients of Equation (7) for Hopf frequency are given below.

$$
\begin{aligned}
b_2 = {} & (- \delta_1^4\delta_2^4\rho_1^2 Y^2 + 2 d_r^2\delta_1^4\delta_2^4) {T_{reg}^{*}}^{8} + (- 2\delta_1^3\delta_2^3\rho_1 (d_a\delta_2\rho_1 + d_i\delta_1\rho_1 - 2 d_r\delta_2\rho_3) I^{*2} + \\
& 8 d_r^2\delta_1^3\delta_2^3 (d_a\delta_2 + d_i\delta_1)) {T_{reg}^{*}}^{7} + (- 2\delta_1^2\delta_2^3\rho_1^2\rho_3 (4\delta_1\rho_1 + \delta_2\rho_1 + 5\delta_2\rho_3) I^{*4} - \delta_1^2\delta_2^2 (d_a^2\delta_2^2\rho_1^2 + \\
& 2 d_a d_i\delta_1\delta_2\rho_1^2 - 8 d_a d_r\delta_2^2\rho_1\rho_3 + d_i^2\delta_1^2\rho_1^2 - 12 d_i d_r\delta_1\delta_2\rho_1\rho_3 + 2 d_r^2\delta_1^2\rho_1^2 + 2 d_r^2\delta_1\delta_2\rho_1^2 + 6 d_r^2\delta_1\delta_2\rho_1\rho_3 + \\
& d_r^2\delta_2^2\rho_1^2 + 8 d_r^2\delta_2^2\rho_3^2) I^{*2} + 2 d_r^2\delta_1^2\delta_2^2 (6 d_a^2\delta_2^2 + 16 d_a d_i\delta_1\delta_2 + 6 d_i^2\delta_1^2 + d_r^2\delta_1^2 + d_r^2\delta_2^2)) {T_{reg}^{*}}^{6} + \\
& (- 2\rho_3\delta_2^2\rho_1\delta_1 (6 d_a\delta_1\delta_2\rho_1^2 + 6 d_a\delta_2^2\rho_1\rho_3 + 9 d_i\delta_1^2\rho_1^2 + 5 d_i\delta_1\delta_2\rho_1^2 + 22 d_i\delta_1\delta_2\rho_1\rho_3 + 6 d_r\delta_1^2\rho_1^2 + \\
& 2 d_r\delta_1\delta_2\rho_1^2 - 4 d_r\delta_2^2\rho_3^2) I^{*4} + 2\delta_1\delta_2 (d_a^2 d_i\delta_1\delta_2^2\rho_1^2 + 2 d_a^2 d_r\delta_2^3\rho_1\rho_3 + d_a d_i^2\delta_1^2\delta_2\rho_1^2 + 10 d_a d_i d_r\delta_1\delta_2^2\rho_1\rho_3 - \\
& 3 d_a d_r^2\delta_1^2\delta_2\rho_1^2 - 2 d_a d_r^2\delta_1\delta_2^2\rho_1^2 - 8 d_a d_r^2\delta_1\delta_2^2\rho_1\rho_3 - 8 d_a d_r^2\delta_2^3\rho_3^2 + 6 d_i^2 d_r\delta_1^2\delta_2\rho_1\rho_3 - d_i d_r^2\delta_1^3\rho_1^2 - \\
& 3 d_i d_r^2\delta_1^2\delta_2\rho_1^2 - 8 d_i d_r^2\delta_1^2\delta_2\rho_1\rho_3 - d_i d_r^2\delta_1\delta_2^2\rho_1^2 - 2 d_i d_r^2\delta_1\delta_2^2\rho_1\rho_3 - 16 d_i d_r^2\delta_1\delta_2^2\rho_3^2 + d_r^3\delta_1^2\delta_2\rho_1\rho_3 + \\
& 2 d_r^3\delta_1\delta_2^2\rho_1\rho_3 + 2 d_r^3\delta_2^3\rho_1\rho_3) I^{*2} + 4 d_r^2\delta_1\delta_2 (2 d_a^3\delta_2^3 + 12 d_a^2 d_i\delta_1\delta_2^2 + 12 d_a d_i^2\delta_1^2\delta_2 + 2 d_a d_r^2\delta_1^2\delta_2 + d_a d_r^2\delta_2^3 + \\
& 2 d_i^3\delta_1^3 + d_i d_r^2\delta_1^3 + 2 d_i d_r^2\delta_1\delta_2^2)) {T_{reg}^{*}}^{5} + (- 2\delta_2^2\rho_1^2\rho_3^2 (\delta_1^2\rho_1^2 + \delta_1\delta_2\rho_1^2 + 3\delta_1\delta_2\rho_1\rho_3 + 2\delta_2^2\rho_3^2) I^{*6} - \\
& 2\rho_3\delta_2\rho_1 (2 d_a^2\delta_1\delta_2^2\rho_1^2 + d_a^2\delta_2^3\rho_1\rho_3 + 9 d_a d_i\delta_1^2\delta_2\rho_1^2 + 24 d_a d_i\delta_1\delta_2^2\rho_1\rho_3 + 12 d_a d_r\delta_1^2\delta_2\rho_1^2 - 4 d_a d_r\delta_2^3\rho_3^2 + \\
& 4 d_i^2\delta_1^3\rho_1^2 + 8 d_i^2\delta_1^2\delta_2\rho_1^2 + 33 d_i^2\delta_1^2\delta_2\rho_1\rho_3 + 9 d_i d_r\delta_1^3\rho_1^2 + 9 d_i d_r\delta_1^2\delta_2\rho_1^2 + 2 d_i d_r\delta_1\delta_2^2\rho_1^2 - 20 d_i d_r\delta_1\delta_2^2\rho_3^2 - \\
& 4 d_r^2\delta_1^2\delta_2\rho_1\rho_3 - 6 d_r^2\delta_1\delta_2^2\rho_1\rho_3 - 4 d_r^2\delta_1\delta_2^2\rho_3^2 + d_r^2\delta_2^3\rho_1\rho_3 - 4 d_r^2\delta_2^3\rho_3^2) I^{*4} + (2 d_a^3 d_i\delta_1\delta_2^3\rho_1^2 + \\
& 6 d_a^2 d_i^2\delta_1^2\delta_2^2\rho_1^2 + 4 d_a^2 d_i d_r\delta_1\delta_2^3\rho_1\rho_3 - 7 d_a^2 d_r^2\delta_1^2\delta_2^2\rho_1^2 - 2 d_a^2 d_r^2\delta_1\delta_2^3\rho_1^2 - 14 d_a^2 d_r^2\delta_1\delta_2^3\rho_1\rho_3 - \\
& 8 d_a^2 d_r^2\delta_2^4\rho_3^2 + 2 d_a d_i^3\delta_1^3\delta_2\rho_1^2 + 12 d_a d_i^2 d_r\delta_1^2\delta_2^2\rho_1\rho_3 - 6 d_a d_i d_r^2\delta_1^3\delta_2\rho_1^2 - 12 d_a d_i d_r^2\delta_1^2\delta_2^2\rho_1^2 - \\
& 42 d_a d_i d_r^2\delta_1^2\delta_2^2\rho_1\rho_3 + 2 d_a d_i d_r^2\delta_1\delta_2^3\rho_1^2 - 8 d_a d_i d_r^2\delta_1\delta_2^3\rho_1\rho_3 - 64 d_a d_i d_r^2\delta_1\delta_2^3\rho_3^2 + 2 d_a d_r^3\delta_1^2\delta_2^2\rho_1\rho_3 + \\
& 8 d_a d_r^3\delta_1\delta_2^3\rho_1\rho_3 + 4 d_i^3 d_r\delta_1^3\delta_2\rho_1\rho_3 - d_i^2 d_r^2\delta_1^4\rho_1^2 - 6 d_i^2 d_r^2\delta_1^3\delta_2\rho_1^2 - 14 d_i^2 d_r^2\delta_1^3\delta_2\rho_1\rho_3 - 2 d_i^2 d_r^2\delta_1^2\delta_2^2\rho_1^2 - \\
& 12 d_i^2 d_r^2\delta_1^2\delta_2^2\rho_1\rho_3 - 48 d_i^2 d_r^2\delta_1^2\delta_2^2\rho_3^2 + 10 d_i d_r^3\delta_1^2\delta_2^2\rho_1\rho_3 + 12 d_i d_r^3\delta_1\delta_2^3\rho_1\rho_3 - 5 d_r^4\delta_1^2\delta_2^2\rho_3^2 - \\
& 4 d_r^4\delta_2^4\rho_3^2) I^{*2} + 2 d_r^2 (d_a^4\delta_2^4 + 16 d_a^3 d_i\delta_1\delta_2^3 + 36 d_a^2 d_i^2\delta_1^2\delta_2^2 + 6 d_a^2 d_r^2\delta_1^2\delta_2^2 + d_a^2 d_r^2\delta_2^4 + 16 d_a d_i^3\delta_1^3\delta_2 + \\
& 8 d_a d_i d_r^2\delta_1^3\delta_2 + 8 d_a d_i d_r^2\delta_1\delta_2^3 + d_i^4\delta_1^4 + d_i^2 d_r^2\delta_1^4 + 6 d_i^2 d_r^2\delta_1^2\delta_2^2)) {T_{reg}^{*}}^{4} + (- 2\delta_2\rho_1^2\rho_3^2 (d_a\delta_1\delta_2\rho_1^2 + \\
& 2 d_a\delta_2^2\rho_1\rho_3 + 2 d_i\delta_1^2\rho_1^2 + 5 d_i\delta_1\delta_2\rho_1^2 + 14 d_i\delta_1\delta_2\rho_1\rho_3 + 2 d_i\delta_2^2\rho_1\rho_3 + 12 d_i\delta_2^2\rho_3^2 - d_r\delta_1\delta_2\rho_1\rho_3 - \\
& 2 d_r\delta_2^2\rho_1\rho_3) I^{*6} + 2\rho_3\rho_1 (2 d_a^2 d_i\delta_1\delta_2^2\rho_1^2 - 2 d_a^2 d_i\delta_2^3\rho_1\rho_3 - 7 d_a^2 d_r\delta_1\delta_2^2\rho_1^2 + 4 d_a d_i^2\delta_1^2\delta_2\rho_1^2 + 2 d_a d_i^2\delta_1\delta_2^2\rho_1^2 - \\
& 30 d_a d_i^2\delta_1\delta_2^2\rho_1\rho_3 - 15 d_a d_i d_r\delta_1^2\delta_2\rho_1^2 - 4 d_a d_i d_r\delta_1\delta_2^2\rho_1^2 + 20 d_a d_i d_r\delta_2^3\rho_3^2 + 8 d_a d_r^2\delta_1\delta_2^2\rho_1\rho_3 + 4 d_a d_r^2\delta_2^3\rho_3^2 - \\
& 4 d_i^3\delta_1^2\delta_2\rho_1^2 - 20 d_i^3\delta_1^2\delta_2\rho_1\rho_3 - 2 d_i^2 d_r\delta_1^3\rho_1^2 - 12 d_i^2 d_r\delta_1^2\delta_2\rho_1^2 - 7 d_i^2 d_r\delta_1\delta_2^2\rho_1^2 + 36 d_i^2 d_r\delta_1\delta_2^2\rho_3^2 + \\
& 6 d_i d_r^2\delta_1^2\delta_2\rho_1\rho_3 + 20 d_i d_r^2\delta_1\delta_2^2\rho_1\rho_3 + 12 d_i d_r^2\delta_1\delta_2^2\rho_3^2 - 2 d_i d_r^2\delta_2^3\rho_1\rho_3 + 20 d_i d_r^2\delta_2^3\rho_3^2 + 3 d_r^3\delta_1\delta_2^2\rho_3^2 - \\
& 4 d_r^3\delta_2^3\rho_3^2) I^{*4} + (2 d_a^3 d_i^2\delta_1\delta_2^2\rho_1^2 - 4 d_a^3 d_i d_r\delta_2^3\rho_1\rho_3 - 4 d_a^3 d_r^2\delta_1\delta_2^2\rho_1^2 - 4 d_a^3 d_r^2\delta_2^3\rho_1\rho_3 + 2 d_a^2 d_i^3\delta_1^2\delta_2\rho_1^2 - \\
& 12 d_a^2 d_i^2 d_r\delta_1\delta_2^2\rho_1\rho_3 - 8 d_a^2 d_i d_r^2\delta_1^2\delta_2\rho_1^2 - 6 d_a^2 d_i d_r^2\delta_1\delta_2^2\rho_1^2 - 36 d_a^2 d_i d_r^2\delta_1\delta_2^2\rho_1\rho_3 - 4 d_a^2 d_i d_r^2\delta_2^3\rho_1\rho_3 - \\
& 32 d_a^2 d_i d_r^2\delta_2^3\rho_3^2 - 2 d_a^2 d_r^3\delta_1\delta_2^2\rho_1\rho_3 + 4 d_a^2 d_r^3\delta_2^3\rho_1\rho_3 - 4 d_a d_i^3 d_r\delta_1^2\delta_2\rho_1\rho_3 - 4 d_a d_i^2 d_r^2\delta_1^3\rho_1^2 - \\
& 12 d_a d_i^2 d_r^2\delta_1^2\delta_2\rho_1^2 - 36 d_a d_i^2 d_r^2\delta_1^2\delta_2\rho_1\rho_3 + 2 d_a d_i^2 d_r^2\delta_1\delta_2^2\rho_1^2 - 24 d_a d_i^2 d_r^2\delta_1\delta_2^2\rho_1\rho_3 - 96 d_a d_i^2 d_r^2\delta_1\delta_2^2\rho_3^2 - \\
& 8 d_a d_i d_r^3\delta_1^2\delta_2\rho_1\rho_3 + 20 d_a d_i d_r^3\delta_1\delta_2^2\rho_1\rho_3 - 4 d_a d_i d_r^3\delta_2^3\rho_1\rho_3 - 10 d_a d_r^4\delta_1\delta_2^2\rho_3^2 - 2 d_i^3 d_r^2\delta_1^3\rho_1^2 - \\
& 4 d_i^3 d_r^2\delta_1^3\rho_1\rho_3 - 2 d_i^3 d_r^2\delta_1^2\delta_2\rho_1^2 - 12 d_i^3 d_r^2\delta_1^2\delta_2\rho_1\rho_3 - 32 d_i^3 d_r^2\delta_1^2\delta_2\rho_3^2 - 2 d_i^2 d_r^3\delta_1^3\rho_1\rho_3 + 8 d_i^2 d_r^3\delta_1^2\delta_2\rho_1\rho_3 + \\
& 12 d_i^2 d_r^3\delta_1\delta_2^2\rho_1\rho_3 - 10 d_i d_r^4\delta_1^2\delta_2\rho_3^2 - 16 d_i d_r^4\delta_2^3\rho_3^2) I^{*2} + 8 d_r^2 (d_a^4 d_i\delta_2^3 + 6 d_a^3 d_i^2\delta_1\delta_2^2 + d_a^3 d_r^2\delta_1\delta_2^2 + \\
& 6 d_i^3\delta_1^2\delta_2 d_a^2 + 3 d_a^2 d_i d_r^2\delta_1^2\delta_2 + d_a^2 d_i d_r^2\delta_2^3 + d_a d_i^4\delta_1^3 + d_a d_i^2\delta_1^3 d_r^2 + 3 d_a d_i^2\delta_1 d_r^2\delta_2^2 + d_i^3\delta_1^2\delta_2 d_r^2)) {T_{reg}^{*}}^{3} + \\
& (2\delta_2^2\rho_1^4\rho_3^4 I^{*8} - \rho_1^2\rho_3^2 (d_a^2\delta_2^2\rho_1^2 + 6 d_a d_i\delta_1\delta_2\rho_1^2 + 18 d_a d_i\delta_2^2\rho_1\rho_3 + 2 d_a d_r\delta_2^2\rho_1\rho_3 + 4 d_i^2\delta_1^2\rho_1^2 + \\
& 16 d_i^2\delta_1\delta_2\rho_1^2 + 40 d_i^2\delta_1\delta_2\rho_1\rho_3 + d_i^2\delta_2^2\rho_1^2 + 20 d_i^2\delta_2^2\rho_1\rho_3 + 52 d_i^2\delta_2^2\rho_3^2 + 4 d_i d_r\delta_1\delta_2\rho_1\rho_3 - 18 d_i d_r\delta_2^2\rho_1\rho_3 + \\
& 5 d_r^2\delta_2^2\rho_3^2) I^{*6} + 2\rho_1\rho_3 (2 d_a^3 d_i\delta_2^2\rho_1^2 - d_a^3 d_r\delta_2^2\rho_1^2 + 12 d_a^2 d_i^2\delta_1\delta_2\rho_1^2 + d_a^2 d_i^2\delta_2^2\rho_1^2 + 3 d_a^2 d_i^2\delta_2^2\rho_1\rho_3 - \\
& 5 d_a^2 d_i d_r\delta_1\delta_2\rho_1^2 - 2 d_a^2 d_i d_r\delta_2^2\rho_1^2 + 4 d_a^2 d_r^2\delta_2^2\rho_1\rho_3 + 4 d_a d_i^3\delta_1^2\rho_1^2 + 6 d_a d_i^3\delta_1\delta_2\rho_1^2 - 12 d_a d_i^3\delta_1\delta_2\rho_1\rho_3 - \\
& 8 d_a d_i^2 d_r\delta_1\delta_2\rho_1^2 + d_a d_i^2 d_r\delta_2^2\rho_1^2 + 36 d_a d_i^2 d_r\delta_2^2\rho_3^2 + 14 d_a d_i d_r^2\delta_1\delta_2\rho_1\rho_3 + 12 d_a d_i d_r^2\delta_2^2\rho_3^2 + 3 d_a d_r^3\delta_2^2\rho_3^2 - \\
& 4 d_i^4\delta_1^2\rho_1\rho_3 - 4 d_i^3 d_r\delta_1^2\rho_1^2 - 7 d_i^3 d_r\delta_1\delta_2\rho_1^2 + 28 d_i^3 d_r\delta_1\delta_2\rho_3^2 + 3 d_i^2 d_r^2\delta_1^2\rho_1\rho_3 + 22 d_i^2 d_r^2\delta_1\delta_2\rho_1\rho_3 + \\
& 12 d_i^2 d_r^2\delta_1\delta_2\rho_3^2 + 3 d_i^2 d_r^2\delta_2^2\rho_1\rho_3 + 36 d_i^2 d_r^2\delta_2^2\rho_3^2 + 9 d_i d_r^3\delta_1\delta_2\rho_3^2 - 12 d_i d_r^3\delta_2^2\rho_3^2) I^{*4} + (- d_a^4 d_i^2\delta_2^2\rho_1^2 - \\
& d_a^4 d_r^2\delta_2^2\rho_1^2 - 2 d_a^3 d_i^3\delta_1\delta_2\rho_1^2 - 12 d_a^3 d_i^2 d_r\delta_2^2\rho_1\rho_3 - 6 d_a^3 d_i d_r^2\delta_1\delta_2\rho_1^2 - 10 d_a^3 d_i d_r^2\delta_2^2\rho_1\rho_3 - 2 d_a^3 d_r^3\delta_2^2\rho_1\rho_3 - \\
& d_a^2 d_i^4\delta_1^2\rho_1^2 - 20 d_a^2 d_i^3 d_r\delta_1\delta_2\rho_1\rho_3 - 7 d_a^2 d_i^2 d_r^2\delta_1^2\rho_1^2 - 6 d_a^2 d_i^2 d_r^2\delta_1\delta_2\rho_1^2 - 30 d_a^2 d_i^2 d_r^2\delta_1\delta_2\rho_1\rho_3 - \\
& 2 d_a^2 d_i^2 d_r^2\delta_2^2\rho_1^2 - 12 d_a^2 d_i^2 d_r^2\delta_2^2\rho_1\rho_3 - 48 d_a^2 d_i^2 d_r^2\delta_2^2\rho_3^2 - 16 d_a^2 d_i d_r^3\delta_1\delta_2\rho_1\rho_3 + 10 d_a^2 d_i d_r^3\delta_2^2\rho_1\rho_3 - \\
& 5 d_a^2 d_r^4\delta_2^2\rho_3^2 - 4 d_a d_i^4 d_r\delta_1^2\rho_1\rho_3 - 4 d_a d_i^3 d_r^2\delta_1^2\rho_1^2 - 10 d_a d_i^3 d_r^2\delta_1^2\rho_1\rho_3 - 2 d_a d_i^3 d_r^2\delta_1\delta_2\rho_1^2 - \\
& 24 d_a d_i^3 d_r^2\delta_1\delta_2\rho_1\rho_3 - 64 d_a d_i^3 d_r^2\delta_1\delta_2\rho_3^2 - 10 d_a d_i^2 d_r^3\delta_1^2\rho_1\rho_3 + 16 d_a d_i^2 d_r^3\delta_1\delta_2\rho_1\rho_3 - 12 d_a d_i^2 d_r^3\delta_2^2\rho_1\rho_3 - \\
& 20 d_a d_i d_r^4\delta_1\delta_2\rho_3^2 - d_i^4 d_r^2\delta_1^2\rho_1^2 - 4 d_i^4 d_r^2\delta_1^2\rho_1\rho_3 - 8 d_i^4 d_r^2\delta_1^2\rho_3^2 + 2 d_i^3 d_r^3\delta_1^2\rho_1\rho_3 + 4 d_i^3 d_r^3\delta_1\delta_2\rho_1\rho_3 - \\
& 5 d_i^2 d_r^4\delta_1^2\rho_3^2 - 24 d_i^2 d_r^4\delta_2^2\rho_3^2) I^{*2} + 2 d_r^2 (6 d_a^4 d_i^2\delta_2^2 + d_a^4 d_r^2\delta_2^2 + 16 d_a^3 d_i^3\delta_1\delta_2 + 8 d_a^3 d_i d_r^2\delta_1\delta_2 +
\end{aligned}
$$

$$6\,da^2di^4\delta_1^2 \;+\; 6\,da^2di^2dr^2\delta_1^2 \;+\; 6\,da^2di^2dr^2\delta_2^2 \;+\; 8\,da\,di^3dr^2\delta_1\delta_2 \;+\; di^4dr^2\delta_1^2\big)\big)T_{reg}^{*\,2} \;+\; \big(8\,di\delta_2\rho_1^4\rho_3^4 I^{*8} \;-$$

$$4\,di\rho_1^2\rho_3^2\big(da^2\delta_2\rho_1^2 \;+\; 2\,da\,di\delta_1\rho_1^2 \;+\; 6\,da\,di\delta_2\rho_1\rho_3 \;+\; 3\,da\,dr\delta_2\rho_1\rho_3 \;+\; 2\,di^2\delta_1\rho_1^2 \;+\; 4\,di^2\delta_1\rho_1\rho_3 \;+\; di^2\delta_2\rho_1^2 \;+$$

$$8\,di^2\delta_2\rho_1\rho_3 \;+\; 12\,di^2\delta_2\rho_3^2 \;+\; 2\,di\,dr\delta_1\rho_1\rho_3 \;-\; 6\,di\,dr\delta_2\rho_1\rho_3 \;+\; 3\,dr^2\delta_2\rho_3^2\big)I^{*6} \;+\; 2\,di\rho_1\rho_3\big(4\,da^3di\delta_2\rho_1^2 \;+\; da^3dr\delta_2\rho_1^2 \;+$$

$$5\,da^2di^2\delta_1\rho_1^2 \;+\; 3\,da^2di^2\delta_2\rho_1^2 \;+\; 8\,da^2di^2\delta_2\rho_1\rho_3 \;+\; 5\,da^2di\,dr\delta_1\rho_1^2 \;-\; 4\,da^2di\,dr\delta_2\rho_1^2 \;+\; 8\,da^2dr^2\delta_2\rho_1\rho_3 \;+\; 4\,da\,di^3\delta_1\rho_1^2 \;-$$

$$2\,da\,di^2dr\delta_1\rho_1^2 \;+\; 3\,da\,di^2dr\delta_2\rho_1^2 \;+\; 28\,da\,di\,dr^2\delta_2\rho_3^2 \;+\; 8\,da\,di\,dr^2\delta_1\rho_1\rho_3 \;+\; 12\,da\,di\,dr^2\delta_2\rho_3^2 \;+\; 9\,da\,dr^3\delta_2\rho_3^2 \;-$$

$$2\,di^3dr\delta_1\rho_1^2 \;+\; 8\,di^3dr\delta_1\rho_3^2 \;+\; 8\,di^2dr\delta_1\rho_1\rho_3 \;+\; 4\,di^2dr^2\delta_1\rho_3^2 \;+\; 8\,di^2dr^2\delta_2\rho_1\rho_3 \;+\; 28\,di^2dr^2\delta_2\rho_3^2 \;+\; 6\,di\,dr^3\delta_1\rho_3^2 \;-$$

$$12\,di\,dr^3\delta_2\rho_3^2\big)I^{*4} \;-\; 2\,di\big(da^4di^2\delta_2\rho_1^2 \;+\; da^4dr^2\delta_2\rho_1^2 \;+\; da^3di^3\delta_1\rho_1^2 \;+\; 6\,da^3di^2dr\delta_2\rho_1\rho_3 \;+\; 3\,da^3di\,dr^2\delta_1\rho_1^2 \;+$$

$$4\,da^3di\,dr^2\delta_2\rho_1\rho_3 \;+\; 4\,da^3dr^3\delta_2\rho_1\rho_3 \;+\; 4\,da^2di^3dr\delta_1\rho_1\rho_3 \;+\; da^2di^2dr^2\delta_1\rho_1^2 \;+\; 4\,da^2di^2dr^2\delta_1\rho_1\rho_3 \;+\; 2\,da^2di^2dr^2\delta_2\rho_1^2 \;+$$

$$6\,da^2di^2dr^2\delta_2\rho_1\rho_3 \;+\; 16\,da^2di^2dr^2\delta_2\rho_3^2 \;+\; 7\,da^2di\,dr^3\delta_1\rho_1\rho_3 \;-\; 4\,da^2di\,dr^3\delta_2\rho_1\rho_3 \;+\; 5\,da^2dr^4\delta_2\rho_3^2 \;+\; da\,di^3dr^2\delta_1\rho_1^2 \;+$$

$$4\,da\,di^3dr^2\delta_1\rho_1\rho_3 \;+\; 8\,da\,di^3dr^2\delta_1\rho_3^2 \;-\; 2\,da\,di^2dr^3\delta_1\rho_1\rho_3 \;+\; 6\,da\,di^2dr^3\delta_2\rho_1\rho_3 \;+\; 5\,da\,di\,dr^4\delta_1\rho_3^2 \;+\; 8\,di^2dr^4\delta_2\rho_3^2\big)I^{*2} \;+$$

$$4\,da\,di\,dr^2\big(2\,da^3di^2\delta_2 \;+\; da^3dr^2\delta_2 \;+\; 2\,da^2di^3\delta_1 \;+\; 2\,da^2di\,dr^2\delta_1 \;+\; 2\,da\,di^2dr^2\delta_2 \;+\; di^3dr^2\delta_1\big)\big)T_{reg}^{*} \;+\; 8\,di^2\rho_1^4\rho_3^4 I^{*8} \;-$$

$$di^2\rho_1^2\rho_3^2\big(5\,da^2\rho_1^2 \;+\; 8\,da\,di\rho_1\rho_3 \;+\; 12\,da\,dr\rho_1\rho_3 \;+\; 4\,di^2\rho_1^2 \;+\; 16\,di^2\rho_1\rho_3 \;+\; 16\,di^2\rho_3^2 \;-\; 8\,di\,dr\rho_1\rho_3 \;+\; 8\,dr^2\rho_3^2\big)I^{*6} \;+$$

$$2\,di^2\rho_1\rho_3\big(da^3di\rho_1^2 \;+\; 3\,da^3dr\rho_1^2 \;+\; 2\,da^2di^2\rho_1^2 \;+\; 4\,da^2di^2\rho_1\rho_3 \;-\; da^2di\,dr\rho_1^2 \;+\; 5\,da^2dr^2\rho_1\rho_3 \;+\; 2\,da\,di^2dr\rho_1^2 \;+$$

$$8\,da\,di^2dr\rho_3^2 \;+\; 4\,da\,di\,dr^2\rho_3^2 \;+\; 6\,da\,dr^3\rho_3^2 \;+\; 4\,di^2dr^2\rho_1\rho_3 \;+\; 8\,di^2dr^2\rho_3^2 \;-\; 4\,di\,dr^3\rho_3^2\big)I^{*4} \;-\; di^2\big(da^4di^2\rho_1^2 \;+$$

$$2\,da^4dr^2\rho_1^2 \;+\; 4\,da^3di^2dr\rho_1\rho_3 \;+\; 2\,da^3di\,dr^2\rho_1\rho_3 \;+\; 6\,da^3dr^3\rho_1\rho_3 \;+\; 2\,da^2di^2dr^2\rho_1^2 \;+\; 4\,da^2di^2dr^2\rho_1\rho_3 \;+\; 8\,da^2di^2dr^2\rho_3^2 \;-$$

$$2\,da^2di\,dr^3\rho_1\rho_3 \;+\; 5\,da^2dr^4\rho_3^2 \;+\; 4\,da\,di^2dr^3\rho_1\rho_3 \;+\; 4\,di^2dr^4\rho_3^2\big)I^{*2} \;+\; 2\,da^2di^2dr^2\big(da^2di^2 \;+\; da^2dr^2 \;+\; di^2dr^2\big).$$

$$b_4 \;=\; T_{reg}^{*\,8}\delta_1^4\delta_2^4 \;+\; 4\,\delta_1^3\delta_2^3\,(da\delta_2 + di\delta_1)\,T_{reg}^{*\,7} \;+\; \Big(\;-\;\delta_1^2\delta_2^2\big(2\,\delta_1^2\rho_1^2 \;+\; 2\,\delta_1\delta_2\rho_1^2 \;+\; 6\,\delta_1\delta_2\rho_1\rho_3 \;+\; \delta_2^2\rho_1^2 \;+$$

$$4\,\delta_2^2\rho_3^2\big)I^{*2} \;+\; 2\,\delta_1^2\delta_2^2\big(3\,da^2\delta_2^2 \;+\; 8\,da\,di\delta_1\delta_2 \;+\; 3\,di^2\delta_1^2 \;+\; 2\,dr^2\delta_1^2 \;+\; 2\,dr^2\delta_2^2\big)\big)T_{reg}^{*\,6} \;+\; \Big(\;-\;2\,\delta_1\delta_2\big(3\,da\delta_1^2\delta_2\rho_1^2 \;+$$

$$2\,da\delta_1\delta_2^2\rho_1^2 \;+\; 8\,da\delta_1\delta_2^2\rho_1\rho_3 \;+\; 4\,da\delta_2^3\rho_3^2 \;+\; di\delta_1^3\rho_1^2 \;+\; 3\,di\delta_1^2\delta_2\rho_1^2 \;+\; 8\,di\delta_1^2\delta_2\rho_1\rho_3 \;+\; di\delta_1\delta_2^2\rho_1^2 \;+\; 2\,di\delta_1\delta_2^2\rho_1\rho_3 \;+$$

$$8\,di\delta_1\delta_2^2\rho_3^2 \;-\; dr\delta_1^2\delta_2\rho_1\rho_3 \;-\; 2\,dr\delta_1\delta_2^2\rho_1\rho_3 \;-\; 2\,dr\delta_2^3\rho_1\rho_3\big)I^{*2} \;+\; 4\,\delta_1\delta_2\big(da^3\delta_2^3 \;+\; 6\,da^2di\delta_1\delta_2^2 \;+\; 6\,da\,di^2\delta_1^2\delta_2 \;+$$

$$4\,da\,dr^2\delta_1^2\delta_2 \;+\; 2\,da\,dr^2\delta_2^3 \;+\; di^3\delta_1^3 \;+\; 2\,di\,dr^2\delta_1^3 \;+\; 4\,di\,dr^2\delta_1\delta_2^2\big)\big)T_{reg}^{*\,5} \;+\; \Big(2\,\rho_3\delta_2\rho_1\big(4\,\delta_1^3\rho_1^2 \;+\; 4\,\delta_1^2\delta_2\rho_1\rho_3 \;-$$

$$2\,\delta_1\delta_2^2\rho_1^2 \;+\; 6\,\delta_1\delta_2^2\rho_1\rho_3 \;+\; 4\,\delta_1\delta_2^2\rho_3^2 \;-\; \delta_2^3\rho_1\rho_3 \;+\; 4\,\delta_2^3\rho_3^2\big)Y^4 \;+\; \Big(\;-\;7\,da^2\delta_1^2\delta_2^2\rho_1^2 \;-\; 2\,da^2\delta_1\delta_2^3\rho_1^2 \;-$$

$$14\,da^2\delta_1\delta_2^3\rho_1\rho_3 \;-\; 4\,da^2\delta_2^4\rho_3^2 \;-\; 6\,da\,di\delta_1^3\delta_2\rho_1^2 \;-\; 12\,da\,di\delta_1^2\delta_2^2\rho_1^2 \;-\; 42\,da\,di\delta_1^2\delta_2^2\rho_1\rho_3 \;+\; 2\,da\,di\delta_1\delta_2^3\rho_1^2 \;-$$

$$8\,da\,di\delta_1\delta_2^3\rho_1\rho_3 \;-\; 32\,da\,di\delta_1\delta_2^3\rho_3^2 \;+\; 2\,da\,dr\delta_1^2\delta_2^2\rho_1\rho_3 \;+\; 8\,da\,dr\delta_1\delta_2^3\rho_1\rho_3 \;-\; di^2\delta_1^4\rho_1^2 \;-\; 6\,di^2\delta_1^3\delta_2\rho_1^2 \;-$$

$$14\,di^2\delta_1^3\delta_2\rho_1\rho_3 \;-\; 2\,di^2\delta_1^2\delta_2^2\rho_1^2 \;-\; 12\,di^2\delta_1^2\delta_2^2\rho_1\rho_3 \;-\; 24\,di^2\delta_1^2\delta_2^2\rho_3^2 \;+\; 10\,di\,dr\delta_1^2\delta_2^2\rho_1\rho_3 \;+\; 12\,di\,dr\delta_1\delta_2^3\rho_1\rho_3 \;-$$

$$dr^2\delta_1^4\rho_1^2 \;-\; 2\,dr^2\delta_1^3\delta_2\rho_1^2 \;-\; 6\,dr^2\delta_1^3\delta_2\rho_1\rho_3 \;-\; 3\,dr^2\delta_1^2\delta_2^2\rho_1^2 \;-\; 2\,dr^2\delta_1^2\delta_2^2\rho_1\rho_3 \;-\; 10\,dr^2\delta_1^2\delta_2^2\rho_3^2 \;-\; 2\,dr^2\delta_1\delta_2^3\rho_1^2 \;-$$

$$6\,dr^2\delta_1\delta_2^3\rho_1\rho_3 \;-\; 8\,dr^2\delta_1\delta_2^4\rho_3^2\big)I^{*2} \;+\; da^4\delta_2^4 \;+\; 16\,da^3di\delta_1\delta_2^3 \;+\; 36\,da^2di^2\delta_1^2\delta_2^2 \;+\; 24\,da^2dr^2\delta_1^2\delta_2^2 \;+\; 4\,da^2dr^2\delta_2^4 \;+$$

$$16\,da\,di^3\delta_1^3\delta_2 \;+\; 32\,da\,di\,dr^2\delta_1^3\delta_2 \;+\; 32\,da\,di\,dr^2\delta_1\delta_2^3 \;+\; di^4\delta_1^4 \;+\; 4\,di^2dr^2\delta_1^4 \;+\; 24\,di^2dr^2\delta_1^2\delta_2^2 \;+\; dr^4\delta_1^4 \;+\; 4\,dr^4\delta_1^2\delta_2^2 \;+$$

$$dr^4\delta_2^4\big)T_{reg}^{*\,4} \;+\; \Big(2\,\rho_3\rho_1\big(10\,da\delta_1^2\delta_2\rho_1^2 \;-\; 2\,da\delta_1\delta_2^2\rho_1^2 \;+\; 8\,da\delta_1\delta_2^2\rho_1\rho_3 \;+\; 4\,da\delta_2^3\rho_3^2 \;+\; 3\,di\delta_1^3\rho_1^2 \;+\; 3\,di\delta_1^2\delta_2\rho_1^2 \;+$$

$$6\,di\delta_1^2\delta_2\rho_1\rho_3 \;-\; 2\,di\delta_1\delta_2^2\rho_1^2 \;+\; 20\,di\delta_1\delta_2^2\rho_1\rho_3 \;+\; 12\,di\delta_1\delta_2^2\rho_3^2 \;-\; 2\,di\delta_2^3\rho_1\rho_3 \;+\; 20\,di\delta_2^3\rho_3^2 \;+\; dr\delta_1^3\rho_1^2 \;-$$

$$2\,dr\delta_1^2\delta_2\rho_1^2 \;-\; 5\,dr\delta_1\delta_2^2\rho_1^2 \;+\; 3\,dr\delta_1\delta_2^2\rho_3^2 \;-\; 4\,dr\delta_2^3\rho_3^2\big)I^{*4} \;+\; \Big(\;-\;4\,da^3\delta_1\delta_2^2\rho_1^2 \;-\; 4\,da^3\delta_2^3\rho_1\rho_3 \;-\; 8\,da^2di\delta_1^2\delta_2\rho_1^2 \;-$$

$$6\,da^2di\delta_1\delta_2^2\rho_1^2 \;-\; 36\,da^2di\delta_1\delta_2^2\rho_1\rho_3 \;-\; 4\,da^2di\delta_2^3\rho_1\rho_3 \;-\; 16\,da^2di\delta_2^3\rho_3^2 \;-\; 2\,da^2dr\delta_1\delta_2^2\rho_1\rho_3 \;+\; 4\,da^2dr\delta_2^3\rho_1\rho_3 \;-$$

$$4\,da\,di^2\delta_1^3\rho_1^2 \;-\; 12\,da\,di^2\delta_1^2\delta_2\rho_1^2 \;-\; 36\,da\,di^2\delta_1^2\delta_2\rho_1\rho_3 \;+\; 2\,da\,di^2\delta_1\delta_2^2\rho_1^2 \;-\; 24\,da\,di^2\delta_1\delta_2^2\rho_1\rho_3 \;-\; 48\,da\,di^2\delta_1\delta_2^2\rho_3^2 \;-$$

$$8\,da\,di\,dr\delta_1^2\delta_2\rho_1\rho_3 \;+\; 20\,da\,di\,dr\delta_1\delta_2^2\rho_1\rho_3 \;-\; 4\,da\,di\,dr\delta_2^3\rho_1\rho_3 \;-\; 4\,da\,dr^2\delta_1^3\rho_1^2 \;-\; 4\,da\,dr^2\delta_1^2\delta_2\rho_1^2 \;-\; 16\,da\,dr^2\delta_1^2\delta_2\rho_1\rho_3 \;-$$

$$4\,da\,dr^2\delta_1\delta_2^2\rho_1^2 \;-\; 4\,da\,dr^2\delta_1\delta_2^2\rho_1\rho_3 \;-\; 20\,da\,dr^2\delta_1\delta_2^2\rho_3^2 \;-\; 4\,da\,dr^2\delta_2^3\rho_1\rho_3 \;-\; 2\,di^3\delta_1^3\rho_1^2 \;-\; 4\,di^3\delta_1^3\rho_1\rho_3 \;-$$

$$2\,di^3\delta_1^2\delta_2\rho_1^2 \;-\; 12\,di^3\delta_1^2\delta_2\rho_1\rho_3 \;-\; 16\,di^3\delta_1^2\delta_2\rho_3^2 \;-\; 2\,di^2dr\delta_1^3\rho_1\rho_3 \;+\; 8\,di^2dr\delta_1^2\delta_2\rho_1\rho_3 \;+\; 12\,di^2dr\delta_1\delta_2^2\rho_1\rho_3 \;-$$

$$2\,di\,dr^2\delta_1^3\rho_1^2 \;-\; 4\,di\,dr^2\delta_1^3\rho_1\rho_3 \;-\; 4\,di\,dr^2\delta_1^2\delta_2\rho_1^2 \;-\; 8\,di\,dr^2\delta_1^2\delta_2\rho_1\rho_3 \;-\; 20\,di\,dr^2\delta_1^2\delta_2\rho_3^2 \;-\; 6\,di\,dr^2\delta_1\delta_2^2\rho_1^2 \;-$$

$$16\,di\,dr^2\delta_1\delta_2^2\rho_1\rho_3 \;-\; 4\,di\,dr^2\delta_2^3\rho_1\rho_3 \;-\; 32\,di\,dr^2\delta_2^3\rho_3^2 \;-\; 2\,dr^3\delta_1^3\rho_1\rho_3 \;+\; 4\,dr^3\delta_1^2\delta_2\rho_1\rho_3 \;+\; 2\,dr^3\delta_1\delta_2^2\rho_1\rho_3 \;+$$

$$4\,dr^3\delta_2^3\rho_1\rho_3\big)I^{*2} \;+\; 4\,da^4di\delta_2^3 \;+\; 24\,da^3di^2\delta_1\delta_2^2 \;+\; 16\,da^3dr^2\delta_1\delta_2^2 \;+\; 24\,di^3\delta_1^2\delta_2da^2 \;+\; 48\,da^2di\,dr^2\delta_1^2\delta_2 \;+$$

$$16\,da^2di\,dr^2\delta_2^3 \;+\; 4\,da\,di^4\delta_1^3 \;+\; 16\,da\,di^2\delta_1^3dr^2 \;+\; 48\,da\,di^2dr^2\delta_1\delta_2^2 \;+\; 4\,da\,dr^4\delta_1^3 \;+\; 8\,da\,dr^4\delta_1\delta_2^2 \;+\; 16\,di^3\delta_1^2\delta_2dr^2 \;+$$

$$8\,di\,dr^4\delta_1^2\delta_2 \;+\; 4\,di\,dr^4\delta_1^3\delta_2\big)T_{reg}^{*\,3} \;+\; \Big(\;-\;\rho_1^2\rho_3^2\big(\delta_1^2\rho_1^2 \;+\; 2\,\delta_1\delta_2\rho_1^2 \;+\; 6\,\delta_1\delta_2\rho_1\rho_3 \;+\; \delta_2^2\rho_1^2 \;+\; 2\,\delta_2^2\rho_1\rho_3 \;+\; 5\,\delta_2^2\rho_3^2\big)I^{*6} \;+$$

$$2\,\rho_3\rho_1\big(8\,da^2\delta_1\delta_2\rho_1^2 \;-\; da^2\delta_2^2\rho_1^2 \;+\; 4\,da^2\delta_2^2\rho_1\rho_3 \;+\; 7\,da\,di\delta_1^2\rho_1^2 \;+\; 2\,da\,di\delta_1\delta_2\rho_1^2 \;+\; 14\,da\,di\delta_1\delta_2\rho_1\rho_3 \;+\; 2\,da\,di\delta_2^2\rho_1^2 \;+$$

$$12\,da\,di\delta_2^2\rho_3^2 \;+\; 3\,da\,dr\delta_1^2\rho_1^2 \;-\; 4\,da\,dr\delta_1\delta_2^2\rho_1^2 \;-\; da\,dr\delta_2^2\rho_1^2 \;+\; 3\,da\,dr\delta_2^2\rho_3^2 \;+\; 4\,di^2\delta_1^2\rho_1^2 \;+\; 3\,di^2\delta_1^2\rho_1\rho_3 \;+$$

$$4\,di^2\delta_1\delta_2\rho_1^2 \;+\; 22\,di^2\delta_1\delta_2\rho_1\rho_3 \;+\; 12\,di^2\delta_1\delta_2\rho_3^2 \;+\; di^2\delta_2^2\rho_1^2 \;+\; 3\,di^2\delta_2^2\rho_1\rho_3 \;+\; 36\,di^2\delta_2^2\rho_3^2 \;-\; di\,dr\delta_1^2\rho_1^2 \;-$$

$$7\,di\,dr\delta_1\delta_2\rho_1^2 \;+\; 9\,di\,dr\delta_1\delta_2\rho_3^2 \;-\; 2\,di\,dr\delta_2^2\rho_1^2 \;-\; 12\,di\,dr\delta_2^2\rho_3^2 \;+\; dr^2\delta_1^2\rho_1\rho_3 \;+\; 2\,dr^2\delta_1\delta_2\rho_1\rho_3 \;+\; 2\,dr^2\delta_1\delta_2\rho_3^2 \;+$$

$$4d_r^2\delta_2^2\rho_1\rho_3 + 3d_r^2\delta_2^2\rho_3^2\big)I^{*4} + \Big(-d_a^4\delta_2^2\rho_1^2 - 6d_a^3d_i\delta_1\delta_2\rho_1^2 - 10d_a^3d_i\delta_2^2\rho_1\rho_3 - 2d_a^3d_r\delta_2^2\rho_1\rho_3 -$$
$$7d_a^2d_i^2\delta_1^2\rho_1^2 - 6d_a^2d_i^2\delta_1\delta_2\rho_1^2 - 30d_a^2d_i^2\delta_1\delta_2\rho_1\rho_3 - 2d_a^2d_i^2\delta_2^2\rho_1^2 - 12d_a^2d_i^2\delta_2^2\rho_1\rho_3 - 24d_a^2d_i^2\delta_2^2\rho_3^2 -$$
$$16d_a^2d_id_r\delta_1\delta_2\rho_1\rho_3 + 10d_a^2d_id_r\delta_2^2\rho_1\rho_3 - 6d_a^2d_r^2\delta_1^2\rho_1^2 - 2d_a^2d_r^2\delta_1\delta_2\rho_1^2 - 14d_a^2d_r^2\delta_1\delta_2\rho_1\rho_3 - 2d_a^2d_r^2\delta_2^2\rho_1^2 -$$
$$2d_a^2d_r^2\delta_2^2\rho_1\rho_3 - 10d_a^2d_r^2\delta_2^2\rho_3^2 - 4d_ad_i^3\delta_1^2\rho_1^2 - 10d_ad_i^3\delta_1^2\rho_1\rho_3 - 2d_ad_i^3\delta_1\delta_2\rho_1^2 - 24d_ad_i^3\delta_1\delta_2\rho_1\rho_3 -$$
$$32d_ad_i^3\delta_1\delta_2\rho_3^2 - 10d_ad_i^2d_r\delta_1^2\rho_1\rho_3 + 16d_ad_i^2d_r\delta_1\delta_2\rho_1\rho_3 - 12d_ad_i^2d_r\delta_2^2\rho_1\rho_3 - 4d_ad_id_r\delta_1^2\rho_1^2 -$$
$$10d_ad_id_r^2\delta_1^2\rho_1\rho_3 - 6d_ad_id_r^2\delta_1\delta_2\rho_1^2 - 16d_ad_id_r^2\delta_1\delta_2\rho_1\rho_3 - 40d_ad_id_r^2\delta_1\delta_2\rho_3^2 - 10d_ad_id_r^2\delta_2^2\rho_1\rho_3 -$$
$$6d_ad_r^3\delta_1^2\rho_1\rho_3 + 8d_ad_r^3\delta_1\delta_2\rho_1\rho_3 - 2d_ad_r^3\delta_2^2\rho_1\rho_3 - d_i^4\delta_1^2\rho_1^2 - 4d_i^4\delta_1^2\rho_1\rho_3 - 4d_i^4\delta_1^2\rho_3^2 + 2d_i^3d_r\delta_1^2\rho_1\rho_3 +$$
$$4d_i^3d_r\delta_1\delta_2\rho_1\rho_3 - 3d_i^2d_r^2\delta_1^2\rho_1^2 - 6d_i^2d_r^2\delta_1^2\rho_1\rho_3 - 10d_i^2d_r^2\delta_1^2\rho_3^2 - 6d_i^2d_r^2\delta_1\delta_2\rho_1^2 - 14d_i^2d_r^2\delta_1\delta_2\rho_1\rho_3 -$$
$$d_i^2d_r^2\delta_2^2\rho_1^2 - 12d_i^2d_r^2\delta_2^2\rho_1\rho_3 - 48d_i^2d_r^2\delta_2^2\rho_3^2 + 2d_id_r^3\delta_1^2\rho_1\rho_3 + 10d_id_r^3\delta_2^2\rho_1\rho_3 - d_r^4\delta_1^2\rho_3^2 -$$
$$5d_r^4\delta_2^2\rho_3^2\big)I^{*2} + 6d_a^4d_i^2\delta_2^2 + 4d_a^4d_r^2\delta_2^2 + 16d_a^3d_i^2\delta_1\delta_2 + 32d_a^3d_id_r^2\delta_1\delta_2 + 6d_a^2d_i^4\delta_1^2 + 24d_a^2d_i^2d_r^2\delta_1^2 +$$
$$24d_a^2d_i^2d_r^2\delta_2^2 + 6d_a^2d_r^4\delta_1^2 + 4d_a^2d_r^4\delta_2^2 + 32d_ad_i^3d_r^2\delta_1\delta_2 + 16d_ad_id_r^4\delta_1\delta_2 + 4d_i^4d_r^2\delta_1^2 + 4d_i^2d_r^4\delta_1^2 +$$
$$6d_i^2d_r^4\delta_2^2\big)T_{reg}^{*2} + \Big(-2\rho_1^2\rho_3^2\big(d_a\delta_1\rho_1^2 + 2d_a\delta_2\rho_1\rho_3 + d_i\delta_1\rho_1^2 + 2d_i\delta_1\rho_1\rho_3 + 2d_i\delta_2\rho_1^2 + 6d_i\delta_2\rho_1\rho_3 +$$
$$6d_i\delta_2\rho_3^2 + d_r\delta_1\rho_1\rho_3 - 2d_r\delta_2\rho_1\rho_3\big)I^{*6} + 2\rho_3\rho_1\big(2d_a^3\delta_2\rho_1^2 + 5d_a^2d_i\delta_1\rho_1^2 + d_a^2d_i\delta_2\rho_1^2 + 8d_a^2d_i\delta_2\rho_1\rho_3 +$$
$$3d_a^2d_r\delta_1\rho_1^2 - 2d_a^2d_r\delta_2\rho_1^2 + 6d_ad_i^2\delta_1\rho_1^2 + 8d_ad_i^2\delta_1\rho_1\rho_3 + 4d_ad_i^2\delta_2\rho_1^2 + 12d_ad_i^2\delta_2\rho_3^2 - 2d_ad_id_r\delta_1\rho_1^2 +$$
$$d_ad_id_r\delta_2\rho_1^2 + 9d_ad_id_r\delta_2\rho_3^2 + 2d_ad_r^2\delta_1\rho_1\rho_3 + 2d_ad_r^2\delta_2\rho_3^2 + 3d_i^3\delta_1\rho_1^2 + 8d_i^3\delta_1\rho_1\rho_3 + 4d_i^3\delta_1\rho_3^2 + 3d_i^3\delta_2\rho_1^2 +$$
$$8d_i^3\delta_2\rho_1\rho_3 + 28d_i^3\delta_2\rho_3^2 - d_i^2d_r\delta_1\rho_1^2 + 6d_i^2d_r\delta_1\rho_3^2 - 4d_i^2d_r\delta_2\rho_1^2 - 12d_i^2d_r\delta_2\rho_3^2 + 2d_id_r^2\delta_1\rho_1\rho_3 + d_id_r\delta_1\rho_3^2 +$$
$$8d_id_r^2\delta_2\rho_1\rho_3 + 9d_id_r^2\delta_2\rho_3^2 + d_r^3\delta_1\rho_3^2 - 2d_r^3\delta_2\rho_3^2\big)I^{*4} + \Big(-2d_a^4\delta_2\rho_1^2 - 6d_a^3d_i^2\delta_1\rho_1^2 - 8d_a^3d_i^2\delta_2\rho_1\rho_3 -$$
$$8d_a^3d_id_r\delta_2\rho_1\rho_3 - 4d_a^3d_r^2\delta_1\rho_1^2 - 4d_a^3d_r^2\delta_2\rho_1\rho_3 - 2d_a^2d_i^3\delta_1\rho_1^2 - 8d_a^2d_i^3\delta_1\rho_1\rho_3 - 4d_a^2d_i^3\delta_2\rho_1^2 -$$
$$12d_a^2d_i^3\delta_2\rho_1\rho_3 - 16d_a^2d_i^3\delta_2\rho_3^2 - 14d_a^2d_i^2d_r\delta_1\rho_1\rho_3 + 8d_a^2d_i^2d_r\delta_2\rho_1\rho_3 - 2d_a^2d_id_r\delta_1\rho_1^2 - 8d_a^2d_id_r\delta_1\rho_1\rho_3 -$$
$$4d_a^2d_id_r^2\delta_2\rho_1^2 - 8d_a^2d_id_r^2\delta_2\rho_1\rho_3 - 20d_a^2d_id_r^2\delta_2\rho_3^2 - 6d_a^2d_r^3\delta_1\rho_1\rho_3 + 4d_a^2d_r^3\delta_2\rho_1\rho_3 - 2d_ad_i^4\delta_1\rho_1^2 -$$
$$8d_ad_i^4\delta_1\rho_1\rho_3 - 8d_ad_i^4\delta_1\rho_3^2 + 4d_ad_i^3d_r\delta_1\rho_1\rho_3 - 12d_ad_i^3d_r\delta_2\rho_1\rho_3 - 6d_ad_i^2d_r^2\delta_1\rho_1^2 - 12d_ad_i^2d_r^2\delta_1\rho_1\rho_3 -$$
$$20d_ad_i^2d_r^2\delta_1\rho_3^2 - 8d_ad_i^2d_r^2\delta_2\rho_1\rho_3 + 4d_ad_id_r^3\delta_1\rho_1\rho_3 - 8d_ad_id_r^3\delta_2\rho_1\rho_3 - 2d_ad_r^4\delta_1\rho_3^2 - 2d_i^3d_r^2\delta_1\rho_1^2 -$$
$$4d_i^3d_r^2\delta_1\rho_1\rho_3 - 2d_i^3d_r^2\delta_2\rho_1^2 - 12d_i^3d_r^2\delta_2\rho_1\rho_3 - 32d_i^3d_r^2\delta_2\rho_3^2 - 2d_i^2d_r^3\delta_1\rho_1\rho_3 + 8d_i^2d_r^3\delta_2\rho_1\rho_3 -$$
$$10d_id_r^4\delta_2\rho_3^2\big)I^{*2} + 4d_a^4d_i^3\delta_2 + 8d_a^4d_id_r^2\delta_2 + 4d_a^3d_i^4\delta_1 + 16d_a^3d_i^2d_r^2\delta_1 + 4d_a^3d_r^4\delta_1 + 16d_a^2d_i^3d_r^2\delta_2 +$$
$$8d_a^2d_id_r^4\delta_2 + 8d_ad_i^4d_r^2\delta_1 + 8d_ad_i^2d_r^4\delta_1 + 4d_i^3d_r^4\delta_2\big)T_{reg}^* + I^{*8}\rho_1^4\rho_3^4 - \rho_1^2\rho_3^2\big(d_a^2\rho_1^2 + 2d_ad_i\rho_1\rho_3 +$$
$$2d_ad_r\rho_1\rho_3 + 5d_i^2\rho_1^2 + 12d_i^2\rho_1\rho_3 + 8d_i^2\rho_3^2 - 2d_id_r\rho_1\rho_3 + d_r^2\rho_3^2\big)I^{*6} + 2\rho_3\rho_1\big(d_a^3d_i\rho_1^2 + d_a^3d_r\rho_1^2 +$$
$$3d_a^2d_i^2\rho_1^2 + 5d_a^2d_i^2\rho_1\rho_3 - d_a^2d_id_r\rho_1^2 + d_a^2d_r^2\rho_1\rho_3 + d_ad_i^3\rho_1^2 + 4d_ad_i^3\rho_3^2 + 3d_ad_i^2d_r\rho_1^2 + 6d_ad_i^2d_r\rho_3^2 +$$
$$d_ad_id_r^2\rho_3^2 + d_ad_r^3\rho_3^2 + 2d_i^4\rho_1^2 + 4d_i^4\rho_1\rho_3 + 8d_i^4\rho_3^2 - d_i^3d_r\rho_1^2 - 4d_i^3d_r\rho_3^2 + 5d_i^2d_r^2\rho_1\rho_3 + 6d_i^2d_r^2\rho_3^2 -$$
$$d_id_r^3\rho_3^2\big)I^{*4} + \Big(-2d_a^4d_i^2\rho_1^2 - d_a^4d_r^2\rho_1^2 - 2d_a^3d_i^3\rho_1\rho_3 - 6d_a^3d_i^2d_r\rho_1\rho_3 - 2d_a^3d_id_r^2\rho_1\rho_3 - 2d_a^3d_r^3\rho_1\rho_3 -$$
$$2d_a^2d_i^4\rho_1^2 - 4d_a^2d_i^4\rho_1\rho_3 - 4d_a^2d_i^4\rho_3^2 + 2d_a^2d_i^3d_r\rho_1\rho_3 - 4d_a^2d_i^2d_r^2\rho_1^2 - 6d_a^2d_i^2d_r^2\rho_1\rho_3 - 10d_a^2d_i^2d_r^2\rho_3^2 +$$
$$2d_a^2d_id_r^3\rho_1\rho_3 - d_a^2d_r^4\rho_3^2 - 4d_ad_i^4d_r\rho_1\rho_3 - 2d_ad_i^3d_r^2\rho_1\rho_3 - 6d_ad_i^2d_r^3\rho_1\rho_3 - d_i^4d_r^2\rho_1^2 - 4d_i^4d_r^2\rho_1\rho_3 -$$
$$8d_i^4d_r^2\rho_3^2 + 2d_i^3d_r^3\rho_1\rho_3 - 5d_i^2d_r^4\rho_3^2\big)I^{*2} + d_a^4d_i^4 + 4d_a^4d_i^2d_r^2 + d_a^4d_r^4 + 4d_a^2d_i^4d_r^2 + 4d_a^2d_i^2d_r^4 + d_i^4d_r^4.$$

$$b_6 = 2\delta_1^2\delta_2^2\big(\delta_1^2 + \delta_2^2\big)T_{reg}^{*6} + 4\delta_1\delta_2\big(2d_a\delta_1^2\delta_2 + d_a\delta_2^3 + d_i\delta_1^3 + 2d_i\delta_1\delta_2^2\big)T_{reg}^{*5} + \Big(\big(-\delta_1^4\rho_1^2 -$$
$$2\delta_1^3\delta_2\rho_1^2 - 6\delta_1^3\delta_2\rho_1\rho_3 - 3\delta_1^2\delta_2^2\rho_1^2 - 2\delta_1^2\delta_2^2\rho_1\rho_3 - 5\delta_1^2\delta_2^2\rho_3^2 - 2\delta_1\delta_2^3\rho_1^2 - 6\delta_1\delta_2^3\rho_1\rho_3 - 4\delta_2^4\rho_3^2\big)I^{*2} +$$
$$12d_a^2\delta_1^2\delta_2^2 + 2d_a^2\delta_2^4 + 16d_ad_i\delta_1^3\delta_2 + 16d_ad_i\delta_1\delta_2^3 + 2d_i^2\delta_1^4 + 12d_i^2\delta_1^2\delta_2^2 + 2d_r^2\delta_1^4 + 8d_r^2\delta_1^2\delta_2^2 +$$
$$2d_r^2\delta_2^4\big)T_{reg}^{*4} + \Big(\big(-4d_a\delta_1^3\rho_1^2 - 4d_a\delta_1^2\delta_2\rho_1^2 - 16d_a\delta_1^2\delta_2\rho_1\rho_3 - 4d_a\delta_1\delta_2^2\rho_1^2 - 4d_a\delta_1\delta_2^2\rho_1\rho_3 - 10d_a\delta_1\delta_2^2\rho_3^2 -$$
$$4d_a\delta_2^3\rho_1\rho_3 - 2d_i\delta_1^3\rho_1^2 - 4d_i\delta_1^3\rho_1\rho_3 - 4d_i\delta_1^2\delta_2\rho_1^2 - 8d_i\delta_1^2\delta_2\rho_1\rho_3 - 10d_i\delta_1^2\delta_2\rho_3^2 - 6d_i\delta_1\delta_2^2\rho_1^2 -$$
$$16d_i\delta_1\delta_2^2\rho_1\rho_3 - 4d_i\delta_2^3\rho_1\rho_3 - 16d_i\delta_2^3\rho_3^2 - 2d_r\delta_1^3\rho_1\rho_3 + 4d_r\delta_1^2\delta_2\rho_1\rho_3 + 2d_r\delta_1\delta_2^2\rho_1\rho_3 + 4d_r\delta_2^3\rho_1\rho_3\big)I^{*2} +$$
$$8d_a^3\delta_1\delta_2^2 + 24d_a^2d_i\delta_1^2\delta_2 + 8d_a^2d_i\delta_2^3 + 8d_ad_i^2\delta_1^3 + 24d_ad_i^2\delta_1\delta_2^2 + 8d_ad_r^2\delta_1^3 + 16d_ad_r^2\delta_1\delta_2^2 + 8d_i^3\delta_1^2\delta_2 +$$
$$16d_id_r^2\delta_1^2\delta_2 + 8d_id_r^2\delta_2^3\big)T_{reg}^{*3} + \Big(2\rho_1\rho_3\big(\delta_1^2\rho_1^2 + \delta_1^2\rho_1\rho_3 + 4\delta_1\delta_2\rho_1^2 + 2\delta_1\delta_2\rho_1\rho_3 + 2\delta_1\delta_2\rho_3^2 - \delta_2^2\rho_1^2 +$$
$$4\delta_2^2\rho_1\rho_3 + 3\delta_2^2\rho_3^2\big)I^{*4} + \Big(-6d_a^2\delta_1^2\rho_1^2 - 2d_a^2\delta_1\delta_2\rho_1^2 - 14d_a^2\delta_1\delta_2\rho_1\rho_3 - 2d_a^2\delta_2^2\rho_1^2 - 2d_a^2\delta_2^2\rho_1\rho_3 -$$
$$5d_a^2\delta_2^2\rho_3^2 - 4d_ad_i\delta_1^2\rho_1^2 - 10d_ad_i\delta_1^2\rho_1\rho_3 - 6d_ad_i\delta_1\delta_2\rho_1^2 - 16d_ad_i\delta_1\delta_2\rho_1\rho_3 - 20d_ad_i\delta_1\delta_2\rho_3^2 - 10d_ad_i\delta_2^2\rho_1\rho_3 -$$
$$6d_ad_r\delta_1^2\rho_1\rho_3 + 8d_ad_r\delta_1\delta_2\rho_1\rho_3 - 2d_ad_r\delta_2^2\rho_1\rho_3 - 3d_i^2\delta_1^2\rho_1^2 - 6d_i^2\delta_1^2\rho_1\rho_3 - 5d_i^2\delta_1^2\rho_3^2 - 6d_i^2\delta_1\delta_2\rho_1^2 -$$
$$14d_i^2\delta_1\delta_2\rho_1\rho_3 - d_i^2\delta_2^2\rho_1^2 - 12d_i^2\delta_2^2\rho_1\rho_3 - 24d_i^2\delta_2^2\rho_3^2 + 2d_id_r\delta_1\rho_1\rho_3 + 10d_id_r\delta_2^2\rho_1\rho_3 - 2d_r^2\delta_1^2\rho_1^2 -$$
$$2d_r^2\delta_1^2\rho_1\rho_3 - 2d_r^2\delta_1^2\rho_3^2 - 2d_r^2\delta_1\delta_2\rho_1^2 - 6d_r^2\delta_1\delta_2\rho_1\rho_3 - d_r^2\delta_2^2\rho_1^2 - 2d_r^2\delta_2^2\rho_1\rho_3 - 10d_r^2\delta_2^2\rho_3^2\big)I^{*2} +$$
$$2d_a^4\delta_2^2 + 16d_a^3d_i\delta_1\delta_2 + 12d_a^2d_i^2\delta_1^2 + 12d_a^2d_i^2\delta_2^2 + 12d_a^2d_r^2\delta_1^2 + 8d_a^2d_r^2\delta_2^2 + 16d_ad_i^3\delta_1\delta_2 + 32d_ad_id_r^2\delta_1\delta_2 +$$
$$2d_i^4\delta_1^2 + 8d_i^2d_r^2\delta_1^2 + 12d_i^2d_r^2\delta_2^2 + 2d_r^4\delta_1^2 + 2d_r^4\delta_2^2\big)T_{reg}^{*2} + \Big(2\rho_1\rho_3\big(2d_a\delta_1\rho_1^2 + 2d_a\delta_1\rho_1\rho_3 + 2d_a\delta_2\rho_1^2 +$$

$$2 d_a \delta_2 \rho_3{}^2 + 3 d_i \delta_1 \rho_1{}^2 + 2 d_i \delta_1 \rho_1 \rho_3 + d_i \delta_1 \rho_3{}^2 + d_i \delta_2 \rho_1{}^2 + 8 d_i \delta_2 \rho_1 \rho_3 + 9 d_i \delta_2 \rho_3{}^2 + d_r \delta_1 \rho_1{}^2 + d_r \delta_1 \rho_3{}^2 - 2 d_r \delta_2 \rho_1{}^2 -$$
$$2 d_r \delta_2 \rho_3{}^2 \big) I^{*4} + \big(- 4 d_a{}^3 \delta_1 \rho_1{}^2 - 4 d_a{}^3 \delta_2 \rho_1 \rho_3 - 2 d_a{}^2 d_i \delta_1 \rho_1{}^2 - 8 d_a{}^2 d_i \delta_1 \rho_1 \rho_3 - 4 d_a{}^2 d_i \delta_2 \rho_1{}^2 - 8 d_a{}^2 d_i \delta_2 \rho_1 \rho_3 -$$
$$10 d_a{}^2 d_i \delta_2 \rho_3{}^2 - 6 d_a{}^2 d_r \delta_1 \rho_1 \rho_3 + 4 d_a{}^2 d_r \delta_2 \rho_1 \rho_3 - 6 d_a d_i{}^2 \delta_1 \rho_1{}^2 - 12 d_a d_i{}^2 \delta_1 \rho_1 \rho_3 - 10 d_a d_i{}^2 \delta_1 \rho_3{}^2 - 8 d_a d_i{}^2 \delta_2 \rho_1 \rho_3 +$$
$$4 d_a d_i d_r \delta_1 \rho_1 \rho_3 - 8 d_a d_i d_r \delta_2 \rho_1 \rho_3 - 4 d_a d_r{}^2 \delta_1 \rho_1{}^2 - 4 d_a d_r{}^2 \delta_1 \rho_1 \rho_3 - 4 d_a d_r{}^2 \delta_1 \rho_3{}^2 - 4 d_a d_r{}^2 \delta_2 \rho_1 \rho_3 - 2 d_i{}^3 \delta_1 \rho_1{}^2 -$$
$$4 d_i{}^3 \delta_1 \rho_1 \rho_3 - 2 d_i{}^3 \delta_2 \rho_1{}^2 - 12 d_i{}^3 \delta_2 \rho_1 \rho_3 - 16 d_i{}^3 \delta_2 \rho_3{}^2 - 2 d_i{}^2 d_r \delta_1 \rho_1 \rho_3 + 8 d_i{}^2 d_r \delta_2 \rho_1 \rho_3 - 2 d_i d_r{}^2 \delta_1 \rho_1{}^2 -$$
$$4 d_i d_r{}^2 \delta_1 \rho_1 \rho_3 - 2 d_i d_r{}^2 \delta_2 \rho_1{}^2 - 8 d_i d_r{}^2 \delta_2 \rho_1 \rho_3 - 20 d_i d_r{}^2 \delta_2 \rho_3{}^2 - 2 d_r{}^3 \delta_1 \rho_1 \rho_3 + 4 d_r{}^3 \delta_2 \rho_1 \rho_3 \big) Y^2 + 4 d_a{}^4 d_i \delta_2 +$$
$$8 d_a{}^3 d_i{}^2 \delta_1 + 8 d_a{}^3 d_r{}^2 \delta_1 + 8 d_a{}^2 d_i{}^3 \delta_2 + 16 d_a{}^2 d_i d_r{}^2 \delta_2 + 4 d_a d_i{}^4 \delta_1 + 16 d_a d_i{}^2 d_r{}^2 \delta_1 + 4 d_a d_r{}^4 \delta_1 + 8 d_i{}^3 d_r{}^2 \delta_2 +$$
$$4 d_i d_r{}^4 \delta_2 \big) T^*_{reg} - \rho_1{}^2 \rho_3{}^2 \big(\rho_1 + \rho_3 \big)^2 I^{*6} + 2 \rho_1 \rho_3 \big(d_a{}^2 \rho_1{}^2 + d_a{}^2 \rho_1 \rho_3 + d_a d_i \rho_1{}^2 + d_a d_i \rho_3{}^2 + d_a d_r \rho_1{}^2 + d_a d_r \rho_3{}^2 +$$
$$3 d_i{}^2 \rho_1{}^2 + 5 d_i{}^2 \rho_1 \rho_3 + 6 d_i{}^2 \rho_3{}^2 - d_i d_r \rho_1{}^2 - d_i d_r \rho_3{}^2 + d_r{}^2 \rho_1 \rho_3 + d_r{}^2 \rho_3{}^2 \big) Y^4 + \big(- d_a{}^4 \rho_1{}^2 - 2 d_a{}^3 d_i \rho_1 \rho_3 -$$
$$2 d_a{}^3 d_r \rho_1 \rho_3 - 4 d_a{}^2 d_i{}^2 \rho_1{}^2 - 6 d_a{}^2 d_i{}^2 \rho_1 \rho_3 - 5 d_a{}^2 d_i{}^2 \rho_3{}^2 + 2 d_a{}^2 d_i d_r \rho_1 \rho_3 - 2 d_a{}^2 d_r{}^2 \rho_1{}^2 - 2 d_a{}^2 d_r{}^2 \rho_1 \rho_3 - 2 d_a{}^2 d_r{}^2 \rho_3{}^2 -$$
$$2 d_a d_i{}^3 \rho_1 \rho_3 - 6 d_a d_i{}^2 d_r \rho_1 \rho_3 - 2 d_a d_i d_r{}^2 \rho_1 \rho_3 - 2 d_a d_r{}^3 \rho_1 \rho_3 - d_i{}^4 \rho_1{}^2 - 4 d_i{}^4 \rho_1 \rho_3 - 4 d_i{}^4 \rho_3{}^2 + 2 d_i{}^3 d_r \rho_1 \rho_3 -$$
$$2 d_i{}^2 d_r{}^2 \rho_1{}^2 - 6 d_i{}^2 d_r{}^2 \rho_1 \rho_3 - 10 d_i{}^2 d_r{}^2 \rho_3{}^2 + 2 d_i d_r{}^3 \rho_1 \rho_3 - d_r{}^4 \rho_3{}^2 \big) I^{*2} + 2 d_a{}^4 d_i{}^2 + 2 d_a{}^4 d_r{}^2 + 2 d_a{}^2 d_i{}^4 + 8 d_a{}^2 d_i{}^2 d_r{}^2 +$$
$$2 d_a{}^2 d_r{}^4 + 2 d_i{}^4 d_r{}^2 + 2 d_i{}^2 d_r{}^4.$$

$$b_8 = \big(\delta_1{}^4 + 4 \delta_1{}^2 \delta_2{}^2 + \delta_2{}^4 \big) T^*_{reg}{}^4 + \big(4 d_a \delta_1{}^3 + 8 d_a \delta_1 \delta_2{}^2 + 8 d_i \delta_1{}^2 \delta_2 + 4 d_i \delta_2{}^3 \big) T^*_{reg}{}^3 + \big(\big(- 2 \delta_1{}^2 \rho_1{}^2 -$$
$$2 \delta_1{}^2 \rho_1 \rho_3 - \delta_1{}^2 \rho_3{}^2 - 2 \delta_1 \delta_2 \rho_1{}^2 - 6 \delta_1 \delta_2 \rho_1 \rho_3 - \delta_2{}^2 \rho_1{}^2 - 2 \delta_2{}^2 \rho_1 \rho_3 - 5 \delta_2{}^2 \rho_3{}^2 \big) I^{*2} + 6 d_a{}^2 \delta_1{}^2 + 4 d_a{}^2 \delta_2{}^2 + 16 d_a d_i \delta_1 \delta_2 +$$
$$4 d_i{}^2 \delta_1{}^2 + 6 d_i{}^2 \delta_2{}^2 + 4 d_r{}^2 \delta_1{}^2 + 4 d_r{}^2 \delta_2{}^2 \big) T^*_{reg}{}^2 + \big(\big(- 4 d_a \delta_1 \rho_1{}^2 - 4 d_a \delta_1 \rho_1 \rho_3 - 2 d_a \delta_1 \rho_3{}^2 - 4 d_a \delta_2 \rho_1 \rho_3 - 2 d_i \delta_1 \rho_1{}^2 -$$
$$4 d_i \delta_1 \rho_1 \rho_3 - 2 d_i \delta_2 \rho_1{}^2 - 8 d_i \delta_2 \rho_1 \rho_3 - 10 d_i \delta_2 \rho_3{}^2 - 2 d_r \delta_1 \rho_1 \rho_3 + 4 d_r \delta_2 \rho_1 \rho_3 \big) I^{*2} + 4 d_a{}^3 \delta_1 + 8 d_a{}^2 d_i \delta_2 + 8 d_a d_i{}^2 \delta_1 +$$
$$8 d_a d_r{}^2 \delta_1 + 4 d_i{}^3 \delta_2 + 8 d_i d_r{}^2 \delta_2 \big) T^*_{reg} + 2 \rho_1 \rho_3 \big(\rho_1{}^2 + \rho_3 \rho_1 + \rho_3{}^2 \big) I^{*4} + \big(- 2 d_a{}^2 \rho_1{}^2 - 2 d_a{}^2 \rho_1 \rho_3 - d_a{}^2 \rho_3{}^2 -$$
$$2 d_a d_i \rho_1 \rho_3 - 2 d_a d_r \rho_1 \rho_3 - 2 d_i{}^2 \rho_1{}^2 - 6 d_i{}^2 \rho_1 \rho_3 - 5 d_i{}^2 \rho_3{}^2 + 2 d_i d_r \rho_1 \rho_3 - d_r{}^2 \rho_1{}^2 - 2 d_r{}^2 \rho_1 \rho_3 - 2 d_r{}^2 \rho_3{}^2 \big) I^{*2} + d_a{}^4 +$$
$$4 d_a{}^2 d_i{}^2 + 4 d_a{}^2 d_r{}^2 + d_i{}^4 + 4 d_i{}^2 d_r{}^2 + d_r{}^4.$$

$$b_{10} = \big(2 \delta_1{}^2 + 2 \delta_2{}^2 \big) T^*_{reg}{}^2 + \big(4 d_a \delta_1 + 4 d_i \delta_2 \big) Treg^* - \big(\rho_1 + \rho_3 \big)^2 I^{*2} + 2 d_a{}^2 + 2 d_i{}^2 + 2 d_r{}^2.$$

References

1. Mason, D. A very high level of crossreactivity is an essential feature of the T-cell receptor. *Immunol. Today* **1998**, *19*, 395–404. [CrossRef]
2. Anderson, A.C.; Waldner, H.; Turchin, V.; Jabs, C.; Prabhu Das, M.; Kuchroo, V.K.; Nicholson, L.B. Autoantigen responsive T cell clones demonstrate unfocused TCR cross-reactivity towards multiple related ligands: Implications for autoimmunity. *Cell. Immunol.* **2000**, *202*, 88–96. [CrossRef] [PubMed]
3. Kerr, E.C.; Copland, D.A.; Dick, A.D.; Nicholson, L.B. The dynamics of leukocyte infiltration in experimental autoimmune uveoretinitis. *Prog. Retin. Eye Res.* **2008**, *27*, 527–535. [CrossRef] [PubMed]
4. Prat, E.; Martin, R. The immunopathogenesis of multiple sclerosis. *J. Rehabil. Res. Dev.* **2002**, *39*, 187–200. [PubMed]
5. Santamaria, P. The long and winding road to understanding and conquering type 1 diabetes. *Immunity* **2010**, *32*, 437–445. [CrossRef] [PubMed]
6. Root-Bernstein, R.; Fairweather, D. Unresolved issues in theories of autoimmune disease using myocarditis as a framework. *J. Theor. Biol.* **2015**, *375*, 101–123. [CrossRef] [PubMed]
7. Caforio, A.L.P.; Iliceto, S. Genetically determined myocarditis: Clinical presentation and immunological characteristics. *Curr. Opin. Cardiol.* **2008**, *23*, 219–226. [CrossRef] [PubMed]
8. Li, H.S.; Ligons, D.L.; Rose, N.R. Genetic complexity of autoimmune myocarditis. *Autoimmun. Rev.* **2008**, *7*, 168–173. [CrossRef] [PubMed]
9. Guilherme, L.; Köhler, K.F.; Postol, E.; Kalil, J. Genes, autoimmunity and pathogenesis of rheumatic heart disease. *Ann. Pediatr. Cardiol.* **2011**, *4*, 13–21. [CrossRef] [PubMed]
10. Germolic, D.; Kono, D.H.; Pfau, J.C.; Pollard, K.M. Animal models used to examine the role of environment in the development of autoimmune disease: Findings from an NIEHS Expert Panel Workshop. *J. Autoimmun.* **2012**, *39*, 285–293. [CrossRef] [PubMed]

11. Mallampalli, M.P.; Davies, E.; Wood, D.; Robertson, H.; Polato, F.; Carter, C.L. Role of environment and sex differences in the development of autoimmune disease: A roundtable meeting report. *J. Women's Health* **2013**, *22*, 578–586. [CrossRef] [PubMed]

12. Manfredo Vieira, S.; Hiltensperger, M.; Kumar, V.; Zegarra-Ruiz, D.; Dehne, C.; Khan, N.; Costa, F.R.C.; Tiniakou, E.; Greiling, T.; et al. Translocation of a gut pathobiont drives autoimmunity in mice and humans. *Science* **2018**, *359*, 1156–1161. [CrossRef] [PubMed]

13. Fujinami, R.S. Can virus infections trigger autoimmune disease? *J. Autoimmun.* **2001**, *16*, 229–234. [CrossRef] [PubMed]

14. Von Herrath, M.G.; Oldstone, M.B.A. Virus-induced autoimmune disease. *Curr. Opin. Immunol.* **1996**, *8*, 878–885. [CrossRef]

15. Ercolini, A.M.; Miller, S.D. The role of infections in autoimmune disease. *Clin. Exp. Immunol.* **2009**, *155*, 1–15. [CrossRef] [PubMed]

16. Segel, L.A.; Jäger, E.; Elias, D.; Cohen, I.R. A quantitative model of autoimmune disease and T-cell vaccination: Does more mean less? *Immunol. Today* **1995**, *16*, 80–84. [CrossRef]

17. Borghans, J.A.M.; De Boer, R.J. A minimal model for T-cell vaccination. *Proc. R. Soc. Lond. B Biol. Sci.* **1995**, *259*, 173–178. [CrossRef] [PubMed]

18. Borghans, J.A.M.; De Boer, R.J.; Sercarz, E.; Kumar, V. T cell vaccination in experimental autoimmune encephalomyelitis: A mathematical model. *J. Immunol.* **1998**, *161*, 1087–1093. [PubMed]

19. León, K.; Perez, R.; Lage, A.; Carneiro, J. Modelling T-cell-mediated suppression dependent on interactions in multicellular conjugates. *J. Theor. Biol.* **2000**, *207*, 231–254. [CrossRef] [PubMed]

20. León, K.; Lage, A.; Carneiro, J. Tolerance and immunity in a mathematical model of T-cell mediated suppression. *J. Theor. Biol.* **2003**, *225*, 107–126. [CrossRef]

21. León, K.; Faro, J.; Lage, A.; Carneiro, J. Inverse correlation between the incidences of autoimmune disease and infection predicted by a model of T cell mediated tolerance. *J. Autoimmun.* **2004**, *22*, 31–42. [CrossRef] [PubMed]

22. Carneiro, J.; Paixão, T.; Milutinovic, D.; Sousa, J.; Leon, K.; Gardner, R.; Faro, J. Immunological self-tolerance: Lessons from mathematical modeling. *J. Comput. Appl. Math.* **2005**, *184*, 77–100. [CrossRef]

23. Alexander, H.K.; Wahl, L.M. Self-tolerance and autoimmunity in a regulatory T cell model. *Bull. Math. Biol.* **2011**, *73*, 33–71. [CrossRef] [PubMed]

24. Iwami, S.; Takeuchi, Y.; Miura, Y.; Sasaki, T.; Kajiwara, T. Dynamical properties of autoimmune disease models: Tolerance, flare-up, dormancy. *J. Theor. Biol.* **2007**, *246*, 646–659. [CrossRef] [PubMed]

25. Iwami, S.; Takeuchi, Y.; Iwamoto, K.; Naruo, Y.; Yasukawa, M. A mathematical design of vector vaccine against autoimmune disease. *J. Theor. Biol.* **2009**, *256*, 382–392. [CrossRef] [PubMed]

26. Burroughs, N.J.; Ferreira, M.; Oliveira, B.M.P.M.; Pinto, A.A. Autoimmunity arising from bystander proliferation of T cells in an immune response model. *Math. Comput. Model.* **2011**, *53*, 1389–1393. [CrossRef]

27. Burroughs, N.J.; Ferreira, M.; Oliveira, B.M.P.M.; Pinto, A.A. A transcritical bifurcation in an immune response model. *J. Differ. Equ. Appl.* **2011**, *17*, 1101–1106. [CrossRef]

28. Burroughs, N.J.; de Oliveira, B.M.P.M.; Pinto, A.A. Regulatory T cell adjustment of quorum growth thresholds and the control of local immune responses. *J. Theor. Biol.* **2006**, *241*, 134–141. [CrossRef] [PubMed]

29. Sakaguchi, S. Naturally arising CD4$^+$ regulatory T cells for immunologic self-tolerance and negative control of immune responses. *Ann. Rev. Immunol.* **2004**, *22*, 531–562. [CrossRef] [PubMed]

30. Josefowicz, S.Z.; Lu, L.F.; Rudensky, A.Y. Regulatory T cells: Mechanisms of differentiation and function. *Ann. Rev. Immunol.* **2012**, *30*, 531–564. [CrossRef] [PubMed]

31. Fontenot, J.D.; Gavin, M.A.; Rudensky, A.Y. Foxp3 programs the development and function of CD4$^+$CD25$^+$ regulatory T cells. *Nat. Immunol.* **2003**, *4*, 330–336. [CrossRef] [PubMed]

32. Khattri, R.; Cox, T.; Yasayko, S.A.; Ramsdell, F. An essential role for Scurfin in CD4$^+$CD25$^+$ T regulatory cells. *Nat. Immunol.* **2003**, *4*, 337–342. [CrossRef] [PubMed]

33. Grossman, Z.; Paul, W.E. Adaptive cellular interactions in the immune system: The tunable activation threshold and the significance of subthreshold responses. *Proc. Natl. Acad. Sci. USA* **1992**, *89*, 10365–10369. [CrossRef] [PubMed]

34. Grossman, Z.; Paul, W.E. Self-tolerance: Context dependent tuning of T cell antigen recognition. *Semin. Immunol.* **2000**, *12*, 197–203. [CrossRef] [PubMed]

35. Grossman, Z.; Singer, A. Tuning of activation thresholds explains flexibility in the selection and development of T cells in the thymus. *Proc. Natl. Acad. Sci. USA* **1996**, *93*, 14747–14752. [CrossRef] [PubMed]
36. Altan-Bonnet, G.; Germain, R.N. Modeling T cell antigen discrimination based on feedback control of digital ERK responses. *PLoS Biol.* **2005**, *3*, e356. [CrossRef] [PubMed]
37. Bitmansour, A.D.; Douek, D.C.; Maino, V.C.; Picker, L.J. Direct ex vivo analysis of human CD4$^+$ memory T cell activation requirements at the single clonotype level. *J. Immunol.* **2002**, *169*, 1207–1218. [CrossRef] [PubMed]
38. Nicholson, L.B.; Anderson, A.C.; Kuchroo, V.K. Tuning T cell activation threshold and effector function with cross-reactive peptide ligands. *Int. Immunol.* **2000**, *12*, 205–213. [CrossRef] [PubMed]
39. Römer, P.S.; Berr, S.; Avota, E.; Na, S.Y.; Battaglia, M.; ten Berge, I.; Einsele, H.; Hünig, T. Preculture of PBMC at high cell density increases sensitivity of T-cell responses, revealing cytokine release by CD28 superagonist TGN1412. *Blood* **2011**, *118*, 6772–6782. [CrossRef] [PubMed]
40. Stefanová, I.; Dorfman, J.R.; Germain, R.N. Self-recognition promotes the foreign antigen sensitivity of naive T lymphocytes. *Nature* **2002**, *420*, 429–434. [CrossRef] [PubMed]
41. Van den Berg, H.A.; Rand, D.A. Dynamics of T cell activation threshold tuning. *J. Theor. Biol.* **2004**, *228*, 397–416. [CrossRef] [PubMed]
42. Scherer, A.; Noest, A.; de Boer, R.J. Activation-threshold tuning in an affinity model for the T-cell repertoire. *Proc. R. Soc. B* **2004**, *271*, 609–616. [CrossRef] [PubMed]
43. Blyuss, K.B.; Nicholson, L.B. The role of tunable activation thresholds in the dynamics of autoimmunity. *J. Theor. Biol.* **2012**, *308*, 45–55. [CrossRef] [PubMed]
44. Blyuss, K.B.; Nicholson, L.B. Understanding the roles of activation threshold and infections in the dynamics of autoimmune disease. *J. Theor. Biol.* **2015**, *375*, 13–20. [CrossRef] [PubMed]
45. Ben Ezra, D.; Forrester, J.V. Fundal white dots: The spectrum of a similar pathological process. *Br. J. Ophthalmol.* **1995**, *79*, 856–860. [CrossRef] [PubMed]
46. Davies, T.F.; Evered, D.C.; Rees Smith, B.; Yeo, P.P.B.; Clark, F.; Hall, R. Value of thyroid-stimulating-antibody determinations in predicting the short-term thyrotoxic relapse in Graves' disease. *Lancet* **1977**, *309*, 1181–1182. [CrossRef]
47. Nylander, A.; Hafler, D.A. Multiple sclerosis. *J. Clin. Investig.* **2012**, *122*, 1180–1188. [CrossRef] [PubMed]
48. Fatehi, F.; Kyrychko, Y.N.; Blyuss, K.B. Interactions between cytokines and T cells with tunable activation thresholds in the dynamics of autoimmunity. 2018, submitted.
49. Fatehi, F.; Kyrychko, S.N.; Ross, A.; Kyrychko, Y.N.; Blyuss, K.B. Stochastic effects in autoimmune dynamics. *Front. Physiol.* **2018**, *9*, 45. [CrossRef] [PubMed]
50. Wu, H.J.; Ivanov, I.I.; Darce, J.; Hattori, K.; Shima, T.; Umesaki, Y.; Littman, D.R.; Benoist, C.; Mathis, D. Gut-residing segmented filamentous bacteria drive autoimmune arthritis via T helper 17 cells. *Immunity* **2010**, *32*, 815–827. [CrossRef] [PubMed]
51. Wolf, S.D.; Dittel, B.N.; Hardardottir, F.; Janeway, C.A. Experimental autoimmune encephalomyelitis induction in genetically B cell-deficient mice. *J. Exp. Med.* **1996**, *184*, 2271–2278. [CrossRef] [PubMed]
52. Baltcheva, I.; Codarri, L.; Pantaleo, G.; Le Boudec, J.Y. Lifelong dynamics of human CD4$^+$CD25$^+$ regulatory T cells: Insights from in vivo data and mathematical modeling. *J. Theor. Biol.* **2010**, *266*, 307–322. [CrossRef] [PubMed]
53. Shevach, E.M.; McHugh, R.S.; Piccirillo, C.A.; Thornton, A.M. Control of T-cell activation by CD4$^+$ CD25$^+$ suppressor T cells. *Immunol. Rev.* **2001**, *182*, 58–67. [CrossRef] [PubMed]
54. Abbas, A.K.; Lichtman, A.H.; Pillai, S. *Cellular and Molecular Immunology*; Elsevier: Philadelphia, PA, USA, 2015.
55. Thornton, A.M.; Shevach, E.M. CD4$^+$CD25$^+$ immunoregulatory T cells suppress polyclonal T cell activation in vitro by inhibiting interleukin 2 production. *J. Exp. Med.* **1998**, *188*, 287–296. [CrossRef] [PubMed]
56. Baccam, P.; Beauchemin, C.; Macken, C.A.; Hayden, F.G.; S, P.A. Kinetics of influenza A virus infection in humans. *J. Virol.* **2006**, *80*, 7590–7599. [CrossRef] [PubMed]
57. Pawelek, K.A.; Huynh, G.T.; Quinlivan, M.; Cullinane, A.; Rong, L.; Perelson, A.S. Modeling within-host dynamics of influenza virus infection including immune responses. *PLoS Comput. Biol.* **2012**, *8*, e1002588. [CrossRef] [PubMed]
58. Kim, P.S.; Lee, P.P.; Levy, D. Modeling regulation mechanisms in the immune system. *J. Theor. Biol.* **2007**, *246*, 33–69. [CrossRef] [PubMed]

59. Li, J.; Zhang, L.; Wang, Z. Two effective stability criteria for linear time-delay systems with complex coefficients. *J. Syst. Sci. Complex.* **2011**, *24*, 835. [CrossRef]
60. Rahman, B.; Blyuss, K.B.; Kyrychko, Y.N. Dynamics of neural systems with discrete and distributed time delays. *SIAM J. Appl. Dyn. Syst.* **2015**, *14*, 2069–2095. [CrossRef]
61. Breda, D.; Maset, S.; Vermiglio, R. Numerical computation of characteristic multipliers for linear time periodic coefficients delay differential equations. *IFAC Proc. Vol.* **2006**, *39*, 163–168. [CrossRef]
62. Kuznetsov, Y.A. *Elements of Applied Bifurcation Theory*; Springer-Verlag: New York, NY, USA, 1995.
63. Cavoretto, R.; Rossi, A.D.; Perracchione, E.; Venturino, E. Reliable approximation of separatrix manifolds in competition models with safety niches. *Int. J. Comput. Math.* **2015**, *92*, 1826–1837. [CrossRef]
64. Cavoretto, R.; De Rossi, A.; Perracchione, E.; Venturino, E. Robust approximation algorithms for the detection of attraction basins in dynamical systems. *J. Sci. Comput.* **2016**, *68*, 395–415. [CrossRef]
65. De Rossi, A.; Perracchione, E.; Venturino, E. Fast strategy for PU interpolation: An application for the reconstruction of separatrix manifolds. *Dolomites Res. Notes Approx.* **2016**, *9*, 3–12.
66. Francomano, E.; Hilker, F.M.; Paliaga, M.; Venturino, E. On basins of attraction for a predator-prey model via meshless approximation. In *AIP Conference Proceedings*; AIP Publishing: Melville, NY, USA, 2016; Volume 1776, p. 070007.
67. Cavoretto, R.; De Rossi, A.; Perracchione, E.; Venturino, E. Graphical representation of separatrices of attraction basins in two and three-dimensional dynamical systems. *Int. J. Comput. Methods* **2017**, *14*, 1750008. [CrossRef]
68. Francomano, E.; Hilker, F.M.; Paliaga, M.; Venturino, E. Separatrix reconstruction to identify tipping points in an eco-epidemiological model. *Appl. Math. Comput.* **2018**, *318*, 80–91. [CrossRef]
69. Skapenko, A.; Leipe, J.; Lipsky, P.E.; Schulze-Koops, H. The role of the T cell in autoimmune inflammation. *Arthritis Res. Ther.* **2005**, *7* (Suppl. 2), S4–S14. [CrossRef] [PubMed]
70. Antia, R.; Ganusov, V.V.; Ahmed, R. The role of models in understanding CD8$^+$ T-cell memory. *Nat. Rev. Immunol.* **2005**, *5*, 101–111. [CrossRef] [PubMed]
71. Schluns, K.S.; Kieper, W.C.; Jameson, S.C.; Lefrancois, L. Interleukin-7 mediates the homeostasis of naïve and memory CD8 T cells in vivo. *Nat. Immunol.* **2000**, *1*, 426–432. [CrossRef] [PubMed]
72. Fatehi Chenar, F.; Kyrychko, Y.N.; Blyuss, K.B. Mathematical model of immune response to hepatitis B. *J. Theor. Biol.* **2018**, *447*, 98–110. [CrossRef] [PubMed]
73. Perelson, A.S.; Weisbuch, G. Immunology for physicists. *Rev. Mod. Phys.* **1997**, *69*, 1219–1267. [CrossRef]

![Σ] *mathematics*

MDPI

Article
Critical Domain Problem for the Reaction–Telegraph Equation Model of Population Dynamics

Weam Alharbi * and Sergei Petrovskii

Department of Mathematics, University of Leicester, University Road, Leicester LE1 7RH, UK; sp237@le.ac.uk
* Correspondence: wsa4@le.ac.uk

Received: 28 February 2018; Accepted: 13 April 2018; Published: 17 April 2018

Abstract: A telegraph equation is believed to be an appropriate model of population dynamics as it accounts for the directional persistence of individual animal movement. Being motivated by the problem of habitat fragmentation, which is known to be a major threat to biodiversity that causes species extinction worldwide, we consider the reaction–telegraph equation (i.e., telegraph equation combined with the population growth) on a bounded domain with the goal to establish the conditions of species survival. We first show analytically that, in the case of linear growth, the expression for the domain's critical size coincides with the critical size of the corresponding reaction–diffusion model. We then consider two biologically relevant cases of nonlinear growth, i.e., the logistic growth and the growth with a strong Allee effect. Using extensive numerical simulations, we show that in both cases the critical domain size of the reaction–telegraph equation is larger than the critical domain size of the reaction–diffusion equation. Finally, we discuss possible modifications of the model in order to enhance the positivity of its solutions.

Keywords: animal movement; fragmented environment; critical size; extinction

---

1. Introduction

Habitat fragmentation due to the climate change and anthropogenic activities is regarded as a major threat to biodiversity worldwide [1]. Understanding the factors affecting species survival in small and fragmented habitats is therefore a problem of high practical importance [2–4]. Although being essentially an ecological problem, it can hardly be studied in full by only traditional ecological methods and tools. Replicated ecological experiments under controlled conditions required for an exhaustive empirical study are often impossible for a variety of reasons, e.g., due to high costs, potential damage to the environment, and irreproducibility of the weather conditions and initial species distribution, to name just a few.

Mathematical modelling has long been recognized as an efficient research tool in theoretical ecology that can be used as a supplement and sometimes even a substitute to a field experiment [5,6]. Mathematical models along with computer simulations create a virtual laboratory where hypotheses can be tested and different scenarios can be followed under safe working conditions and at relatively low costs. In particular, the problem of population persistence in small habitats, often referred to as the problem of critical domain, has been studied almost exhaustively in terms of reaction–diffusion models [7–11].

The choice of an adequate model, however, is a subtle issue. Whilst reaction–diffusion models have been used extensively in mathematical ecology for several decades [8,12–15] helping to make a number of important insights and inferences, they have their limits and their relevance may sometimes become questionable. One point of their criticism is that diffusion (Brownian motion) as a baseline animal movement pattern is not entirely realistic as the diffusively moving 'animal' can change its movement direction with a high frequency (i.e., with any two consequent turns infinitely close in time) and choose the new direction uniformly distributed over the circle. However, this is at odds with many observations as well as with common sense [16]: since the body of animals of all species has the front end and the rear end, they are more likely to choose the new movement direction (following re-orientation) close to the movement direction at the preceding moment. The corresponding movement pattern is known as the correlated random walk (CRW) [17] and the corresponding microscopic stochastic process as the telegraph process [18], and their mean-field counterpart is known as a telegraph equation [19–23]:

$$\tau \frac{\partial^2 u}{\partial t^2} + \frac{\partial u}{\partial t} = D \frac{\partial^2 u}{\partial x^2}, \tag{1}$$

where $u = u(x, t)$ is the population density at location x and time t, D is the diffusion coefficient, and τ is a characteristic time of the microscopic movement that quantifies directional persistence in individual animal movement. The precise meaning of parameter τ can be slightly different depending on the details of the microscopic model; for instance, in the telegraph movement process, it is the time over which the animal moves without changing its movement direction [18–20]. Brownian motion thus corresponds to the limit $\tau \to 0$ when Equation (1) turns into the diffusion equation.

Equation (1) describes the evolution of the population density because of animal movement. It does not take into account the contribution of births and deaths. Interestingly, in order to include them into the model, it is not enough to add the 'reaction' term (i.e., the population growth rate)—say, $F(u)$—into the right-hand side of Equation (1) (as is done in the case of diffusive animal movement [14]). An accurate account for birth and death events also modifies the factor in front of the first order derivative on the left-hand side [20,23]. The corresponding model is known as the reaction–telegraph equation:

$$\tau \frac{\partial^2 u}{\partial t^2} + [1 - \tau F'(u)] \frac{\partial u}{\partial t} = D \frac{\partial^2 u}{\partial x^2} + F(u). \tag{2}$$

Here and below, the prime denotes the ordinary derivative of a function with regard to its argument.

In this paper, we consider the properties of the reaction–telegraph equation in a bounded domain. In doing this, we are motivated by the ecological problem of habitat fragmentation and its effect on species survival. Our main goal is therefore to understand what the conditions are of species survival (i.e., when $u(x, t)$ does not tend to zero uniformly over the domain in the large-time limit), how they may differ from the predictions of the corresponding reaction–diffusion model, and how they may depend on the type of density dependence in the population growth, e.g., the existence of the Allee effect.

2. Non-Conservative Property of the Telegraph Equation

In order to demonstrate that the telegraph equation has properties significantly different from the diffusion equation, we begin with a simple example. Consider Equation (1) in a bounded domain $0 < x < L$ with the Neumann-type boundary conditions at the domain boundaries:

$$\frac{\partial u(0, t)}{\partial x} = 0, \qquad \frac{\partial u(L, t)}{\partial x} = 0. \tag{3}$$

Integrating Equation (1) over space and taking Equation (3) into account, we readily obtain the equation for the total population size:

$$\tau M''(t) + M'(t) = 0, \tag{4}$$

where $M(t) = \int_0^L u(x, t) dx$. Assuming for simplicity that $M'(0) = 0$, the generic solution of Equation (4) is $M(t) = M_0 \exp(-t/\tau)$ so that, obviously, $M(t) \to 0$ for $t \to \infty$. Therefore, the population goes extinct in the large time limit regardless of its initial population size and regardless of the domain size. Note that, in the corresponding diffusion model, boundary condition (3) correspond to impenetrable boundaries and the total mass is conserved, $M(t) = M_0 = const$, so that species extinction is impossible. This drastic difference between the predictions of the diffusion equation and telegraph equation is explained by the observation that, for the telegraph movement process, the flux is described differently from diffusion [23]; the flux is not proportional to the gradient of the population density (as given by the Fick law). Therefore, the boundary conditions as given by Equation (3) that make the boundary impenetrable in case the diffusion process does not block the population flux in the case of the telegraph movement process.

In the context of the habitat fragmentation problem, which is the main focus of this paper, the domain is open rather than closed. As we will show in the next section, the difference between diffusion and the telegraph movement process becomes somewhat less pronounced when the boundary conditions are of the Dirichlet-type and when the population reproduction is taken into account.

3. Telegraph Equation with Linear Growth

Once population multiplication is taken into account, telegraph equation becomes the reaction–telegraph equation; see Equation (2). In case the per capita growth rate of the population can be regarded as density independent, i.e., $F(u) = \alpha u$, Equation (2) takes the following form:

$$\tau \frac{\partial^2 u}{\partial t^2} + (1 - \alpha\tau)\frac{\partial u}{\partial t} = D\frac{\partial^2 u}{\partial x^2} + \alpha u, \tag{5}$$

where α is a coefficient. The linear growth in Equation (5) assumes that the individuals of the given species (u) do not interact; collective behaviours such as competition or cooperation are neglected. Although it is clearly not true in a general case, if the population density u is sufficiently small (e.g., because of the initial conditions), the linear growth may provide a reasonable approximation to the true (nonlinear) growth rate; see [14] and also Section 2 in [24].

For the convenience of notation, we define new coefficients $\omega_1 = \tau$ and $\omega_2 = 1 - \tau F'(u) = 1 - \alpha\tau$. Equation (5) then becomes

$$\omega_1 \frac{\partial^2 u}{\partial t^2} + \omega_2 \frac{\partial u}{\partial t} = D\frac{\partial^2 u}{\partial x^2} + \alpha u. \tag{6}$$

We consider Equation (6) in a bounded domain of length L, i.e., $0 < x < L$. The environment outside of the domain is assumed to be unfavorable for the given species, in fact very harsh, so that an animal that leaves the domain will die very soon without any chance of coming back. This is described by the following Dirichlet-type boundary conditions:

$$\text{(a)} \quad u(0,t) = 0, \qquad \text{(b)} \quad u(L,t) = 0. \tag{7}$$

Equation (6) is further complemented by the initial conditions:

$$\text{(a)} \quad u(x,0) = f(x), \qquad \text{(b)} \quad \frac{\partial u(x,0)}{\partial t} = g(x), \qquad 0 \le x \le L, \tag{8}$$

where $f(x)$ and $g(x)$ are certain functions, and $f(x) \ge 0$.

An analytical solution of the boundary problem given by Equations (6)–(8) can be found the method of variable separation [25–27], which we briefly revisit below. We look for a solution using the following anzatz:

$$u(x,t) = X(x)T(t), \tag{9}$$

where X and T are certain functions to be determined. Having substituted it into Equation (6), we obtain:

$$\frac{\omega_1 T'' + \omega_2 T' - \alpha T}{T} = \frac{DX''}{X} = -\lambda, \tag{10}$$

where λ is a certain constant (since the left-hand side of Equation (10) is a function of time t and the middle part is a function of x). From Equation (10), we obtain:

$$DX'' + \lambda X = 0, \tag{11}$$

$$w_1 T'' + w_2 T' + (\lambda - \alpha) T = 0. \tag{12}$$

It is readily seen that Equation (11) does not have non-trivial solutions for $\lambda \le 0$ to satisfy boundary conditions given by Equation (7), i.e., $X(0) = X(L) = 0$. For $\lambda > 0$, the solution for Equation (11) is as follows:

$$X(x) = A \cos\left(x\sqrt{\frac{\lambda}{D}}\right) + B \sin\left(x\sqrt{\frac{\lambda}{D}}\right), \tag{13}$$

where coefficients A and B are to be found. From $X(0) = 0$, we obtain $A = 0$, and from $X(L) = 0$ we obtain (assuming $B \ne 0$) that $\sin\left(L\sqrt{\frac{\lambda}{D}}\right) = 0$. The latter leads to the following expression for λ:

$$\lambda = \lambda_n = \left(\frac{\pi n}{L}\right)^2 D, \quad n = 1,2,3,\dots. \tag{14}$$

Obviously, solution $X_n(x)$ corresponding to a given $\lambda = \lambda_n$ is

$$X(x) = X_n(x) = B_n \sin\left(\frac{\pi n x}{L}\right), \quad n \ge 1. \tag{15}$$

Correspondingly, the equation for $T_n(t)$ is

$$w_1 T_n'' + w_2 T_n' + (\lambda_n - \alpha) T_n = 0, \tag{16}$$

with the characteristic polynomial

$$w_1 r^2 + w_2 r + (\lambda_n - \alpha) = 0. \tag{17}$$

The roots of Equation (17) are

$$r_{1,2} = \frac{-w_2 \pm \sqrt{w_2^2 - 4w_1(\lambda_n - \alpha)}}{2w_1}. \tag{18}$$

The properties of the solution essentially depend on the sign of the expression under the square root, i.e., on the following quantity:

$$R_n = w_2^2 - 4w_1(\lambda_n - \alpha) = \left(w_2^2 + 4w_1\alpha\right) - 4w_1\lambda_n. \tag{19}$$

Since $\lambda_n < \lambda_{n+1}$, $n = 1,2,\dots$, we observe that, if

$$\lambda_1 > \frac{(w_2^2 + 4w_1\alpha)}{4w_1}, \tag{20}$$

then $R_n < 0$ for any n. In this case,

$$r_{1,2} = \frac{-w_2 \pm i\sqrt{|R_n|}}{2w_1}, \tag{21}$$

and the solution of Equation (12) is

$$T_n(t) = e^{\left(-\frac{\omega_2}{2\omega_1}\right)t}\left[A_n \cos\left(\frac{\sqrt{|R_n|}}{2\omega_1}t\right) + B_n \sin\left(\frac{\sqrt{|R_n|}}{2\omega_1}t\right)\right]. \qquad (22)$$

The product $X_n(x)T_n(t)$ gives a partial solution of Equation (6). Since Equation (6) is linear, a linear combination of solutions is also a solution. Therefore, the general solution of Equation (6) allowing for the boundary conditions given by Equation (7) is:

$$u(x,t) = \sum_{n=1}^{\infty} \sin\left(\frac{\pi n x}{L}\right) e^{\left(-\frac{\omega_2}{2\omega_1}\right)t}\left[A_n \cos\left(\frac{\sqrt{|R_n|}}{2\omega_1}t\right) + B_n \sin\left(\frac{\sqrt{|R_n|}}{2\omega_1}t\right)\right], \qquad (23)$$

where coefficients A_n and B_n are found from the initial conditions given by Equation (8); see Appendix A for details of calculations.

Obviously, regardless the choice of the initial population distribution $f(x)$, the solution given by Equation (23) describes a population decay eventually leading, in the large-time limit, to population extinction. This is in agreement with intuitive expectations: since $\lambda_n \sim L^{-2}$, the condition given by Equation (20) means that the domain is 'too small' to support sustainable population dynamics for the given growth rate.

The situation changes if the size L of the domain is sufficiently large (or the growth rate α is sufficiently large), i.e., the condition given by Equation (20) does not hold. The solution of Equation (16) then always contains oscillatory (trigonometric) and non-oscillatory (exponential) terms. A general study of the properties of the solution in this case is complicated and will be considered in detail elsewhere [28]. For the purposes of this paper, it is sufficient to consider the situation where only $R_1 > 0$ and $R_n < 0$ for $n = 2, 3, \ldots$. In this case, the time-dependent part of the first term in Equation (23) changes to

$$T_1(t) = A_n e^{r_1 t} + B_n e^{r_2 t}, \qquad (24)$$

where

$$r_{1,2} = \frac{\omega_2}{2\omega_1}\left[-1 \pm \left(1 + \frac{2\omega_1}{\omega_2^2}(\alpha - \lambda_1)\right)^{1/2}\right]. \qquad (25)$$

Let, respectively, r_1 correspond to plus and r_2 correspond to minus on the right-hand side of Equation (25). Then, $r_1 < 0$ for any parameter values (as long as $R_1 > 0$) but r_2 can change its sign, so that

$$r_2 < 0 \text{ for } \lambda_1 > \alpha \qquad \text{but} \qquad r_2 > 0 \text{ for } \lambda_1 < \alpha. \qquad (26)$$

Whilst $r_2 < 0$ corresponds to an exponential decay similar to that predicted by the solution given by Equation (23), in case $r_2 > 0$, the first term in the solution of the problem in Equations (6)–(8) grows exponentially. Therefore, the properties of the solution change when the following critical relation between the parameters takes place:

$$\lambda_1 = \alpha. \qquad (27)$$

From Equations (14) and (27), we arrive at the expression for the critical size of the domain:

$$L_{cr} = \pi\sqrt{\frac{D}{\alpha}}. \qquad (28)$$

Interestingly, the expression of Equation (28) for the critical domain size for the population dynamics described by the reaction–telegraph equation coincides with the critical domain size obtained for the linear reaction–diffusion equation, e.g., see Section 3.1 in [24]. Note that the critical size does not depend on the choice of the initial conditions.

The above result is further confirmed by numerical simulations. Figure 1 shows the critical domain size of the reaction–telegraph equation obtained for the initial conditions as in Equation (8) with $f(x) \equiv 1$ and $g(x) \equiv 0$. To perform the simulations, the telegraph equation is solved by finite differences using an implicit method (for details, see Appendix B). Simulations were performed for a broad range of values for parameters ω_1, ω_2 and α. It is readily seen that the simulation results are in a perfect agreement with the analytical result of Equation (28). Although hardly surprising as such, this agreement between the numerical and analytical results indicates high robustness of our numerical method that produces accurate results even in the case where ω_1 and ω_2 differ by two orders of magnitude (see green squares in Figure 1).

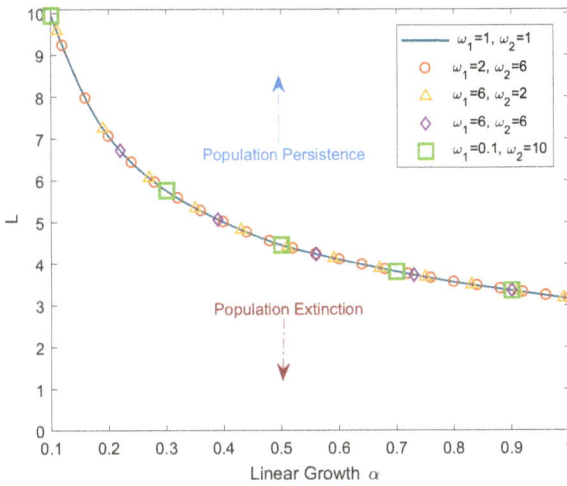

Figure 1. Critical domain size of the reaction–telegraph equation obtained numerically at various values of linear growth rate α and various values of ω_1 and ω_2 as shown by different symbols and different colors. The solid blue curve corresponds to Equation (28). For parameters from below the curve, the population goes to extinction; for parameters from above the curve, the population exhibits unbounded growth.

Note that, in case the size of the domain is overcritical, i.e., $L > L_{cr}$ (or $\lambda_1 < \alpha$), the solution given by Equation (23) of the problem (where $T_1(t)$ is given by Equation (24)) predicts an unbounded growth. This is an artefact of the linear population growth. In real population dynamics, as soon as the population density is large enough, nonlinear feedback (e.g., competition) would slow down the growth and eventually stabilize the dynamics. Understanding the effects of such feedback requires a more realistic model where the corresponding term in the reaction–telegraph equation is a nonlinear function of u. The critical domain problem for the nonlinear model is considered in the next section.

4. Telegraph Equation with Nonlinear Growth

In a more realistic case, the growth rate F is a nonlinear function of the population density u. In this case, the reaction–telegraph equation has the following form:

$$\omega_1 \frac{\partial^2 u}{\partial t^2} + \omega_2 \frac{\partial u}{\partial t} = D \frac{\partial^2 u}{\partial x^2} + F(u). \tag{29}$$

See Section 1 for details.

Our goal here is, on the one hand, to apply the reaction–telegraph equation to determine the critical population size, and, on the other hand, to establish whether the reaction–telegraph equation and the reaction–diffusion equation have the same critical size and, if their critical size is different, to reveal how large the difference can be. In case of a nonlinear growth rate, the method of separation of variables cannot be used to analytically determine the critical size. Therefore, in order to address the above matters, computer simulations have been employed. In simulations, we use the same Dirichlet-type boundary condition as above (see Equation (7)). The initial conditions are used as in Equation (8), where we now consider $f(x) \equiv K$ where parameter K is the carrying capacity (see below). This initial condition can be interpreted ecologically that, prior to the habitat fragmentation, the population was in a spatially uniform steady state. Note that, whilst the meaning of $f(x)$ is straightforward (i.e., the initial distribution of the population density), the meaning of $\partial u(x,0)/\partial t$ is not obvious at all. Therefore, in order to avoid the ambiguity of interpretation, we set $g(x) \equiv 0$.

We consider two different types of the population growth. One of them is given by the logistic growth:

$$F(u) = \alpha u \left(1 - \frac{u}{k}\right). \tag{30}$$

Below, we assume $K = 1$ without any loss of generality as K plays the role of a scaling parameter for the population density. In terms of the notations ω_1 and ω_2 introduced in Section 3, we observe that $\omega_1 = \tau$ and $\omega_2 = 1 - \omega_1\alpha(1 - 2u)$.

Simulations were performed for different values of α varying it in a broad range. For any given set of parameters α and ω_1, Equation (29) was solved numerically by finite differences (using the implicit method, see Appendix B) in domains of different size L. By distinguishing between the simulation runs resulting in species extinction and those resulting in species persistence, the critical value of the domain size was established. The results are shown in Figure 2. In any panel of the figure, parameter values from below the curve(s) correspond to species extinction in the large-time limit (the domain is not large enough to support sustainable population dynamics), and parameter values from above the curve(s) correspond to species survival. It is readily seen that, for a small value of ω_1, the critical domain size of the nonlinear reaction–telegraph equation nearly coincides with the critical domain size of the reaction–diffusion equation. This is perhaps not surprising as, intuitively, Equation (29) with $\omega_1 \ll 1$ may be expected to approximate the diffusion equation. However, the difference between the two models become significant when either ω_1 is sufficiently large or ω_2 is sufficiently small. The diffusion equation and the telegraph equation then have a clearly different critical size. Since ω_2 is a decreasing function of α, the largest difference between the models is reached for larger values of α.

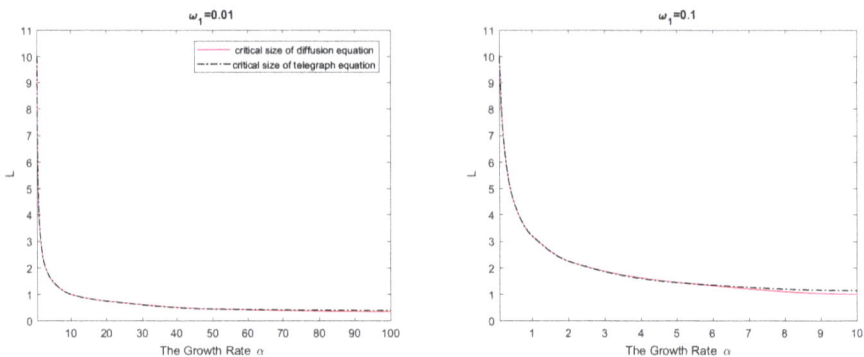

Figure 2. *Cont.*

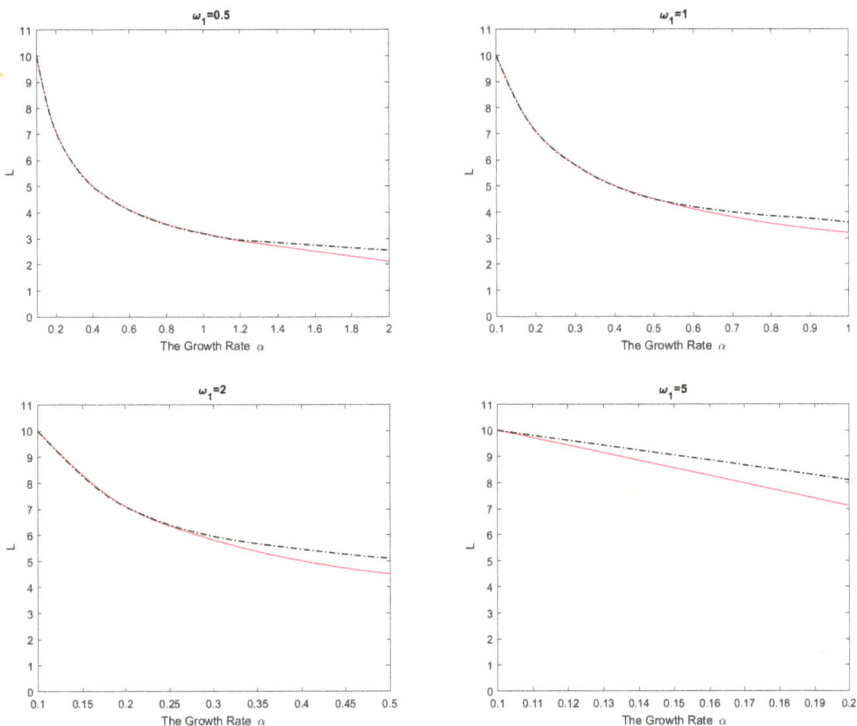

Figure 2. Dependence of the critical domain size on parameter α in case the population growth is logistic (see Equation (30)). The dashed black curve shows the results obtained for the reaction–telegraph Equation (29), and the solid red curve shows the results obtained for the corresponding reaction–diffusion equation, i.e., Equation (29) with $\omega_1 = 0$. Other parameters are $K = 1, D = 1$.

We therefore observe that, in all cases, the critical domain size of the reaction–telegraph equation appears to be larger than the critical size of the reaction–diffusion equation. In ecological terms, it indicates that survival in small domains is more problematic for species where animals consistently employ the movement pattern with a preferred movement direction (such as the CRW or the telegraph process).

We also consider the population growth with the strong Allee effect:

$$F(u) = \alpha u (1 - u)(u - \beta),\qquad(31)$$

where β is a parameter, $0 < \beta < 1$. Correspondingly, we observe that now $\omega_1 = \tau$ and $\omega_2 = 1 - \omega_1 \alpha(-3u^2 + 2\beta u - \beta)$. Simulation results are shown in Figure 3. Similarly to the above, we observe that the difference between the two models is small if $\omega_1 \ll 1$ but may become significant if ω_1 is not small, and the difference tends to be larger for larger values of α. As well as above, the critical size of the reaction–telegraph equation is always larger than the critical size of the reaction–diffusion equation.

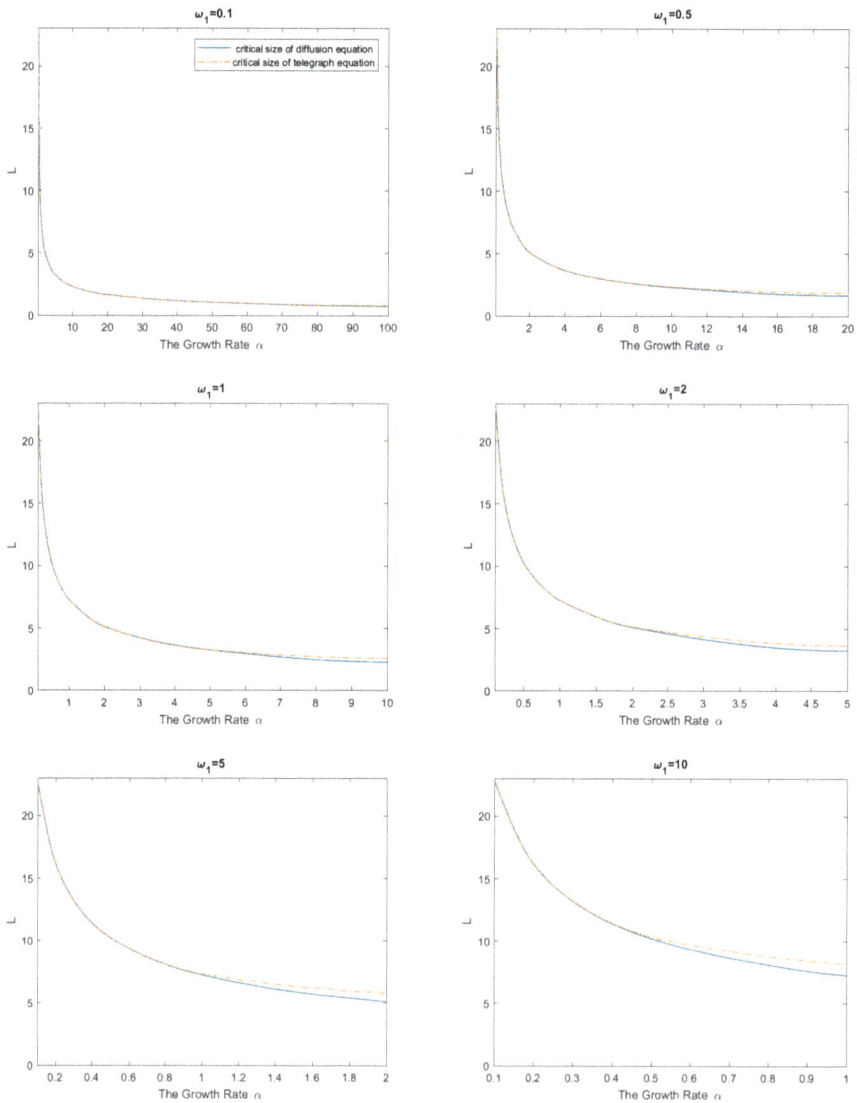

Figure 3. Dependence of the critical domain size on parameter α in case the population growth is subject to the strong Allee effect (see Equation (31)). The dashed orange curve shows the results for the reaction–telegraph Equation (29), and the solid blue curve shows the results obtained for the corresponding reaction–diffusion equation, i.e., Equation (29) with $\omega_1 = 0$. Other parameters are $K = 1, D = 1, \beta = 0.1$.

5. Empirical Model: Telegraph Equation with a Cutoff

One property of the telegraph equation that can make its application to real-world population dynamics somewhat questionable is that it is not positively defined [29]. It means that, for a certain combination of parameters and for certain initial conditions, its solution can become negative, at least in some areas of the domain and for some intervals of time. In our simulations of the reaction–telegraph equation, we also observed this property. Although it does not seem to be a frequent phenomenon, it may indeed happen for some parameter values. By way of example, the left column in Figure 4 shows the solutions of the reaction–telegraph equation with the logistic growth, where u becomes negative at some locations close to the domain boundary. After being negative for a short time (several time steps in the finite-difference procedure), the solution becomes positive again. We mention here that this negativity of the solution is not a numerical artefact, as essentially the same behaviour is observed for smaller values of the mesh steps.

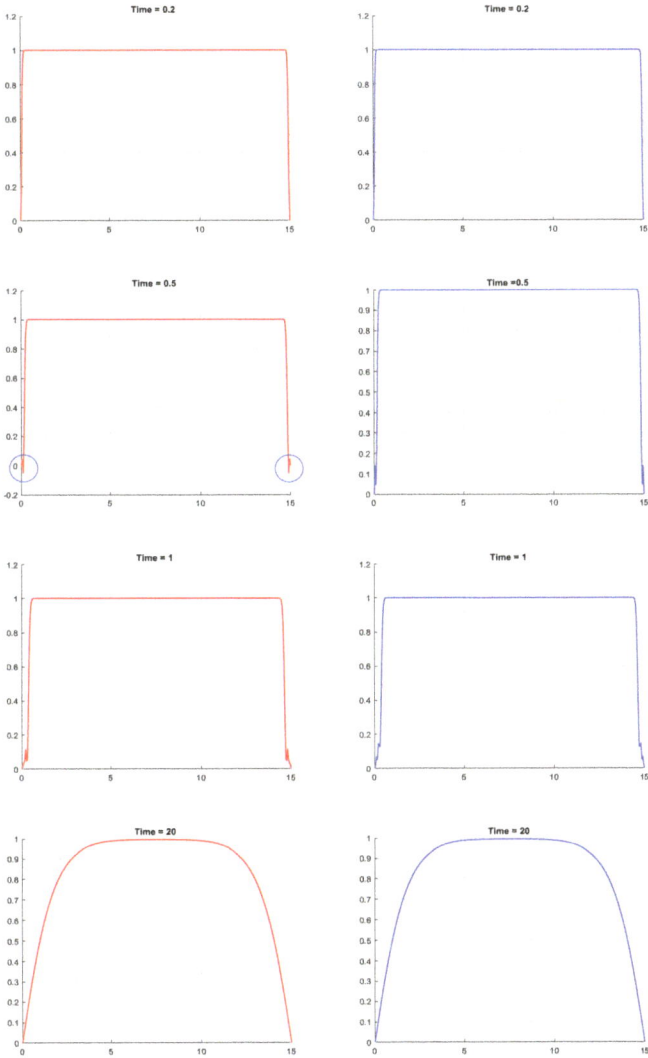

Figure 4. Solution of the reaction–telegraph equation with the logistic growth without cutoff (**left**, red curves) and with the cutoff of negative values (**right**, blue curves). Parameters are $w_1 = 6$ and $w_2 = 2$. The circles mark the location of the areas where the solution becomes negative.

Questions arise here as to what may be the reason for the non-positivity and whether the model can possibly be amended to avoid this unrealistic behaviour. We mention here that the corresponding microscopic model that considers movement of individual 'particles' (e.g., animals) is positively defined [20]. The non-positivity of the solution is therefore an artefact of its mean-field counterpart rather than a genuine property of the movement–reproduction dynamics as such.

One empirical way to keep the solution nonnegative is to introduce a cutoff into the finite-difference method: as soon as $u_i^n = u(x_i, t_n) < 0$, we set $u_i^n = 0$. Since we have observed in our simulations that, if the solution becomes negative, it only attains small values, one can expect that the perturbation introduced to the solution by the cutoff is likely to remain small. In order to look into this issue, for the same set of parameter values and initial conditions, we performed simulations with and without cutoff. We obtained that solution of the reaction–telegraph equation was robust to the cutoff procedure and eventually produced almost the same distribution of the population density. The left column in Figure 4 shows the solution without cutoff and the right column shows the solution after a cutoff was implemented at $t = 0.5$. There is no any visual distinction between the corresponding panels in the left and the right columns of the figure. Similar results are obtained in the case of the strong Allee effect; we do not show them here for sake of brevity.

In order to make a more quantitative insight into the evolution of the perturbation introduced by the cutoff, we calculated the 'error'—the difference between the solutions with and without cutoff—as a function of time and space. Figure 5 shows the distribution of error over the domain at time $t = 0.5$ (i.e., immediately after the cutoff was done), as corresponds to the second row in Figure 4. The error is non-zero only in small areas close to the domain boundaries where the solution became negative. Figure 6 left shows how the maximum error over the domain (i.e., the maximum of functions shown in Figure 5) evolves with time. It is readily seen that the maximum error promptly decays with time, eventually reaching very small values. Figure 6 right shows the distribution of error over space at time $t = 20$, i.e., as corresponds to the bottom row in Figure 4.

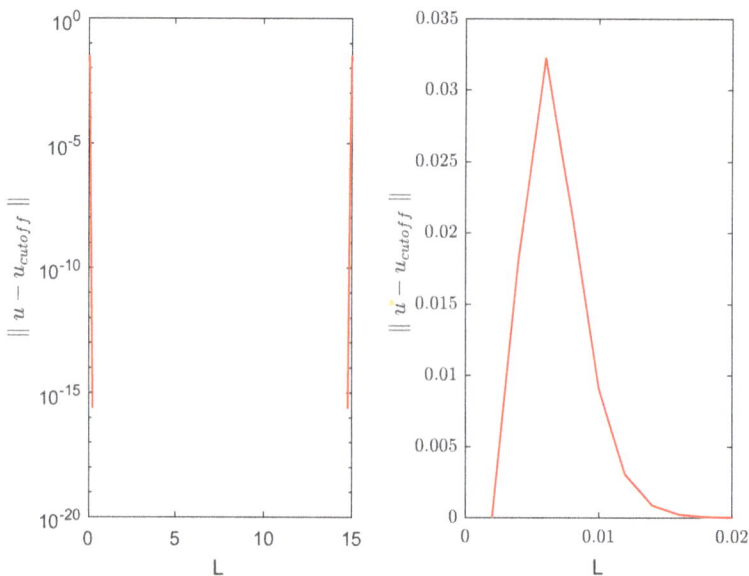

Figure 5. Distribution of the error $\| u - u_{cutoff} \|$ over space at $t = 0.5$, i.e., immediately after the cutoff was implemented. (**left**) distribution over the whole domain $0 < x < L$; (**right**) a magnified view of the part of the domain close to the left-hand side boundary.

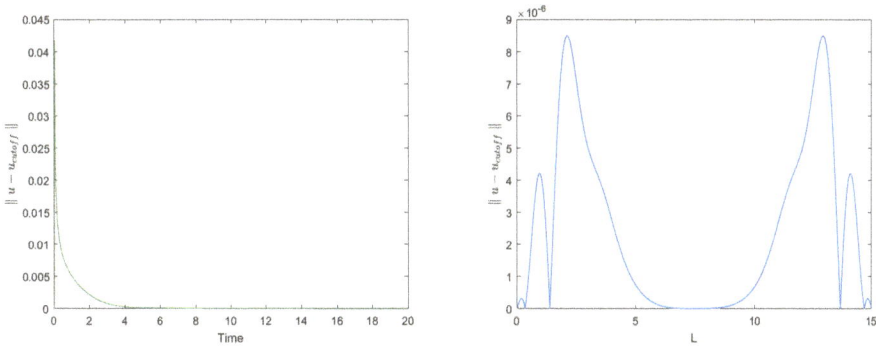

Figure 6. (left) maximum error as a function of time; **(right)** distribution of error $\| u - u_{cutoff} \|$ over the domain $0 < x < L$ at $t = 20$.

6. Discussion and Conclusions

For several decades, the main mathematical framework for modelling animal movement and dispersal was based on the simple isotropic random walk model (SRW) which is unbiased in the sense that, at any moment, the walker can move in any possible direction [13]. This means that the movement direction taken at any given time is independent (uncorrelated) of the movement direction at the preceding moment. The CRW is an extension of the SRW that accounts for the persistence in the walker's direction of movement [17,19,20,23,30]. Indeed, in most real-world systems, the walker is more likely to move in a direction similar to the previous one. This is known as persistence and it can be combined with a fixed speed of movement in the random walk process, resulting in the telegraph or velocity-jump processes [19,31–33]. Our baseline model (1), i.e., the 1D telegraph equation is the mean-field counterpart for the simplest one-dimensional CRW; it includes SRW (the diffusive movement) as the limiting case when the persistence vanishes, i.e., for $\tau \rightarrow 0$.

Being thus a model presumably more adequate than the reaction–diffusion equation, questions arise as to what the properties of the reaction–telegraph equation can be in a relevant ecological context and how different they may be from those of the corresponding reaction–diffusion equation. In this paper, we addressed these questions in the context of habitat fragmentation and species survival in small domains. Having considered the critical domain problem for the reaction–telegraph equation, we arrive at the following conclusions:

- In the case of a linear reaction term (i.e., linear population growth), we found that the critical domain size for the reaction–telegraph equation coincides with that of the corresponding reaction–diffusion equation. This seems to be a surprising result as intuitively the more directional animal movement described by the reaction–telegraph equation should result in a larger critical size.
- In the case of a nonlinear growth (either logistic or with a strong Allee effect), we found that the critical size of the reaction–telegraph equation is indeed somewhat larger than that of the corresponding reaction–diffusion equation. Thus, the difference between the two models arise as a result of a subtle interplay between the movement pattern and the nonlinearity of the population growth.

Our study has potentially important implications for ecology and nature conservation, in particular by means of providing better understanding of the effect that biologically meaningful factors (such as the animal movement pattern and the type of the growth rate) may have on the minimum domain size required to prevent extinctions. However, since the reaction–telegraph equation is widely used in various branches of science and engineering, our findings are likely to have broader range of applications. Traditionally, the telegraph equation was used to describe the voltage and current that can be found on electrical transmission line. In particular, the telegraphic equation was studied by Lord Kelvin in relation to the signals that propagate across the transatlantic cable. The telegraph equation has also been used to model various problems in wave propagation and signal analysis more generally [34,35], random walk theory [31], transport in heterogeneous porous media [32] and pulse transmission through a nerve axon [36,37]. It also has applications in other fields, in particular, it has been used to model transport processes in physical, biological, social, and ecological systems [34,38–43]. Understanding the problem of critical size in those wide-ranging specific contexts is likely to result in exciting new research.

Acknowledgments: An illuminating discussion of the problem with Paulo F. C. Tilles is acknowledged and appreciated.

Author Contributions: Sergei Petrovskii designed the research. Sergei Petrovskii and Weam Alharbi performed the linear analysis. Weam Alharbi wrote the numerical code and performed simulations. Sergei Petrovskii and Weam Alharbi wrote the manuscript.

Conflicts of Interest: The authors declare no conflicts of interest.

Appendix A. Calculation of the Coefficients in the Fourier Series

Setting $t = 0$ in Equation (23) and using the initial condition $f(x) = u(x,0) \equiv 0$ results in the requirement that

$$u(x,0) = \sum_{n=1}^{\infty} A_n \sin\left(\frac{\pi n x}{L}\right) \equiv f(x), \qquad 0 \le x \le L. \tag{A1}$$

Recognizing that this is just a Fourier Sine Series for the function $f(x)$, the constants A_n must be the coefficients of that Sine Series; therefore:

$$A_n = \frac{2}{L} \int_0^L f(x) \sin\left(\frac{\pi n x}{L}\right) dx. \tag{A2}$$

In order to make use of the second initial condition, we have to differentiate Equation (23) with respect to time:

$$\frac{\partial u(x,t)}{\partial t} = \sum_{n=1}^{\infty} \sin\left(\frac{\pi n x}{L}\right) \left[A_n \cos\left(\frac{\sqrt{|R_n|}}{2\omega_1} t\right) + B_n \sin\left(\frac{\sqrt{|R_n|}}{2\omega_1} t\right) \right] e^{-\left(\frac{\omega_2}{2\omega_1}\right)t} \left(\frac{-\omega_2}{2\omega_1}\right)$$
$$+ \sin\left(\frac{\pi n x}{L}\right) \left[B_n \cos\left(\frac{\sqrt{|R_n|}}{2\omega_1} t\right) - A_n \sin\left(\frac{\sqrt{|R_n|}}{2\omega_1} t\right) \right] \frac{\sqrt{|R_n|}}{2\omega_1} e^{-\left(\frac{\omega_2}{2\omega_1}\right)t},$$

where R_n is defined by Equation (19). For $t = 0$, from the above equation, we obtain:

$$\frac{\partial u(x,0)}{\partial t} = \sum_{n=1}^{\infty} \sin\left(\frac{\pi n x}{L}\right) \left[\frac{-\omega_2}{2\omega_1} A_n + \frac{\sqrt{|R_n|}}{2\omega_1} B_n \right] \equiv g(x). \tag{A3}$$

This is a Fourier Sine Series for $g(x)$ with coefficients as

$$\frac{-\omega_2}{2\omega_1} A_n + \frac{\sqrt{|R_n|}}{2\omega_1} B_n = \frac{2}{L} \int_0^L g(x) \sin\left(\frac{\pi n x}{L}\right) dx. \tag{A4}$$

Therefore,

$$B_n = \frac{2\omega_1}{\sqrt{|R_n|}} \left[\frac{2}{L} \int_0^L g(x) \sin\left(\frac{\pi n x}{L}\right) dx + \frac{\omega_2}{2\omega_1} A_n \right]. \tag{A5}$$

Appendix B. Numerical Scheme

The implicit scheme at the point (x_i, t_n) for solving Equation (6) is:

$$-\frac{D}{(\Delta x)^2} u_{i+1}^{n+1} + \left(\frac{\omega_1}{(\Delta t)^2} + \frac{\omega_2}{2\Delta t} + \frac{2D}{(\Delta x)^2} \right) u_i^{n+1} - \frac{D}{(\Delta x)^2} u_{i-1}^{n+1} \tag{A6}$$

$$= \frac{2\omega_1}{(\Delta t)^2} u_i^n + \left(\frac{\omega_2}{2\Delta t} - \frac{\omega_1}{(\Delta t)^2} \right) u_i^{n-1} + F\left(u_i^{n+1} \right),$$

where $u_{i+1}^{n\pm1} = u(x_i, t_n \pm \Delta t)$ and $u_{i\pm1}^n = u(x_i \pm \Delta x, t_n)$.

In case growth rate $F(u)$ is linear, at each time step, Equation (A6) (where $i = 1, \ldots, N$ and N is the number of nodes in the spatial grid) is a linear algebraic system and hence is straightforward to solve (e.g., using Thomas algorithm). In case $F(u)$ is nonlinear, it can be linearized, e.g., it can be interpolated using the first two term of the Taylor expansion:

$$F\left(u_i^{n+1} \right) \approx F\left(u_i^n \right) + \left(F'(u) \right)_i^n \left(u_i^{n+1} - u_i^n \right) \approx F\left(u_i^n \right) + \left(F'(u) \right)_i^n \left(\frac{\partial u}{\partial t} \right)_i^n \Delta t, \tag{A7}$$

where $F'(u)$ is found analytically. Note that, in Equation (A7), we have made use of the fact that, for a smooth function, its left-hand side derivative coincides with its right-hand side derivative, so that $(u_i^{n+1} - u_i^n)/\Delta t \approx (\partial u/\partial t)_i^n \approx (u_i^n - u_i^{n-1})/\Delta t$.

The simulation results presented in the main text were obtained for $\Delta x = 0.05$ and $\Delta t = 0.001$.

References

1. Fahrig, L. Effects of habitat fragmentation on biodiversity. *Annu. Rev. Ecol. Evol. Syst.* **2003**, 34, 487–515. [CrossRef]
2. Ewers, R.M.; Didham, R.K. Confounding factors in the detection of species responses to habitat fragmentation. *Biol. Rev.* **2006**, 81, 117–142. [CrossRef]
3. Fischer, J.; Lindenmayer, D.B. Landscape modification and habitat fragmentation: A synthesis. *Glob. Ecol. Biogeogr.* **2007**, 16, 265–280. [CrossRef]
4. Kirk, R.V.; Lewis, M.A. Edge permeability and population persistence in isolated habitat patches. *Nat. Resour. Model.* **1999**, 12, 37–64. [CrossRef]
5. Mangel, M. *The Theoretical Biologist's Toolbox: Quantitative Methods for Ecology and Evolutionary Biology*; Cambridge University Press: Cambridge, UK, 2006.
6. Maynard Smith, J. *Models in Ecology*; Cambridge University Press: Cambridge, UK, 1974.
7. Kierstead, H.; Slobodkin, L.B. The size of water masses containing plankton blooms. *J. Mar. Res.* **1953**, 12, 141–147.
8. Lewis, M.A.; Kareiva, P. Allee dynamics and the spread of invading organisms. *Theor. Popul. Biol.* **1993**, 43, 141–158. [CrossRef]
9. Petrovskii, S.V. Plankton front waves accelerated by marine turbulence. *J. Mar. Syst.* **1999**, 21, 179–188. [CrossRef]
10. Petrovskii, S.V.; Shigesada, N. Some exact solutions of a generalized Fisher equation related to the problem of biological invasion. *Math. Biosci.* **2001**, 172, 73–94. [CrossRef]
11. Alharbi, W.G.; Petrovskii, S.V. The impact of fragmented habitat's size and shape on populations with Allee effect. *Math. Modell. Nat. Phenom.* **2016**, 11, 5–15. [CrossRef]
12. Fisher, R.A. The wave of advance of advantageous genes. *Ann. Eugen.* **1937**, 7, 355–369. [CrossRef]
13. Skellam, J.G. Random dispersal in theoretical populations. *Biometrika* 1951, 38, 196–218. [CrossRef]
14. Murray, J.D. *Mathematical Biology*; Springer: Berlin, Germany, 1989.
15. Malchow, H.; Petrovskii, S.V.; Venturino, E. *Spatiotemporal Patterns in Ecology and Epidemiology: Theory, Models, and Simulation*; CRC Press: Boca Raton, FL, USA, 2008.
16. Pyke, G.H. Understanding movements of organisms: It's time to abandon the Levy foraging hypothesis. *Methods Ecol. Evol.* **2015**, 6, 1–16. [CrossRef]

17. Kareiva, P.M.; Shigesada, N. Analyzing insect movement as a correlated random walk. *Oecologia* **1983**, *56*, 234–238. [CrossRef]
18. Kac, M. A stochastic model related to the telegrapher equation. *Rocky Mt. J. Math.* **1974**, *4*, 497–509. [CrossRef]
19. Goldstein, S. On diffusion by discontinuous movements and on the telegraph equation. *Quart. J. Mech. Appl. Math.* **1951**, *4*, 129–156. [CrossRef]
20. Holmes, E.E. Are diffusion models too simple? A comparison with telegraph models of invasion. *Am. Nat.* **1993**, *142*, 779–795. [CrossRef]
21. Mainardi, F. Signal velocity for transient waves in linear dissipative media. *Wave Motion* **1983**, *5*, 33–41. [CrossRef]
22. Di Crescenzo, A.; Martinucci, B.; Zacks, S. Telegraph process with elastic boundary at the origin. *Methodol. Comput. Appl. Probabil.* **2018**, *20*, 333–352. [CrossRef]
23. Mendez, V.; Fedotov, S.; Horsthemke, W. *Reaction–Transport Systems: Mesoscopic Foundations, Fronts, and Spatial Instabilities*; Springer: Berlin, Germany, 2010.
24. Lewis, M.A.; Petrovskii, S.V.; Potts, J. *The Mathematics Behind Biological Invasions. Interdisciplinary Applied Mathematics*; Springer: Berlin, Germany, 2016; Volume 44.
25. Renardy, M.; Rogers, R.C. *An Introduction to Partial Differential Equations*; Springer: New York, NY, USA, 2006.
26. Farlow, S.J. *Partial Differential Equations for Scientists and Engineers*; Dover: New York, NY, USA, 1993.
27. Miller, W. The technique of variable separation for partial differential equations. In *Nonlinear Phenomena*; Wolf, K.B., Ed.; Springer: Berlin, Germany, 1983; pp. 184–208.
28. Tilles, P.F.C.; Petrovskii, S.V. On the consistency of the reaction–telegraph process in finite domains. *J. Math. Biol.* **2018**, in preperation.
29. Hillen, T. Existence theory for correlated random walks on bounded domains. *Canad. Appl. Math. Q.* **2010**, *18*, 1–40.
30. Turchin, P. *Quantitative Analysis of Movement*; Sinauer: Sunderland, UK, 1998.
31. Banasiak, J.; Mika, J.R. Singularly perturbed telegraph equations with applications in the random walk theory. *Int. J. Stoch. Anal.* **1998**, *11*, 9–28. [CrossRef]
32. Berkowitz, B.; Kosakowski, G.; Margolin, G.; Scher, H. Application of continuous time random walk theory to tracer test measurements in fractured and heterogeneous porous media. *Groundwater* **2001**, *39*, 593–604. [CrossRef]
33. Van Gorder, R.A.; Vajravelu, K. Analytical and numerical solutions of the density dependent Nagumo telegraph equation. *Nonlinear Anal.* **2010**, *11*, 3923–3929. [CrossRef]
34. Roussy, G.; Pearce, J.A. *Foundations and Industrial Applications of Microwaves and Radio Frequency Fields*; John Wiley & Sons: Chichester, UK, 1995.
35. Jordan, P.M.; Puri, A. Digital signal propagation in dispersive media. *J. Appl. Phys.* **1999**, *85*, 1273–1282. [CrossRef]
36. FitzHugh, R. Impulses and physiological states in theoretical models of nerve membrane. *Biophys. J.* **1961**, *1*, 445–466. [CrossRef]
37. Nagumo, J.; Arimoto, S.; Yoshizawa, S. An active pulse transmission line simulating nerve axon. *Proc. IRE* **1962**, *50*, 2061–2070. [CrossRef]
38. Ahmed, E.; Hassanb, S.Z. On diffusion in some biological and economic systems. *Z. Naturforsch. A* **2000**, *55*, 669–672. [CrossRef]
39. Ahmed, E.; Abdusalam, H.A.; Fahmy, E.S. On telegraph reaction diffusion and coupled map lattice in some biological systems. *Int. J. Modern Phys. C* **2001**, *12*, 717–726. [CrossRef]
40. Giusti, A. Dispersion relations for the time-fractional Cattaneo-Maxwell heat equation. *J. Math. Phys.* **2018**, *59*, 013506. [CrossRef]
41. Harris, P.A.; Garra, R. Nonlinear heat conduction equations with memory: Physical meaning and analytical results. *J. Math. Phys.* **2017**, *58*, 063501. [CrossRef]
42. Mohebbi, A.; Dehghan, M. High order compact solution of the one-space-dimensional linear hyperbolic equation. *Numer. Methods Part. Differ. Equ.* **2008**, *24*, 1222–1235. [CrossRef]
43. Sobolev, S.L. On hyperbolic heat-mass transfer equation. *Int. J. Heat Mass Transfer* **2018**, *122*, 629–630. [CrossRef]

![Σ mathematics logo]

MDPI

Article

Prey-Predator Model with a Nonlocal Bistable Dynamics of Prey

Malay Banerjee [1,*,†], Nayana Mukherjee [1,†] and Vitaly Volpert [2,†]

[1]   Department of Mathematics and Statistics, IIT Kanpur, Kanpur 208016, India
[2]   Institut Camille Jordan, UMR 5208 CNRS, University Lyon 1, 69622 Villeurbanne, France;
       volpert@math.univ-lyon1.fr
*   Correspondence: malayb@iitk.ac.in; Tel.: +91-0512-259-6157
†   These authors contributed equally to this work.

Received: 5 February 2018; Accepted: 5 March 2018; Published: 8 March 2018

Abstract: Spatiotemporal pattern formation in integro-differential equation models of interacting populations is an active area of research, which has emerged through the introduction of nonlocal intra- and inter-specific interactions. Stationary patterns are reported for nonlocal interactions in prey and predator populations for models with prey-dependent functional response, specialist predator and linear intrinsic death rate for predator species. The primary goal of our present work is to consider nonlocal consumption of resources in a spatiotemporal prey-predator model with bistable reaction kinetics for prey growth in the absence of predators. We derive the conditions of the Turing and of the spatial Hopf bifurcation around the coexisting homogeneous steady-state and verify the analytical results through extensive numerical simulations. Bifurcations of spatial patterns are also explored numerically.

Keywords: prey-predator; nonlocal consumption; Turing bifurcation; spatial Hopf bifurcation; spatio-temporal pattern

---

1. Introduction

Investigation of spatiotemporal pattern formation leads to understanding of the interesting and complex dynamics of prey-predator populations. Reaction-diffusion systems of equations are conventionally used to study such dynamics. Various forms of reaction kinetics in the spatiotemporal model give rise to a wide variety of Turing patterns as well as non-Turing patterns including traveling wave [1–5] and spatiotemporal chaos [6,7]. Such patterns can be justified ecologically with the help of the field data and experiments which confirm the presence of patches in the prey-predator distributions. For example, Gause [8] has shown the importance of spatial heterogeneity for the stabilization and long term survival of species in the laboratory experiment on growth of *paramecium* and *didinium*. Luckinbill [9,10] has also studied the effect of dispersal on stability as well as persistence/extinction of population over a longer period of time. Based on these data, works are done where the prey-predator models with spatial distribution are considered for various ecological processes [11], such as plankton patchiness [12–14], semiarid vegetation patterns [15], invasion by exotic species [16,17] etc. (see also [18–21]). Such models have been successful in proving long term coexistence of both prey and predator populations along with formation of stationary or time dependent localized patches with periodic, quasi-periodic and chaotic dynamics [6,7].

The classical representation of two species interacting populations including the spatial aspect, consists of a reaction-diffusion system of equations in the form of two nonlinear coupled partial differential equations,

$$\frac{\partial u}{\partial t} = d_u \frac{\partial^2 u}{\partial x^2} + F_1(u,v), \tag{1}$$

$$\frac{\partial v}{\partial t} = d_v \frac{\partial^2 v}{\partial x^2} + F_2(u,v) \tag{2}$$

with non-negative initial conditions and appropriate boundary conditions. Population densities of prey and predator species at the spatial location x and time t are denoted by $u(x,t)$ and $v(x,t)$, respectively. The nonlinear functions F_1 and F_2 represent the interactions among individuals of the two species. The diffusion coefficients d_u and d_v represent the rate of random movement of individuals of the two species within the considered domain.

A wide variety of spatiotemporal patterns are described by these models, namely, traveling wave, periodic traveling wave, modulated traveling wave, wave of invasion, spatiotemporal chaos, stationary patchy patterns etc. [20–23]. Among all these, only stationary patchy pattern results in due to Turing instability, represents a stationary in time but non-homogeneous in space distribution. A stable co-existence of both species occurs due to formation of localized patches where the average population of each species remains unaltered in time. Whereas the other patterns are time dependent with the individuals of both species following continuously changing resources.

The general assumption for consumption of resources in the spatiotemporal models of interacting populations is taken to be local in space. In other words, it is supposed that the individuals consume resources in some areas surrounding their average location. Whereas nonlocal consumption of resources is more general since it incorporates the interspecific competition for food [24–26]. Such modifications enables the explanation of emergence and evolution of biological species as well as speciation in a more appropriate manner [27–31]. The models with nonlocal consumption of resources present complex dynamics for the single species models [28,29,32–35] as well as for competition models including two or more species [32,36–38]. Furthermore, such complex dynamics cannot be found in the corresponding local models.

Interesting results are obtained due to the introduction of nonlocal consumption of prey by predator in a reaction-diffusion system with Rosenzweig-McArthur type reaction kinetics [39]. Contrary to the local model where Turing patterns are not observed, this model satisfies the Turing instability conditions and gives rise to Turing patterns under proper assumptions on parameters. Other than this, existence of non-Turing patterns like traveling wave, modulated traveling wave, oscillatory pattern and spatiotemporal chaos are also observed for the nonlocal model. Some of the non-Turing patterns are reported for the nonlocal model with the modified Lotka-Volterra reaction kinetics [39,40].

In order to introduce the prey-predator model with a nonlocal bistable dynamics of prey, let us recall the classical models for the single population. Single species population model with the logistic growth law is described by the following ordinary differential equation, assuming homogeneous distribution of the species over their habitat,

$$\frac{du}{dt} = ru(k-u), \tag{3}$$

where r and k denote the intrinsic growth rate and carrying capacity, respectively. Introducing multiplicative Allee effect in this single population growth model, the above equation becomes

$$\frac{du}{dt} = ru(k-u)(u-l), \tag{4}$$

where l is the Allee effect threshold satisfying the restriction $0 < l < k$ [41–45]. This equation accounts for two significant feedback effects: positive feedback due to cooperation at low population density and negative feedback arising through the competition for limited resources at high population density [46]. In the framework of this formulation, the cooperation and competition mechanisms are described by the linear factors $(u-l)$ and $(k-u)$, respectively. Introduction of the Allee effect through a multiplicative term has a significant drawback since it represents a product of cooperation at the low population density and competition at the high population density. In this case, cooperation and competition influence each other, and their effects cannot be considered independently (see [46,47] for detailed discussion). The per capita growth rate is described by the factor $r(k - u)(u - l)$ which is positive for $l < u < k$, it is an increasing function for $l < u < \frac{k+l}{2}$, and a decreasing function for $\frac{k+l}{2} < u < k$. To overcome such situations, an additive form of the per capita growth rate function, proposed by Petrovskii et al. [47], is given by

$$\frac{du}{dt} = u\left(f(u) - \sigma - g(u)\right), \tag{5}$$

where the functions $f(u)$ and $g(u)$ describe population growth due to the reproduction and density dependent enhanced mortality rate, respectively. Here σ is the natural mortality rate independent of population density. Depending upon appropriate parametrization and assumption for the functional forms, the above model describes various types of single species population growth. In particular, if we choose $f(u) = \mu u$ and $g(u) = \eta u^2$ then we get the growth Equation (4) from (5) with appropriate relations between two sets of parameters (r, k, l) and (μ, σ, η). With a different type of parametrization, $f(u) = abu$ and $g(u) = au^2$ we can obtain the single species population growth model with sexual reproduction [34,35,48] as follows

$$\frac{du}{dt} = au^2(b - u) - \sigma u, \tag{6}$$

where a is the intrinsic growth rate, b is the environmental carrying capacity and σ is the density independent natural death rate. Introducing the nonlocal consumption of resources and random motion of the population, we get the following integro-differential equation model,

$$\frac{\partial u(x,t)}{\partial t} = d\frac{\partial^2 u(x,t)}{\partial x^2} + au^2(x,t)\left(b - \int_{-\infty}^{\infty} \phi(x - y)u(y,t)dt\right) - \sigma u(x,t), \tag{7}$$

where $\phi(z)$ is an even function with a bounded support and d is the diffusion coefficient. The kernel function is normalized to satisfy the condition $\int_{-\infty}^{\infty} \phi(z)dz = 1$. It shows the efficacy of consumption of resources as a function of distance $(x - y)$. The integral describes the total consumption of resources at the point x by the individuals located at $y \in (-\infty, \infty)$. This model shows bistability since the corresponding temporal model has two stable steady-states 0 and u_+ separated by an unstable steady-state u_- [34].

Based on this model, we are interested to study pattern formation described by the nonlocal reaction-diffusion system of prey-predator interaction with the bistable reaction kinetics of prey in the absence of predators which are specialist in nature following Holling type-II functional response. We will obtain the conditions of the Turing instability and of the spatial Hopf bifurcation in Section 2. Section 3 describes spatiotemporal pattern formation observed in numerical simulations. Here we also present bifurcation diagrams for the model with nonlocal consumption. Main outcomes of this investigation are summarized in the discussion section.

2. Stability Analysis

In this section, we will introduce the prey-predator models without and with nonlocal consumption term and will study stability of the positive homogeneous stationary solution.

2.1. Local Model

We consider the following reaction-diffusion system for the prey-predator interaction:

$$\frac{\partial u}{\partial t} = d_1\frac{\partial^2 u}{\partial x^2} + au^2(b - u) - \sigma_1 u - \frac{\alpha uv}{\kappa + u}, \tag{8}$$

$$\frac{\partial v}{\partial t} = d_2\frac{\partial^2 v}{\partial x^2} + \frac{\beta uv}{\kappa + u} - \sigma_2 v, \tag{9}$$

subjected to a non-negative initial condition and the periodic boundary condition. The consumption of prey by the predator follows the Holling type-II functional response, α is the rate of consumption of prey by an individual predator, κ is the half-saturation constant and β is the rate of conversion of prey to predator biomass. Furthermore, β/α is the conversion efficiency with the value between 0 and 1, consequently $\beta < \alpha$. The reproduction of prey is proportional to the second power of the population density specific for sexual reproduction. In the absence of predator ($\alpha = 0$) dynamics of prey is described by a bistable reaction-diffusion equation.

The coexistence (positive) equilibrium $E_*(u_*, v_*)$ of the corresponding temporal model

$$\frac{du}{dt} = au^2(b - u) - \sigma_1 u - \frac{\alpha uv}{\kappa + u} \equiv f(u, v), \tag{10}$$

$$\frac{dv}{dt} = \frac{\beta uv}{\kappa + u} - \sigma_2 v \equiv g(u, v), \tag{11}$$

is given by the equalities

$$u_* = \frac{\kappa \sigma_2}{\beta - \sigma_2}, \quad v_* = \frac{\kappa + u_*}{\alpha}\left(au_*(b - u_*) - \sigma_1\right), \tag{12}$$

and associated feasibility conditions

$$0 < \sigma_2 < \beta, \; 0 < \sigma_1 < \frac{a\kappa\sigma_2}{\beta - \sigma_2}\left(b - \frac{\kappa\sigma_2}{\beta - \sigma_2}\right)$$

which provide the positiveness of solutions.

Here we briefly present the local asymptotic stability condition of E_* for the temporal model (10)–(11) that will be required afterwards. Linearizing the nonlinear system (10)–(11) around E_* we can find the associated eigenvalue equation

$$\lambda^2 - a_{11}\lambda - a_{12}a_{21} = 0,$$

where

$$a_{11} = f_u(u_*, v_*), \; a_{12} = f_v(u_*, v_*) < 0, \; a_{21} = g_u(u_*, v_*) > 0, \; a_{22} = g_v(u_*, v_*) = 0.$$

Two eigenvalues of the above characteristic equation have negative real parts if $a_{11} < 0$ and hence E_* is locally asymptotically stable for $a_{11} < 0$. The stationary point E_* loses its stability through the super-critical Hopf bifurcation if $a_{11} = 0$.

It is well known that the models of the form (8)–(9), that is for which $a_{22} = 0$, are unable to produce any Turing pattern as the Turing instability conditions cannot be satisfied [40]. However these type of models are capable to produce non-Turing pattern if the temporal parameter values are well inside the Hopf-bifurcation domain [6]. The spatiotemporal prey-predator models with a specialist predator and linear death rate for predator population can produce spatiotemporal chaos, wave of chaos, modulated traveling wave, wave of invasion and their combinations if the spatial domain is large enough [7].

2.2. Nonlocal Model

Under the assumption that prey can move from one location to another one to access the resources, model (8)–(9) can be extended to the model with nonlocal consumption of resources:

$$\frac{\partial u}{\partial t} = d_1 \frac{\partial^2 u}{\partial x^2} + au^2(b - J(u)) - \sigma_1 u - \frac{\alpha uv}{\kappa + u}, \tag{13}$$

$$\frac{\partial v}{\partial t} = d_2 \frac{\partial^2 v}{\partial x^2} + \frac{\beta uv}{\kappa + u} - \sigma_2 v, \tag{14}$$

subjected to a non-negative initial condition and the periodic boundary condition. Here

$$J(u) = \int_{-\infty}^{\infty} \phi(x - y)u(y, t)dy, \quad \phi(y) = \begin{cases} \frac{1}{2M} & , \quad |y| \le M \\ 0 & , \quad |y| > M \end{cases}.$$

Various forms of kernel functions are considered in literature. Here we consider the step function for simplicity of mathematical calculations [49]. This step function means that the nonlocal consumption is confined within the range $2M$, and the efficacy of consumption inside this range is constant.

We will analyze stability of the homogeneous steady-state (u_*, v_*). We consider the perturbation around it in the form

$$u(x, t) = u_* + \epsilon_1 e^{\lambda t + ikx}, \quad v(x, t) = v_* + \epsilon_2 e^{\lambda t + ikx}, \quad |\epsilon_1|, |\epsilon_2| \ll 1.$$

The characteristic equation writes as $|H - \lambda I| = 0$ where

$$H = \begin{bmatrix} a_1 - au_*^2 \frac{\sin kM}{kM} - d_1 k^2 & -a_2 \\ b_1 & -d_2 k^2 \end{bmatrix} \tag{15}$$

and

$$a_1 = abu_* - au_*^2 + \frac{au_* v_*}{(\kappa + u_*)^2}, \quad a_2 = \frac{au_*}{\kappa + u_*}, \quad b_1 = \frac{\beta u_* v_*}{(\kappa + u_*)^2}. \tag{16}$$

Therefore, the characteristic equation becomes as follows:

$$\lambda^2 - \Gamma(k, M)\lambda + \Delta(k, M) = 0, \tag{17}$$

where

$$\Gamma(k, M) = a_1 - au_*^2 \frac{\sin kM}{kM} - (d_1 + d_2)k^2, \tag{18}$$

$$\Delta(k, M) = \left(au_*^2 \frac{\sin kM}{kM} - a_1 + d_1 k^2 \right) d_2 k^2 + a_2 b_1. \tag{19}$$

The homogeneous steady-state is stable under space dependent perturbations if the following two conditions are satisfied:

$$\Gamma(k, M) < 0, \quad \Delta(k, M) > 0 \tag{20}$$

for all positive real k and M. The homogeneous steady-state loses its stability through the spatial Hopf bifurcation if $\Gamma(k_H, M) = 0, \Delta(k_H, M) > 0$ for some k_H, and through the Turing bifurcation if $\Gamma(k_T, M) < 0, \Delta(k_T, M) = 0$ for some k_T.

2.3. Spatial Hopf Bifurcation

First, we find the spatial Hopf bifurcation threshold in terms of the parameter d_2. It is important to note that $\Gamma(k, M) < 0$ and $\Delta(k, M) > 0$ as $M \to 0+$ if we assume that (u_*, v_*) is locally asymptotically stable for the temporal model (10)–(11). One can easily verify that $\lim_{M \to 0+} \Gamma(k, M) = a_{11}$ and $\lim_{M \to 0+} \Delta(k, M) = -a_{12}a_{21}$. For some suitable M if one can find a unique value $k \equiv k_H$ such that $\Gamma(k, M) = 0$ then k_H is the critical wavenumber for the spatial Hopf bifurcation. This critical wavenumber can be obtained by solving the following two equations simultaneously:

$$\Gamma(k, M) = 0, \quad \frac{\partial}{\partial k}\Gamma(k, M) = 0. \tag{21}$$

Using the expression of $\Gamma(k, M)$, we find d_2 from the equation $\Gamma(k, M) = 0$:

$$d_2(k) = \frac{1}{k^2}\left[a_1 - au_*^2\frac{\sin kM}{kM} - d_1 k^2\right]. \tag{22}$$

Substituting this expression into the second equation in (21) we get:

$$2a_1 + au_*^2\cos kM - 3au_*^2\frac{\sin kM}{kM} = 0. \tag{23}$$

Equation (23) can have more than one positive real root depending upon the values of parameters. It is necessary to verify that the corresponding values of $d_2(k)$ are positive. We choose the root k_H for which $d_2(k_H)$ is the minimal positive number, and $\Delta(k_H, M) > 0$.

Consider, as example, the following values of parameters:

$$a = 1,\ b = 1,\ \sigma_1 = 0.1,\ \alpha = 0.335,\ \kappa = 0.4,\ \beta = 0.335,\ \sigma_2 = 0.2,\ d_1 = 1. \tag{24}$$

Then $u_* = 0.593$, $v_* = 0.419$, and Equation (23) possesses only one positive root $k = 0.297$ for $M = 6$. From (22), we find $d_2 = 0.51$. Since $\Delta(0.51, 6) = 0.0094$, these values of k and d_2 correspond to the desired spatial Hopf bifurcation thresholds, $k_H = 0.297$, $d_{2H} = 0.51$.

Furthermore, $\Gamma(k, 6) > 0$ for $d_2 < d_{2H}$. Hence the spatial Hopf bifurcation takes place as d_2 crosses the critical threshold d_{2H} from higher to lower values. Therefore, oscillatory in space and time patterns emerging due to the spatial Hopf bifurcation are observed below the stability boundary. The spatial Hopf bifurcation curve in the (M, d_2)-parameter space is shown in Figure 1. Spatiotemporal patterns for parameter values lying in the spatial Hopf domain is presented in Figure 3a.

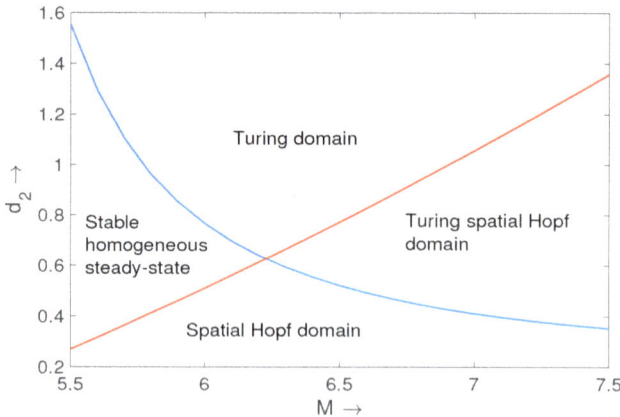

Figure 1. Turing and Hopf stability boundaries in the (M, d_2)-parameter plane.

2.4. Turing Pattern for Nonlocal Prey-Predator Model

Next, we discuss the Turing bifurcation condition and we assume that $\lim_{M\to 0+}\Gamma(k, M) < 0$ and $\lim_{M\to 0+}\Delta(k, M) > 0$. These conditions provide stability of the homogeneous steady-state under space independent perturbations. The critical wavenumber and the corresponding Turing bifurcation threshold in terms of d_2 can be obtained as a solution of the following two equations:

$$\Delta(k, M) = 0,\ \frac{\partial}{\partial k}\Delta(k, M) = 0. \tag{25}$$

From $\Delta(k, M) = 0$, we get

$$d_2(k) = \frac{a_2 b_1}{k^2 \left[a_1 - d_1 k^2 - a u_*^2 \frac{\sin kM}{kM} \right]}. \tag{26}$$

Substituting this expression into the second equation in (25), we obtain

$$4 d_1 k^2 - 2 a_1 + a u_*^2 \left(\cos kM + \frac{\sin kM}{kM} \right) = 0. \tag{27}$$

This equation can have more than one positive root. From now on, we assume that all parameter values are fixed except for M and d_2. Suppose that for a chosen value of M, (27) admits a finite number of positive roots, $k_1, k_2, \cdots, k_m$. The corresponding values of $d_2(k_i)$ in (26) are not necessarily positive. The Turing bifurcation threshold d_{2T} is given by the minimal positive value $d_2(k_i)$, and k_T is the value of k_i for which the minimum is reached.

Let us consider the same set of parameters as in the previous subsection except for $d_1 = 0.4$. An interesting feature of the Turing bifurcation curve is that it is not smooth when plotted in the (M, d_2)-parameter plane (Figure 2). The point of non-differentiability arises around $M = 12.65$. For $M = 12.5$, we find four positive roots of (27), $k_1 = 0.038$, $k_2 = 0.470$, $k_3 = 0.638$ and $k_4 = 0.786$. The corresponding values $d_2(k_{j_r})$ are positive for the last three roots, $d_2(k_2) = 0.198$, $d_2(k_3) = 0.235$ and $d_2(k_4) = 0.199$. Hence we find the Turing bifurcation threshold $d_{2T} = 0.198$, corresponding to k_2 (Figure 2b). Next, if we choose $M = 12.7$, then we find $k_1 = 0.0378$, $k_2 = 0.465$, $k_3 = 0.620$, $k_4 = 0.781$, and $d_2(k_2) = 0.201$, $d_2(k_3) = 0.232$, $d_2(k_4) = 0.189$ are positive d_2-values. Hence, the Turing bifurcation threshold $d_{2T} = 0.189$ corresponds to k_4 (Figure 2c). Hence, the point where the stability boundary is not smooth correspond to the sudden change of the location of the feasible root of Equation (27).

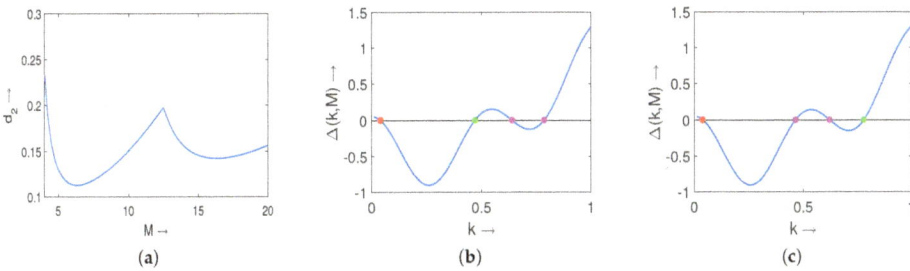

Figure 2. Turing bifurcation curve on the (M, d_2)-parameter plane (**a**). Stationary Turing patterns exist above the bifurcation curve. The functions $\Delta(k, 12.5)$ (**b**) and $\Delta(k, 12.7)$ (**c**). The root corresponding to the Turing instability is shown in green.

Finally, it is important to mention that the choice of parameters (24) leads to an interesting scenario for which the spatial Hopf and Turing bifurcation curves intersect. The two curves are shown in Figure 1 for $d_1 = 1$, and they divide the parametric domain into four different regions. Spatial patterns produced by the prey population for parameter values taken from spatial Hopf domain and Turing domain are presented in Figure 3a,b. In order to emphasize the fact that the stationary Turing pattern can be obtained for equal diffusion coefficients, we set $d_1 = d_2 = 1$ and observe a periodic in space and stationary in time solution (see Figure 3b). Various spatio-temporal patterns are observed for the values of parameters at the intersection of Turing and Hopf instability regions. Some of them are described in Section 3.

(a)   (b)

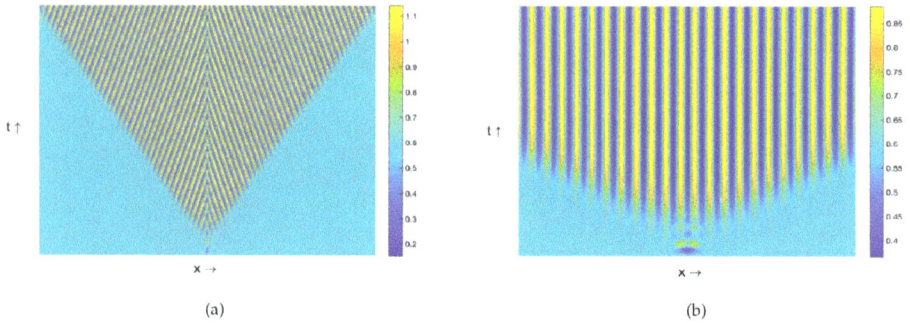

Figure 3. (a) Resulting spatio-temporal patterns produced by the nonlocal model (13)–(14) for $d_2 = 0.3$, $M = 6.5$ and other parameter values as mentioned in the text. **(b)** Stationary pattern produced by the prey population for $d_1 = 1, d_2 = 1$ and $M = 6$, other parameters are mentioned at text.

3. Spatiotemporal Patterns

In this section, we study nonlinear dynamics of the prey-predator model without and with nonlocal term in the equation for prey population. We present the results of numerical simulations performed with a finite difference approximation of systems (8)–(9) and (13)–(14).

3.1. Patterns Produced by the Model (8)–(9)

In this subsection, we consider the non-Turing patterns described by system (8)–(9) in the interval $-L \leq x \leq L$ with non-negative initial condition and periodic boundary condition. Results presented here are obtained for $L = 200$. We consider a small perturbation around the homogeneous steady-state at the center of the domain as initial condition. The values of parameters are as follows

$$a = 1, b = 1, \sigma_1 = 0.1, \alpha = 0.4, \kappa = 0.4, \sigma_2 = 0.2, d_1 = 1, d_2 = 1, \tag{1}$$

and the value of β will vary.

It is known [6,7,16,26,50] that the prey-predator models with specialist predator can manifest time dependent spatial patterns if the parameters of the reaction kinetics are far inside the temporal Hopf domain. In this case, the temporal Hopf-bifurcation threshold is $\beta_* = 0.339$, that is E_* is stable for $\beta < \beta_*$, and it is unstable otherwise.

Solutions homogeneous in space and oscillatory in time are observed for $\beta > \beta_*$ but close to it. The spatiotemporal pattern presented in Figure 4a is almost homogeneous in space but oscillatory in time for the value of β close to the temporal Hopf bifurcation threshold. For larger values of β we find spatiotemporal patterns periodic both in space and time (Figure 4b) and symmetric around $x = 0$. This symmetry is maintained due to the choice of symmetric initial condition. With the increase of β we observe various complex aperiodic spatiotemporal regimes (Figure 4c). They are characterized by specific triangular patterns resulting from the merging of two peaks in the population density moving towards each other.

This model is capable to produce other type of spatiotemporal patterns, such as the traveling wave, periodic travelling wave, wave of invasion, wave of chaos similar to the prey-predator model with Rosenzweig-MacArthur reaction kinetics [26] but those results are beyond the scope of this work and will be addressed in the future.

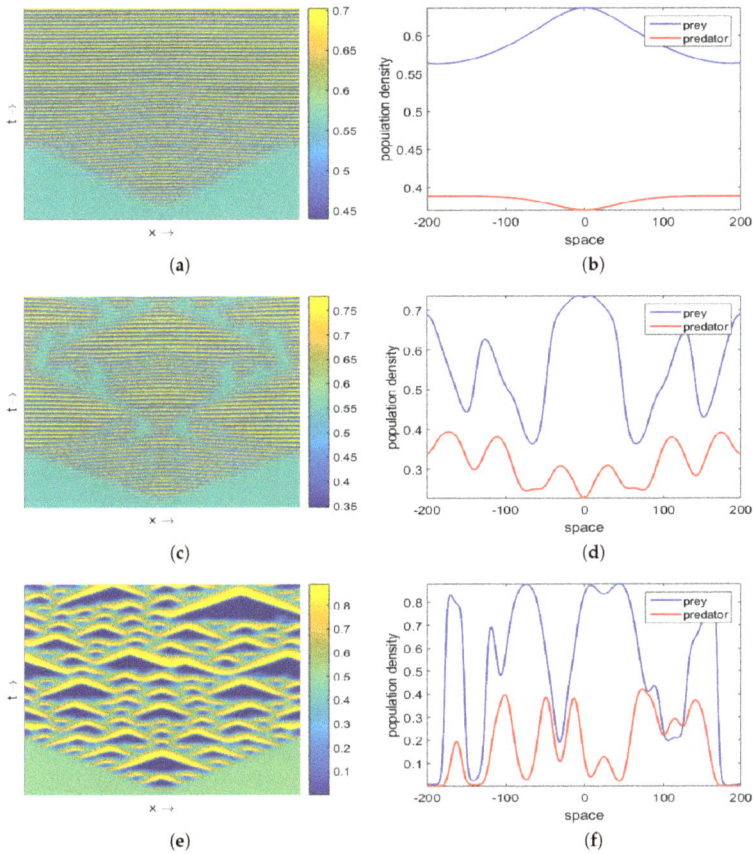

Figure 4. Spatiotemporal patterns (prey density) produced by the model (8)–(9) are presented in the left column for the parameter values as mentioned at the text and (**a**) $\beta = 0.342$; (**c**) $\beta = 0.3445$; (**e**) $\beta = 0.36$. Corresponding distribution of prey and predator population over space at $t = 1000$ are presented in the right column, see (**b**), (**d**), (**f**).

3.2. Effect of Nonlocal Consumption

We will now analyze how the nonlocal term influences dynamics of the prey-predator model. Nonlinear dynamics of prey-predator system with nonlocal consumption of resources by prey is summarized in the diagram in Figure 5a. Parameter regions with different regimes are shown on the (M, β)-plane for the values of other parameters given in (1). For small β, predator disappears while the population of prey is either homogeneous in space or it forms a spatially periodic distribution. For large β, both population go to extinction. More interesting behavior is observed for the intermediate values of β. This can be homogeneous or inhomogeneous in space, stationary or non-stationary in time solutions. Some of the spatiotemporal patterns are shown in Figure 6. For M sufficiently small these patterns become similar to those presented in Figure 4. For M large enough, both prey and predator densities represent stationary periodic in space distributions (similar to Figure 3b). Figure 5b shows a similar diagram in the case of different diffusion coefficients of prey and predator, $d_1 = 0.7, d_2 = 0.5$, with the same values of other parameters. The region of spatiotemporal patterns exists here for a narrower interval of β while the regions of stationary patterns and of extinction change their shape.

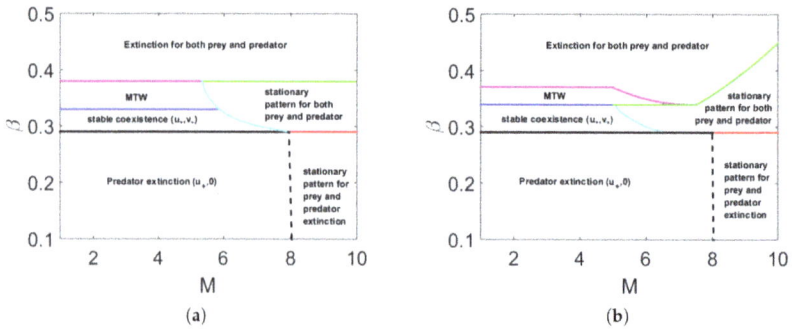

(a)

(b)

Figure 5. Bifurcation diagram in (M, β)-parameter space for fixed parameters $a = 1, b = 1, \sigma_1 = 0.1$, $\alpha = 0.4, \kappa = 0.4$ and $\sigma_2 = 0.2$. (**a**) $d_1 = 1, d_2 = 1$; (**b**) $d_1 = 0.7, d_2 = 0.5$.

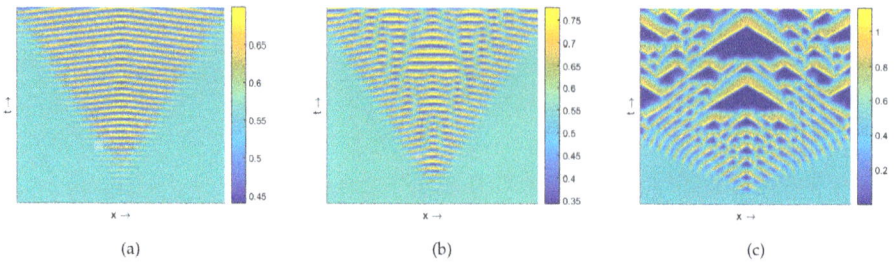

(a)

(b)

(c)

Figure 6. Spatio-temporal patterns produced by the nonlocal model with parameter values as mentioned at (1) with $\beta = 0.342$ and (**a**) $M = 0$; (**b**) $M = 4$; (**c**) $M = 6$.

Multiplicity of Stationary Solutions

Another interesting aspect of the stationary patterns arising through the Turing bifurcation for the spatiotemporal model with nonlocal interaction term is the existence of multiple stationary solutions for a particular value of M. We have used forward and backward numerical continuation method to determine the range of M for the stationary patterns with different periodicity (Figure 7). Fixed parameter values are same as (1) except $d_1 = 0.4$ and $d_2 = 0.2$. For example, stationary pattern with 33 patches (over a spatial domain of size $L = 200$) exists for $2 \leq M \leq 4.5$, with 32 patches for $2.5 \leq M \leq 5$, and so on.

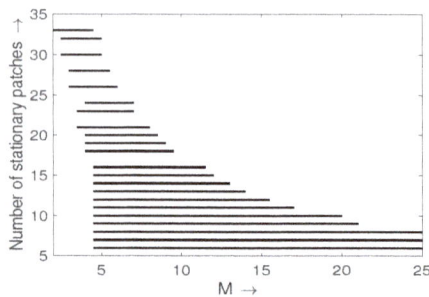

Figure 7. Stationary patterns with various number of patches exist for a range of nonlocal consumption (M) is plotted. Parameter values are same as in (24) except $d_1 = 0.4$ and $d_2 = 0.2$.

4. Discussion

In this work we study a prey-predator model with a bistable nonlocal dynamics of prey. Without the interaction with predator, the prey density is described by a bistable reaction-diffusion equation taking into account the Allee effect or the sexual reproduction of population. In this case, the reproduction rate is proportional to the second power of the population density. The dynamics of the prey population changes due to introduction of nonlocal consumption of resources. The main difference compared to the results of conventional local consumption is that the positive stable equilibrium may become unstable resulting in the appearance of stationary in time but periodic in space solutions.

The interaction with predator provides an additional factor that influences the dynamics of solutions. If we characterize this interaction by the parameter β, which determines the reproduction rate in the equation of predator density, then we can identify three main types of behavior depending on its value (Figure 5). If it is sufficiently small then the predator population vanishes since the reproduction is not enough to overcome the mortality. If this parameter is too large, then both populations go to extinction, particularly due to the bistability of the prey dynamics. Both populations coexist in a relatively narrow interval of the interconnection parameter. There are three different types of patterns inside this parameter domain. The homogeneous in space equilibrium can be stable or it can lose its stability resulting in the emergence of spatiotemporal patterns. They are observed for limited values of the parameter M which determines the range of nonlocal consumption. If the range of nonlocal consumption is sufficiently large, then both populations represent a periodic in space distribution. Such solutions are specific for nonlocal consumption with a large range, in particular for the single prey population. Thus, nonlocal consumption takes over spatiotemporal oscillations specific for the local prey-predator dynamics. Let us note that there is multiplicity of stationary patterns for the same values of parameters. This effect is specific for the problems with nonlocal interaction [39].

Spatiotemporal oscillations are specific for the prey-predator dynamics [6,7,16,26,40,50]. Here we observe the dynamics with "triangular" patterns (Figure 4) appearing when two pulses move towards each other and merge. Nonlocal consumption of resources modifies these patterns.

Some questions related to the prey-predator dynamics with nonlocal bistable model for prey remain beyond the scope of this work. We have not discussed here travelling waves and pulses that can also be observed in such models. We will study them in the subsequent work.

Author Contributions: M.B. and V.V. proposed the formulation of the problem; N.M. performed the numerical simulations; M.B. and N.M. analyzed the results; all authors participated in the preparation of the manuscript.

Conflicts of Interest: The authors declare no conflict of interest.

References

1. Dunbar, S.R. Travelling wave solutions of diffusive Lotka-Volterra equations. *J. Math. Biol.* **1983**, *17*, 11–32.
2. Dunbar, S.R. Travelling waves in diffusive predator-prey equations: Periodic orbits and point-to-periodic heteroclinic orbits. *SIAM J. Appl. Math.* **1986**, *46*, 1057–1078.
3. Sherratt, J.A. Periodic travelling waves in cyclic predator-prey systems. *Ecol. Lett.* **2001**, *4*, 30–37.
4. Sherratt, J.A.; Smith, M. Periodic travelling waves in cyclic populations: Field studies and reaction diffusion models. *J. R. Soc. Interface* **2008**, *5*, 483–505.
5. Volpert, V.; Petrovskii, S.V. Reaction-diffusion waves in biology. *Phys. Life Rev.* **2009**, *6*, 267–310.
6. Petrovskii, S.V.; Malchow, H. A minimal model of pattern formation in a prey-predator system. *Math. Comp. Model.* **1999**, *29*, 49–63.
7. Petrovskii, S.V.; Malchow, H. Wave of chaos: New mechanism of pattern formation in spatio-temporal population dynamics. *Theor. Pop. Biol.* **2001**, *59*, 157–174.
8. Gause, G.F. *The Struggle for Existence*; Williams and Wilkins: Baltimore, MD, USA, 1935.
9. Luckinbill, L.L. Coexistence in laboratory populations of *Paramecium aurelia* and its predator *Didinium nasutum*. *Ecology* **1973**, *54*, 1320–1327.

10. Luckinbill, L.L. The effects of space and enrichment on a predator-prey system. *Ecology* **1974**, *55*, 1142–1147.
11. Fasani, S.; Rinaldi, S. Factors promoting or inhibiting Turing instability in spatially extended prey-predator systems. *Ecol. Model.* **2011**, *222*, 3449–3452.
12. Huisman, J.; Weissing, F.J. Biodiversity of plankton by oscillations and chaos. *Nature* **1999**, *402*, 407–410.
13. Levin, S.A.; Segel, L.A. Hypothesis for origin of planktonic patchiness. *Nature* **1976**, *259*, 659.
14. Segel, L.A.; Jackson, J.L. Dissipative structure: An explanation and an ecological example. *J. Theor. Biol.* **1972**, *37*, 545–559.
15. Klausmeier, C.A. Regular and irregular patterns in semiarid vegetation. *Science* **1999**, *284*, 1826–1828.
16. Medvinsky, A.; Petrovskii, S.; Tikhonova, I.; Malchow, H.; Li, B.L. Spatiotemporal complexity of plankton and fish dynamics. *SIAM Rev.* **2002**, *44*, 311–370.
17. Shigesada, N.; Kawasaki, K. *Biological Invasions: Theory and Practice*; Oxford University Press: Oxford, UK, 1997.
18. Baurmann, M.; Gross, T.; Feudel, U. Instabilities in spatially extended predator-prey systems: Spatio-temporal patterns in the neighborhood of Turing-Hopf bifurcations. *J. Theor. Biol.* **2007**, *245*, 220–229.
19. Cantrell, R.S.; Cosner, C. *Spatial Ecology via Reaction-Diffusion Equations*; Wiley: London, UK, 2003.
20. Murray, J.D. *Mathematical Biology II*; Springer: Berlin/Heidelberg, Germany, 2002.
21. Okubo, A.; Levin, S. *Diffusion and Ecological Problems: Modern Perspectives*; Springer: Berlin/Heidelberg, Germany, 2001.
22. Banerjee, M.; Banerjee, S. Turing instabilities and spatio-temporal chaos in ratio-dependent Holling-Tanner model. *Math. Biosci.* **2012**, *236*, 64–76.
23. Banerjee, M.; Petrovskii, S. Self-organized spatial patterns and chaos in a ratio-dependent predator-prey system. *Theor. Ecol.* **2011**, *4*, 37–53.
24. Gourley, S.A.; Britton, N.F. A predator-prey reaction-diffusion system with nonlocal effects. *J. Math. Biol.* **1996**, *34*, 297–333.
25. Gourley, S.A.; Ruan, S. Convergence and travelling fronts in functional differential equations with nonlocal terms: A competition model. *SIAM J. Appl. Math.* **2003**, *35*, 806–822.
26. Sherratt, J.A.; Eagan, B.T.; Lewis, M.A. Oscillations and chaos behind predator-prey invasion: Mathematical artifact or ecological reality? *Phils. Trans. R. Soc. Lond. B* **1997**, *352*, 21–38.
27. Bessonov, N.; Reinberg, N.; Volpert, V. Mathematics of Darwin's diagram. *Math. Model. Nat. Phenom.* **2014**, *9*, 5–25.
28. Genieys, S.; Bessonov, N.; Volpert, V. Mathematical model of evolutionary branching. *Math. Comp. Model.* **2009**, *49*, 2109–2115.
29. Genieys, S.; Volpert, V.; Auger, P. Pattern and waves for a model in population dynamics with nonlocal consumption of resources. *Math. Model. Nat. Phenom.* **2006**, *1*, 63–80.
30. Genieys, S.; Volpert, V.; Auger, P. Adaptive dynamics: Modelling Darwin's divergence principle. *Comp. Ren. Biol.* **2006**, *329*, 876–879.
31. Volpert, V. Branching and aggregation in self-reproducing systems. *ESAIM Proc. Surv.* **2014**, *47*, 116–129.
32. Apreutesei, N.; Bessonov, N.; Volpert, V.; Vougalter, V. Spatial structures and generalized travelling waves for an integro-differential equation. *DCDS B* **2010**, *13*, 537–557.
33. Aydogmus, O. Patterns and transitions to instability in an intraspecific competition model with nonlocal diffusion and interaction. *Math. Model. Nat. Phenom.* **2015**, *10*, 17–19.
34. Volpert, V. Pulses and waves for a bistable nonlocal reaction-diffusion equation. *Appl. Math. Lett.* **2015**, *44*, 21–25.
35. Volpert, V. *Elliptic Partial Differential Equations*; Reaction-diffusion equations; Birkhäuser: Basel, Switzerland, 2014; Volume 2.
36. Apreutesei, N.; Ducrot, A.; Volpert, V. Competition of species with intra-specific competition. *Math. Model. Nat. Phenom.* **2008**, *3*, 1–27.
37. Apreutesei, N.; Ducrot, A.; Volpert, V. Travelling waves for integro-differential equations in population dynamics. *DCDS B* **2009**, *11*, 541–561.
38. Bayliss, A.; Volpert, V.A. Patterns for competing populations with species specific nonlocal coupling. *Math. Model. Nat. Phenom.* **2015**, *10*, 30–47.
39. Banerjee, M.; Volpert, V. Prey-predator model with a nonlocal consumption of prey. *Chaos* **2016**, *26*, 083120.

40. Banerjee, M.; Volpert, V. Spatio-temporal pattern formation in Rosenzweig-McArthur model: Effect of nonlocal interactions. *Ecol. Complex.* **2016**, *30*, 2–10.
41. Amarasekare, P. Interactions between local dynamics and dispersal: Insights from single species models. *Theor. Popul. Biol.* **1998**, *53*, 44–59.
42. Amarasekare, P. Allee effects in metapopulation dynamics. *Am. Nat.* **1998**, *152*, 298–302.
43. Courchamp, F.; Berec, L.; Gascoigne, J. *Allee Effects in Ecology and Conservation*; Oxford University Press: Oxford, UK, 2008.
44. Courchamp, F.; Clutton-Brock, T.; Grenfell, B. Inverse density dependence and the Allee effect. *Trends. Ecol. Evol.* **1999**, *14*, 405–410.
45. Lewis, M.A.; Kareiva, P. Allee dynamics and the spread of invading organisms. *Theor. Popul. Biol.* **1993**, *43*, 141–158.
46. Jankovic, M.; Petrovskii, S. Are time delays always destabilizing? Revisiting the role of time delays and the Allee effect. *Theor. Ecol.* **2014**, *7*, 335–349.
47. Petrovskii, S.; Blackshaw, R.; Li, B.L. Consequences of the Allee effect and intraspecific competition on population persistence under adverse environmental conditions. *Bull. Math. Biol.* **2008**, *70*, 412–437.
48. Banerjee, M.; Vougalter, V.; Volpert, V. Doubly nonlocal reaction diffusion equations and the emergence of species. *Appl. Math. Model.* **2017**, *42*, 591–599.
49. Segal, B.L.; Volpert, V.A.; Bayliss, A. Pattern formation in a model of competing populations with nonlocal interactions. *Physics D* **2013**, *253*, 12 – 23.
50. Petrovskii, S.V.; Li, B.L.; Malchow, H. Quantification of the spatial aspect of chaotic dynamics in biological and chemical systems. *Bull. Math. Biol.* **2003**, *65*, 425–446.

mathematics

MDPI

Article

Role of Bi-Directional Migration in Two Similar Types of Ecosystems

Nikhil Pal [1], Sudip Samanta [2], Maia Martcheva [3],*,[†] and Joydev Chattopadhyay [4]

[1] Department of Mathematics, Visva-Bharati University, Santiniketan 731235, India;
nikhil.pal@visva-bharati.ac.in

[2] Department of Mathematics, Faculty of Science & Arts-Rabigh, King Abdulaziz University,
Rabigh 25732, Saudi Arabia; samanta.sudip.09@gmail.com

[3] Department of Mathematics, University of Florida, Gainesville, FL 32611, USA

[4] Agricultural and Ecological Research Unit, Indian Statistical Institute 203, B. T. Road, Kolkata 700108, India;
joydev@isical.ac.in

* Correspondence: maia@ufl.edu; Tel.: +352-294-2319

† Work supported by NSF grant DMS-1515661.

Received: 15 January 2018; Accepted: 20 February 2018; Published: 2 March 2018

Abstract: Migration is a key ecological process that enables connections between spatially separated populations. Previous studies have indicated that migration can stabilize chaotic ecosystems. However, the role of migration for two similar types of ecosystems, one chaotic and the other stable, has not yet been studied properly. In the present paper, we investigate the stability of ecological systems that are spatially separated but connected through migration. We consider two similar types of ecosystems that are coupled through migration, where one system shows chaotic dynamics, and other shows stable dynamics. We also note that the direction of the migration is bi-directional and is regulated by the population densities. We propose and analyze the coupled system. We also apply our proposed scheme to three different models. Our results suggest that bi-directional migration makes the coupled system more regular. We have performed numerical simulations to illustrate the dynamics of the coupled systems.

Keywords: food web; dispersal; bifurcation; chaos; stability

1. Introduction

In mathematical biology, population theory plays an important role. Historically, the first model of population dynamics was formulated by Malthus [1] and was later on adapted for more realistic situations by Verhulst [2]. Lotka and Volterra [3,4] first modeled oscillations occurring in natural populations. Subsequently, the Lotka–Volterra model was modified by several researchers, and many of them observed chaotic dynamics [5–9]. The occurrence of chaos in a simple ecological system motivated researchers to investigate complex dynamical behaviors of ecological systems, such as bi-stability, bifurcation and chaos. However, in real-world populations, the evidence of chaos is rare. In ecology, until now, many researchers have investigated three-species food chain/web models with the aim of controlling the chaos by incorporating several biological phenomena [10–12].

Spatial structure is an important factor in ecological systems. Natural systems are rarely isolated but rather interact among themselves as well as with their natural surroundings, and the dynamics of ecological systems connected by migration are very different from the dynamics of the individual systems. The concept of a metapopulation is a formalism to describe spatially separated interacting populations [13,14]. A metapopulation consists of a group of spatially separated populations living in patches; individuals are allowed to migrate to surrounding patches. Levins (1969) [13] proposed a metapopulation theory and applied it in a pest-control situation. In landscape ecology and conservation biology, the idea of a metapopulation plays an important role [15,16].

In population biology, two systems can be coupled through migration, which is a common biological phenomenon and plays a vital role in the stability of ecosystems. Migration has been studied in a variety of taxa [17–19]. In the stability of an ecosystem, migration can have a stabilizing effect [20–23]. Holt (1985) [23] observed that passive dispersal between sink and source habitats can stabilize an otherwise unstable system. MacCullum observed that immigration could stabilize a chaotic system of the crown of thorns starfish *Acanthaster planci* and its associated larval recruitment patterns [21]. Stone and Hart [22] observed that a discrete-time chaotic system could be stabilized by constant immigration. Silva et al. [24] also synchronized chaotic oscillations of uncoupled populations through migration. Furthermore, it has been established that unstable equilibria of a single-patch predator–prey model cannot be stabilized by coupling with identical patches [25]. The persistence of coupled locally unstable systems depends on asynchronous behaviors between the populations [25–27]. Ruxton [28] showed that weak coupling between two chaotic systems exhibited simple cycles or remained at a stable level and reduced populations' extinction probabilities. Recently, Pal et al. [29] investigated the effect of bi-directional migration on the stability of two non-identical ecosystems, which were connected through migration. They observed that an increase in the rate of migration could stabilize the non-identical coupled ecosystem. The above observations clearly indicate that migration has a major role in stabilizing chaotic ecosystems. However, the role of bi-directional migration for two similar types of ecosystems, where one is chaotic and the other is stable in nature, has not yet been investigated properly.

In the present paper, we consider metapopulation dynamics of spatially separated food webs that are connected through bi-directional migrations. Our aim of the present study is to investigate the role of migration on the stability of a coupled ecosystem for which one system shows chaotic dynamics, and the other system shows stable dynamics. In the next section, we formulate the model and analyze its behavior regarding the interior equilibrium point. In Section 3, we show the applications of the present scheme in three different models. Finally, the paper ends with a brief conclusion.

2. General Model Formulation and Stability Analysis

Two isolated systems can be coupled via migration. We consider the general case of two coupled ecological systems:

$$
\begin{aligned}
\frac{dX}{dt} &= f(X) + F(X, Y) \\
\frac{dY}{dt} &= g(Y) + G(X, Y)
\end{aligned}
\tag{1}
$$

where X and Y are the variables in the vector notation. The individual systems are described by the functions $f(X)$ and $g(Y)$; $F(X, Y)$ and $G(X, Y)$ are coupling functions. The equilibrium solutions of the uncoupled system are given by $f(X) = 0$ and $g(Y) = 0$. When coupling occurs, the equilibrium points of the system given by Equation (1) are given by $f(X) + F(X, Y) = g(Y) + G(X, Y) = 0$.

Now we consider two three-species food-chain ecological systems that are coupled through bi-directional migrations. In bi-directional migration, a population can migrate from one patch to another depending on the population densities. The flow of the migration is from higher to lower density. Therefore, in bi-directional migration, the migration depends on the relative density difference between two patches. Then Equation (1) with bi-directional migration can be written as

$$\frac{dx_1}{dt} = f_1(x_1, y_1, z_1) + k_1(x_2 - x_1)$$

$$\frac{dy_1}{dt} = f_2(x_1, y_1, z_1) + k_2(y_2 - y_1)$$

$$\frac{dz_1}{dt} = f_3(x_1, y_1, z_1) + k_3(z_2 - z_1)$$

$$\frac{dx_2}{dt} = g_1(x_2, y_2, z_2) + k_1(x_1 - x_2)$$ (2)

$$\frac{dy_2}{dt} = g_2(x_2, y_2, z_2) + k_2(y_1 - y_2)$$

$$\frac{dz_2}{dt} = g_3(x_2, y_2, z_2) + k_3(z_1 - z_2)$$

where x_1, y_1, and z_1 are the populations of system 1 and x_2, y_2, and z_2 are the populations of system 2; $f_i(i = 1, 2, 3)$ and $g_i(i = 1, 2, 3)$ are the functions describing systems 1 and 2, respectively; k_1, k_2, and k_3 are the migration coefficients of the three different populations. To study the stability behavior of the coupled system around the interior equilibrium point $E^*(x_1{}^*, y_1{}^* z_1{}^*, x_2{}^*, y_2{}^*, z_2{}^*)$, we have to calculate the Jacobian matrix of the system given by Equation (2) at the interior equilibrium point. The Jacobian matrix at $E^*(x_1{}^*, y_1{}^* z_1{}^*, x_2{}^*, y_2{}^*, z_2{}^*)$ is

$$J(E^*) = \begin{pmatrix} V_1 & V_2 & V_3 & k_1 & 0 & 0 \\ V_4 & V_5 & V_6 & 0 & k_2 & 0 \\ V_7 & V_8 & V_9 & 0 & 0 & k_3 \\ k_1 & 0 & 0 & M_1 & M_2 & M_3 \\ 0 & k_2 & 0 & M_4 & M_5 & M_6 \\ 0 & 0 & k_3 & M_7 & M_8 & M_9 \end{pmatrix}$$

where the
$V_1 = f_{1_{x1}} - k_1$, $V_2 = f_{1_{y1}}$, $V_3 = f_{1_{z1}}$, $V_4 = f_{2_{x1}}$, $V_5 = f_{2_{y1}} - k_2$, $V_6 = f_{2_{z1}}$, $V_7 = f_{3_{x1}}$, $V_8 = f_{3_{y1}}$,
$V_9 = f_{3_{z1}} - k_3$, $M_1 = g_{1_{x2}} - k_1$, $M_2 = g_{1_{y2}}$, $M_3 = g_{1_{z2}}$, $M_4 = g_{2_{x2}}$, $M_5 = g_{2_{y2}} - k_2$, $M_6 = g_{2_{z2}}$,
$M_7 = g_{3_{x2}}$, $M_8 = g_{3_{y2}}$, and $M_9 = g_{3_{z2}} - k_3$
suffixes denote the partial derivatives with respect to the corresponding variable.

The characteristic equation of the above Jacobian matrix is

$$\lambda^6 + \sigma_1 \lambda^5 + \sigma_2 \lambda^4 + \sigma_3 \lambda^3 + \sigma_4 \lambda^2 + \sigma_5 \lambda + \sigma_6 = 0,$$

where
$\sigma_1 = -(A_1 + B_1)$

$\sigma_2 = -(k_1{}^2 + k_2{}^2 + k_3{}^2) - A_2 - B_2 + A_1 B_1$

$\sigma_3 = -A_3 - B_3 + A_2 B_1 + B_2 A_1 + k_1{}^2(V_5 + V_9 + M_5 + M_9) + k_2{}^2(V_1 + V_9 + M_1 + M_9) + k_3{}^2(V_1 + V_5 + M_1 + M_5)$

$\sigma_4 = A_1 B_3 + A_2 B_2 + A_3 B_1 - k_1{}^2((V_5 + V_9)(M_5 + M_9) + (V_5 V_9 - V_6 V_8) + (M_5 M_9 - M_6 M_8))$
$-k_2{}^2((V_1 + V_9)(M_1 + M_9) + (V_1 V_9 - V_3 V_7) + (M_1 M_9 - M_3 M_7)) - k_1 k_2(V_2 M_4 + V_4 M_2) - k_1 k_3(V_3 M_7 + V_7 M_3)$
$-k_3{}^2((V_1 + V_5)(M_1 + M_5) + (V_1 V_5 - V_2 V_4) + (M_1 M_5 - M_2 M_4)) - k_2 k_3(V_8 M_6 + V_6 M_8) + k_1{}^2 k_2{}^2 + k_2{}^2 k_3{}^2 + k_3{}^2 k_1{}^2$

$\sigma_5 = (A_2 B_3 + A_3 B_2) + k_1{}^2((V_5 + V_9)(M_5 M_9 - M_6 M_8) + (M_5 + M_9)(V_5 V_9 - V_6 V_8))$
$+k_2{}^2((V_1 + V_9)(M_1 M_9 - M_3 M_7) + (M_1 + M_9)(V_1 V_9 - V_3 V_7)) - k_1{}^2 k_2{}^2(V_9 + M_9)$
$+k_3{}^2((V_1 + V_5)(M_1 M_5 - M_2 M_4) + (M_1 + M_5)(V_1 V_5 - V_2 V_4)) - k_2{}^2 k_3{}^2(V_1 + M_1) - k_1{}^2 k_3{}^2(V_5 + M_5)$
$+k_1 k_2 (V_2(M_4 M_9 - M_6 M_7) + M_4(V_2 V_9 - V_3 V_8) + M_2(V_4 V_9 - V_6 V_7) + V_4(M_2 M_9 - M_3 M_8))$
$+k_2 k_3 (V_8(M_1 M_6 - M_3 M_4) + M_6(V_1 V_8 - V_2 V_7) + M_8(V_1 V_6 - V_3 V_4) + V_6(M_1 M_8 - M_2 M_7))$
$+k_1 k_3 (V_3(M_5 M_7 - M_4 M_8) + M_7(V_3 V_5 - V_2 V_6) + M_3(V_5 V_7 - V_4 V_8) + V_7(M_3 M_5 - M_2 M_6))$

$\sigma_6 = A_3 B_3 - k_1{}^2(V_5 V_9 - V_6 V_8)(M_5 M_9 - M_6 M_8) - k_2{}^2(V_1 V_9 - V_3 V_7)(M_1 M_9 - M_3 M_7)$

$$-k_3{}^2(V_1V_5 - V_2V_4)(M_1M_5 - M_2M_4) + k_1{}^2k_2{}^2V_9M_9 + k_2{}^2k_3{}^2V_1M_1 + k_1{}^2k_3{}^2V_5M_5 - k_1{}^2k_2{}^2k_3{}^2$$
$$+k_1{}^2k_2k_3(V_6M_8 + M_6V_8) + k_1k_2{}^2k_3(V_7M_3 + M_7V_3) + k_1k_2k_3{}^2(V_2M_4 + M_2V_4)$$
$$+k_1k_2\left((V_3V_8 - V_2V_9)(M_4M_9 - M_6M_7) + (M_3M_8 - M_2M_9)(V_4V_9 - V_6V_7)\right)$$
$$+k_2k_3\left((V_2V_7 - V_1V_8)(M_1M_6 - M_3M_4) + (M_2M_7 - M_1M_8)(V_1V_6 - V_3V_4)\right)$$
$$+k_1k_3\left((V_5V_7 - V_4V_8)(M_2M_6 - M_3M_5) + (M_5M_7 - M_4M_8)(V_2V_6 - V_3V_5)\right)$$

with $A_1 = (V_1 + V_5 + V_9)$, $A_2 = (V_6V_8 - V_5V_9) + (V_3V_7 - V_1V_9) + (V_2V_4 - V_1V_5)$,
$A_3 = V_1(V_5V_9 - V_6V_8) + V_2(V_6V_7 - V_4V_9) + V_3(V_4V_8 - V_5V_7)$,
$B_1 = (M_1 + M_5 + M_9)$, $B_2 = (M_6M_8 - M_5M_9) + (M_3M_7 - M_1M_9) + (M_2M_4 - M_1M_5)$,
and $B_3 = M_1(M_5M_9 - M_6M_8) + M_2(M_6M_7 - M_4M_9) + M_3(M_4M_8 - M_5M_7)$.

Now, the eigenvalues of the characteristic equation are negative or have negative real parts if all Routh–Hurwitz (RH) determinants ($RH_i, i = 1, 2, ..., 6$) are positive, where $RH_1 = |\sigma_1|$, $RH_2 = \begin{vmatrix} \sigma_1 & 1 \\ \sigma_3 & \sigma_2 \end{vmatrix}$ and $RH_n = \begin{vmatrix} \sigma_1 & 1 & 0 & 0 & ... & 0 \\ \sigma_3 & \sigma_2 & \sigma_1 & 1 & ... & 0 \\ ... & ... & ... & ... & ... & ... \\ 0 & 0 & 0 & 0 & 0 & \sigma_n \end{vmatrix}$, where $\sigma_j = 0$ if $j > n$.

3. Applications

Migration within a population with spatial subdivision is important in some species and systems. It is observed that, if two identical patches are coupled through migration, then the coupled system acts exactly as a single-patch system. The persistence of coupled locally unstable systems depends on asynchrony behaviors among populations [25–27]. It is to be noted that two identical chaotic systems cannot be stabilized by diffusive migration. Here we consider two tri-trophic food-chain systems of the same type with different parameter values, where one system shows chaotic dynamics and the other system shows stable dynamics. We also note that the two tri-trophic food web systems are spatially separated but are connected through bi-directional migrations. In this section, we describe the application of the above scheme developed in Section 2 to three different models, namely, the Hastings–Powell (HP) model, the Upadhyay–Rai (UR) model and the Priyadarshi–Gakkhar (PG) model, which are able to produce stable dynamics as well as chaotic dynamics for different sets of parameter values.

3.1. Hastings–Powell Model

In 1991, Hastings and Powell [6] proposed and analyzed a three-species food-chain model with a Holling type II functional response. The model is known for exhibiting chaotic dynamics in a continuous-time food-chain model. The non-dimensional HP model is governed by the following equations:

$$\begin{aligned}
\frac{dx}{dt} &= x(1-x) - \frac{a_1xy}{1+b_1x} \\
\frac{dy}{dt} &= \frac{a_1xy}{1+b_1x} - \frac{a_2yz}{1+b_2y} - d_1y \\
\frac{dz}{dt} &= \frac{a_2yz}{1+b_2y} - d_2z
\end{aligned} \tag{3}$$

where x, y and z are the densities of the prey, middle-predator and top-predator populations, respectively; a_1, a_2, b_1, b_2, d_1 and d_2 are the non-negative parameters that have the usual meanings [6]. Hastings and Powell [6] studied the model given by Equation (3) and observed switching of the dynamics of the system between stable focus, limit cycle oscillations and chaos by changing the parameter b_1.

Coupling between Chaotic HP Model and Stable HP Model

The HP model shows different dynamical behaviors, including chaos. In the present section, we investigate the dynamics of the coupled ecosystem, for which one HP system shows chaotic dynamics and the other HP system shows stable dynamics. Here, we assume that the two different systems are connected by migration and that the direction of the migration is bi-directional. Further, all populations are free to migrate from one system to another. We denote the chaotic HP system with subscript 1 and the stable HP system with subscript 2. The coupled system is governed by the following equations:

$$
\begin{aligned}
\frac{dx_1}{dt} &= x_1(1-x_1) - \frac{a_1 x_1 y_1}{1+b_{11}x_1} + k_1(x_2 - x_1)\\
\frac{dy_1}{dt} &= \frac{a_1 x_1 y_1}{1+b_{11}x_1} - \frac{a_2 y_1 z_1}{1+b_2 y_1} - d_1 y_1 + k_2(y_2 - y_1)\\
\frac{dz_1}{dt} &= \frac{a_2 y_1 z_1}{1+b_2 y_1} - d_2 z_1 + k_3(z_2 - z_1)\\
\frac{dx_2}{dt} &= x_2(1-x_2) - \frac{a_1 x_2 y_2}{1+b_{21}x_2} + k_1(x_1 - x_2)\\
\frac{dy_2}{dt} &= \frac{a_1 x_2 y_2}{1+b_{21}x_2} - \frac{a_2 y_2 z_2}{1+b_2 y_2} - d_1 y_2 + k_2(y_1 - y_2)\\
\frac{dz_2}{dt} &= \frac{a_2 y_2 z_2}{1+b_2 y_2} - d_2 z_2 + k_3(z_1 - z_2)
\end{aligned}
\tag{4}
$$

where k_1, k_2, and k_3 are the migration coefficients of the prey, middle-predator and top-predator populations, respectively. We assume that two systems differ only in the parameter b_1 in Equation (3); b_{11} and b_{21} are the parameters corresponding to systems 1 and 2, respectively.

Non-Negativity of the Solutions: We let $R_+^6 = [0, \infty)^6$ be the non-negative octant in R^6. Then the interaction functions of the system given by Equation (4) are continuously differentiable and locally satisfy Lipschitz conditions in R_+^6. Thus, any solution of the system given by Equation (4) with non-negative initial conditions satisfies the non-negativity condition and exists uniquely in the interval $[0, M)$ for some $M > 0$ ([30], Theorem A.4).

Boundedness of the Solutions:
We define a function

$$
P(t) = x_1(t) + y_1(t) + z_1(t) + x_2(t) + y_2(t) + z_2(t).
\tag{5}
$$

The time derivative of Equation (5) along with the solutions of Equation (4) are

$$
\frac{dP}{dt} = \frac{dx_1}{dt} + \frac{dy_1}{dt} + \frac{dz_1}{dt} + \frac{dx_2}{dt} + \frac{dy_2}{dt} + \frac{dz_2}{dt} = x_1(1-x_1) - d_1 y_1 - d_2 z_1 + x_2(1-x_2) - d_1 y_2 - d_2 z_2
$$
$$
\Rightarrow \frac{dP}{dt} + \mu P = x_1(1-x_1+\mu) - (d_1-\mu)(y_1+y_2) - (d_2-\mu)(z_1+z_2) + x_2(1-x_2+\mu) \le \frac{(1+\mu)^2}{2} = Q \text{ (say)},
$$

where $\mu \le min\{d_1, d_2\}$.

Applying the theorem of differential inequality [31], we obtain $P(x_1, y_1, z_1, x_2, y_2, z_2) \le \frac{Q}{\mu}(1 - e^{-\mu t}) + P(x_1(0), y_1(0), z_1(0), x_2(0), y_2(0), z_2(0))e^{-\mu t}$, which implies that $P \le Q/\mu + \epsilon$ for all $t \ge t_0$. Therefore, all the solutions of the system given by Equation (4) are bounded.

Hence, all the solutions of the system given by Equation (4), which are initiated in R_+^6, are positively invariant in the region $B = \{(x_1, y_1, z_1, x_2, y_2, z_2) \in R_+^6 : P \le Q/\mu + \epsilon\}$.

Now we describe the numerical simulations for the system given by Equation (3) and the coupled system given by Equation (4) by considering the following parameter values:

$$a_1 = 5, a_2 = 0.1, b_2 = 2, d_1 = 0.4, d_2 = 0.01, \tag{6}$$

which were taken from [6]. Choosing $b_{11} = b_{21} = 3$, then the coupled system given by Equation (4) remained chaotic for any coupling strength (migration rate). We then chose two different values of $b_1 (b_1 = 3$ and $b_1 = 2)$, and the HP model of Equation (3) showed chaotic dynamics and stable dynamics. For system 1, we set $b_{11} = 3$, so that the system showed chaotic dynamics, and for system 2, we set $b_{21} = 2$, so that the system showed stable dynamics (Figure 1). The initial condition for the simulation of the coupled system given by Equation (4) was $(x_1(0), y_1(0), z_1(0), x_2(0), y_2(0), z_2(0)) = (0.7, 0.6, 12, 0.75, 0.5, 11)$. We could then investigate the effect of bi-directional migration between the two systems. For simplicity, we considered $k_1 = k_2 = k_3 = k$ and drew the bifurcation diagram of the coupled system of Equation (4) with respect to the rate of migration k (Figure 2). It is to be noted that in the absence of migration ($k = 0$), system 1 showed chaotic dynamics and system 2 showed stable dynamics. When we introduced migration between these two systems, then the coupled system became stable through a Hopf bifurcation when the migration rate (k) crossed a threshold value, ($k_{HP}^* = 0.0145$) (Figure 2). We observed that a small migration destabilized the stable system, and the coupled system showed higher periodic and chaotic oscillations, but if the strength of migration was increased gradually, then the coupled system became stable. We also observed that for $k = 0.25$, the coupled system of Equation (4) had a unique positive interior equilibrium $E_{HP}^*(0.837058, 0.0841652, 12.2809, 0.692788, 0.171415, 12.4183)$. We also obtained the RH determinants, $RH_1 = 2.4080 > 0$, $RH_2 = 4.2563 > 0$, $RH_3 = 2.7726 > 0$, $RH_4 = 0.3547 > 0$, $RH_5 = 0.0058 > 0$, and $RH_6 = 7.4751 \times 10^{-6} > 0$, which satisfied the RH stability criterion of order 6. The eigenvalues of the coupled system given by Equation (4) were $(-0.9721, -0.1091 + 0.1388i, -0.1091 - 0.1388i, -0.1544, -0.5317 + 0.0778i, -0.5317 - 0.0778i)$. Hence, the system given by Equation (4) was stable around the positive interior equilibrium E_{HP}^* (Figure 3).

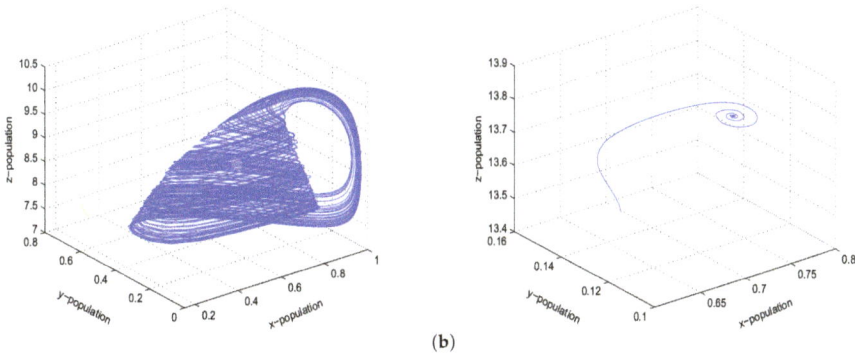

(a)

(b)

Figure 1. (**a,b**) Chaotic oscillations and stable focus of the system given by Equation (3) for $b_1 = 3$ and $b_1 = 2$, respectively.

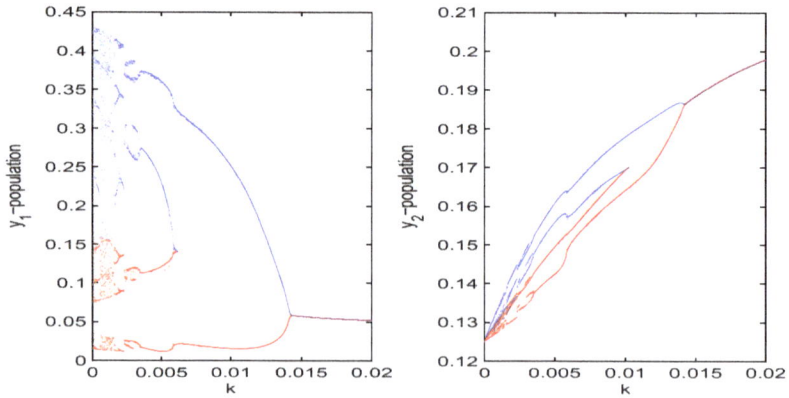

Figure 2. Bifurcation diagram for y_1 and y_2 populations of the coupled Hastings–Powell (HP)–HP system corresponding to the bifurcating parameter k, where $k \in [0, 0.02]$.

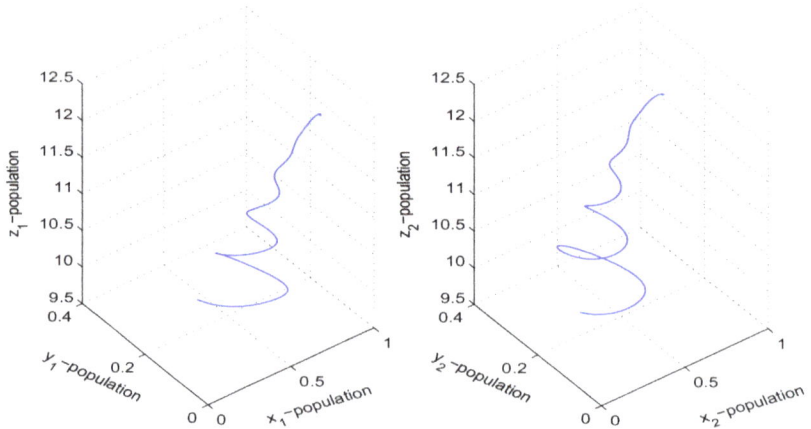

Figure 3. Figure shows stable dynamics of the coupled system given by Equation (4) for $k = 0.25$.

Further, we performed numerical simulations of the coupled system for the realistic parameter values considered by McCann and Yodzis [32]. McCann and Yodzis [32] considered the modified HP model [6]; they produced a range of more "plausible" parameter values and demonstrated the existence of chaos for a wide range of these values. We considered the following parameter values:

$$x_c = 0.4, y_c = 2.01, x_p = 0.08, y_p = 5, c_0 = 0.5, \tag{7}$$

which were taken from [32]. The model and the meaning of the parameter values are given in [32]. For system 1, we set $r_0 = 0.161$, so that the system showed chaotic dynamics, and for system 2, we set $r_0 = 0.75$, so that the system showed stable dynamics (Figure 4). The initial condition for the simulations of the coupled system was $(x_1(0), y_1(0), z_1(0), x_2(0), y_2(0), z_2(0)) = (0.35, 0.5, 0.9, 0.35, 0.5, 0.9)$. If we introduced migration between the two systems (chaotic system and stable system), then the coupled system showed limit cycle oscillations via period-halving bifurcations (Figure 5). We observed that a gradual increase in migration made the coupled system switch its stability from chaotic dynamics to limit cycle oscillations (Figure 6). Therefore, migration could stabilize the coupled system by producing stable focus or more regular oscillations.

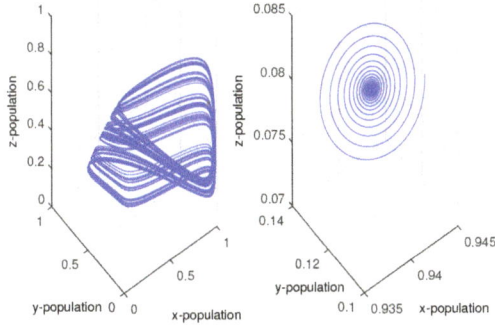

Figure 4. Figure shows chaotic oscillations and stable focus for $r_0 = 0.161$ and $r_0 = 0.75$, respectively.

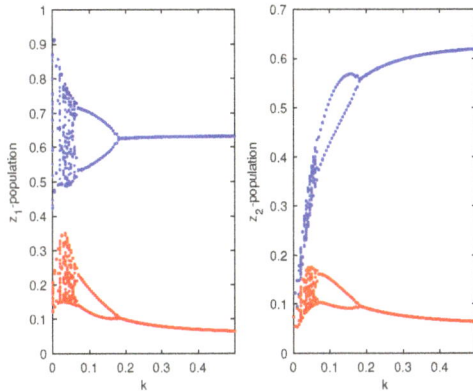

Figure 5. Bifurcation diagram of the top-predator populations of the coupled system corresponding to the bifurcating parameter k, where $k \in [0, 0.5]$.

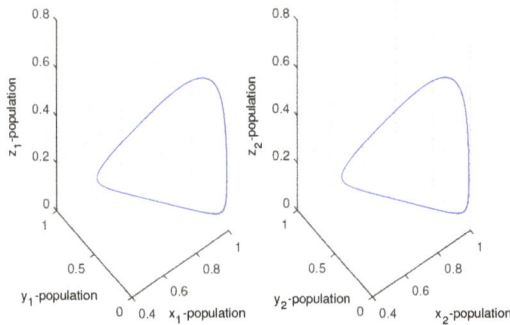

Figure 6. Figure shows limit cycle oscillations of the coupled system for $k = 5$.

3.2. Upadhyay–Rai Model

Upadhyay and Rai [7,33] proposed and analyzed a tri-trophic food-chain model by considering the middle predator as a specialist predator and the top predator as a generalist predator.

The prey–specialist predator–generalist system is governed by the following equations [7,33]:

$$
\begin{aligned}
\frac{dx}{dt} &= m_1 x - n_1 x^2 - \frac{wxy}{x + D_1} \\
\frac{dy}{dt} &= -n_2 y + \frac{w_1 xy}{x + D_1} - \frac{w_2 yz}{y + D_2} \\
\frac{dz}{dt} &= cz^2 - \frac{w_3 z^2}{y + D_3}
\end{aligned}
\tag{8}
$$

where x, y and z are the densities of the prey, specialist predator, and generalist predator populations, respectively; $m_1, n_1, n_2, w, w_1, w_2, w_3, D, D_1, D_2$ and c are the non-negative parameters that have the usual meanings [7,33]. In the above model, the prey population (x) grows logistically; the specialist predator (y) predates prey (only food item available to the specialist predator) via a Holling type II functional response; the generalist predator (z) sexually reproduces, its population growing quadratically (cz^2) and decaying as a result of intraspecific competition ($-\frac{w_3 z^2}{y + D_3}$). Additionally, males and females in the generalist predator population are assumed to be equal in terms of numbers, and the mating frequency is directly proportional to the number of males as well as the number of females. The interaction between the generalist predator and specialist predator follows a modified Leslie–Gower scheme. Here, the specialist middle predator is the favourite food choice of the generalist top predator and the generalist predator feeds on other food items (alternative food resources), in case of a short supply of the middle predator.

It is to be noted here that the system given by Equation (8) is not always dissipative, and the solutions may blow-up in finite time (explosive instability) depending on the parameter values and initial conditions [34]. In recent literature, few researchers have investigated different models that show finite-time blow-up in the solutions [34–40]. However, the above system given by Equation (8) shows very rich dynamics when $\frac{w_3}{y+D_3} < c < \frac{w_3}{D_3}$. Upadhyay and Rai [7,33] explored chaotic dynamics in the system by increasing the intrinsic growth rate m_1.

Coupling between Chaotic UR Model and Stable UR Model

In this section, we denote the chaotic UR system with the subscript 1 and the stable UR system with the subscript 2. The coupled system is governed by the following equations:

$$
\begin{aligned}
\frac{dx_1}{dt} &= m_{11} x_1 - n_1 x_1^2 - \frac{wx_1 y_1}{x_1 + D_1} + k_1(x_2 - x_1) \\
\frac{dy_1}{dt} &= -n_2 y_1 + \frac{w_1 x_1 y_1}{x_1 + D_1} - \frac{w_2 y_1 z_1}{y_1 + D_2} + k_2(y_2 - y_1) \\
\frac{dz_1}{dt} &= cz_1^2 - \frac{w_3 z_1^2}{y_1 + D_3} + k_3(z_2 - z_1) \\
\frac{dx_2}{dt} &= m_{21} x_2 - n_1 x_2^2 - \frac{wx_2 y_2}{x_2 + D_1} + k_1(x_1 - x_2) \\
\frac{dy_2}{dt} &= -n_2 y_2 + \frac{w_1 x_2 y_2}{x_2 + D_1} - \frac{w_2 y_2 z_2}{y_2 + D_2} + k_2(y_1 - y_2) \\
\frac{dz_2}{dt} &= cz_2^2 - \frac{w_3 z_2^2}{y_2 + D_3} + k_3(z_1 - z_2)
\end{aligned}
\tag{9}
$$

where $k_1, k_2,$ and k_3 are the migration coefficients of the prey, specialist predator and generalist predator populations, respectively. We assume that two systems differ only in the parameter m_1 in Equation (8); m_{11} and m_{21} are the parameters corresponding to systems 1 and 2, respectively.

Now we describe the numerical simulations of the system given by Equation (8) and the coupled system given by Equation (9). The set of parameters were as follows:

$$n_1 = 0.06, w = 1, D_1 = 10, n_2 = 1, w_1 = 2,$$
$$w_2 = 0.405, D_2 = 10, c = 0.03, w_3 = 1, D_3 = 20,$$

(10)

which were taken from [33]. Choosing $m_{11} = m_{21} = 1.93$, then the coupled system given by Equation (8) remained chaotic for any coupling strength (migration rate). We then chose two different values of m_1 ($m_1 = 1.93$ and $m_1 = 1.2$), and the UR model given by Equation (8) showed chaotic dynamics and stable dynamics, respectively (Figure 7). For system 1, we set $m_{11} = 1.93$, so that the system showed chaotic dynamics, and for system 2, we set $m_{21} = 1.2$, so that the system showed stable dynamics. The initial condition for the simulations of the coupled system given by Equation (9) was $(x_1(0), y_1(0), z_1(0), x_2(0), y_2(0), z_2(0)) = (0.7, 0.5, 7, 0.7, 0.4, 6)$. We then investigated the effect of bi-directional migration on the two systems. For simplicity, we considered $k_1 = k_2 = k_3 = k$ and drew the bifurcation diagram of the coupled system given by Equation (9) with respect to the rate of migration k (Figure 8). We observed that the coupled system given by Equation (9) became stable through a Hopf bifurcation when the migration coefficient crossed a threshold value $k_{UR}^* = 0.21$. We observed that when the migration was weak (k small), the stable system became unstable and the coupled system showed higher periodic and chaotic oscillations, but if the strength of migration was increased gradually, then the coupled system became stable. Further, we observed that for $k = 0.25$, the coupled system given by Equation (9) had a unique positive interior equilibrium $E_{UR}^*(22.7980, 15.6757, 19.5396, 15.1274, 10.5329, 16.5322)$. We also obtained the RH determinants $RH_1 = 2.9322 > 0$, $RH_2 = 7.7900 > 0$, $RH_3 = 9.8214 > 0$, $RH_4 = 3.1712 > 0$, $RH_5 = 0.0300 > 0$, and $RH_6 = 8.2866 \times 10^{-4} > 0$, which satisfied the RH stability criterion of order 6. The eigenvalues of the coupled system given by Equation (9) were $(-1.1705, -0.0096 + 0.3009i, -0.0096 - 0.3009i, -0.6209, -0.5608 + 0.3227i, -0.5608 - 0.3227i)$. Hence, the coupled system given by Equation (9) was stable around the positive interior equilibrium E_{UR}^* (Figure 9).

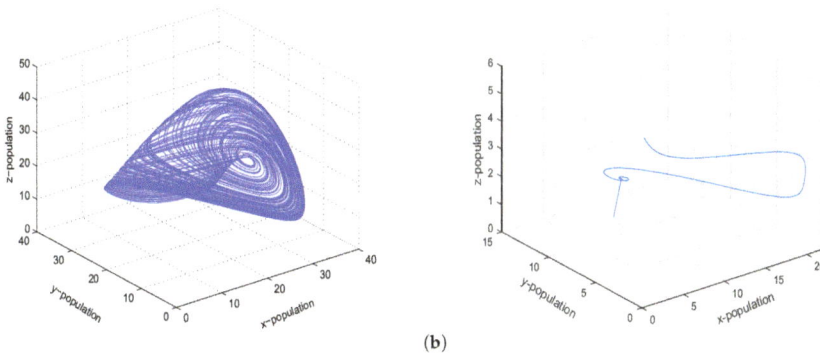

(a)   (b)

Figure 7. (**a**,**b**) Chaotic oscillations and stable focus of the system given by Equation (8) for $m_1 = 1.93$ and $m_1 = 1.2$, respectively.

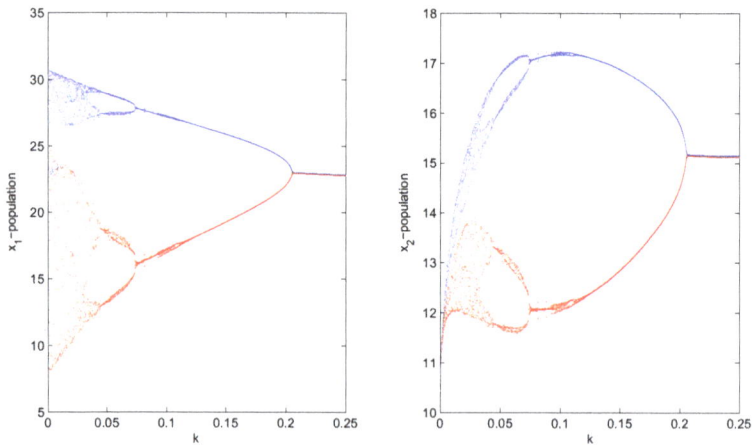

Figure 8. Bifurcation diagram for x_1 and x_2 populations of the coupled Upadhyay–Rai (UR)–UR system corresponding to the bifurcating parameter k, where $k \in [0, 0.25]$.

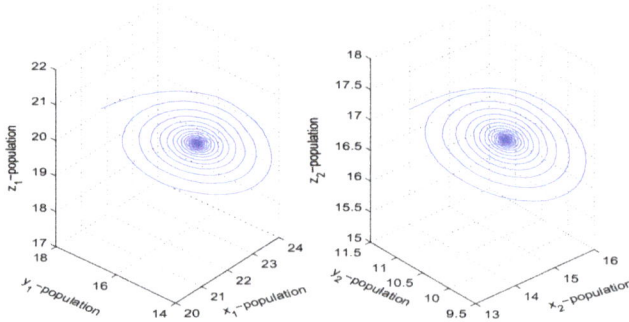

Figure 9. The figure shows stable dynamics of the coupled system given by Equation (9) for $k = 0.25$.

3.3. Priyadarshi–Gakkhar Model

Priyadarshi and Gakkhar [9] proposed and analyzed a tri-trophic food-web model consisting of a Leslie–Gower-type generalist predator, where the middle predator is a specialist predator and the top predator is a generalist predator.

The prey–specialist predator–generalist predator system is governed by the following equations [9]:

$$
\begin{aligned}
\frac{dx}{dt} &= x(1-x) - \frac{xy}{1+w_1 x} - \frac{xz}{1+w_2 x + w_3 y} \\
\frac{dy}{dt} &= -w_5 y + \frac{w_4 xy}{1+w_1 x} - \frac{w_6 yz}{1+w_2 x + w_3 y} \\
\frac{dz}{dt} &= w_7 z^2 - \frac{w_8 z^2}{1+w_9 x + w_{10} y}
\end{aligned}
\tag{11}
$$

where x, y and z are the densities of the prey, specialist predator, and generalist predator populations, respectively. The parameters $w_1, w_2, w_3, w_4, w_5, w_6, w_7, w_8, w_9$ and w_{10} are non-negative parameters that have the usual meanings [9]. The formulation of the above model is similar to that of the UR model. However, in the above model, the specialist predator predates prey according to a Holling type II functional response, whereas the generalist predator predates prey and the specialist predator following a modified Holling type II functional response. It is to be noted that the system given by Equation (11) may not be dissipative and shows the blow-up phenomenon depending on the parameter values and initial conditions [34]. Priyadarshi and Gakkhar [9] explored a "snail-shell" chaotic attractor in the system.

Coupling between Chaotic PG Model and Stable PG Model

In this section, we investigate the dynamics of the coupled ecosystem, where one PG system shows chaotic dynamics and the other PG system shows stable dynamics. Here, we consider that two different systems are connected by bi-directional migration. We assume that all populations are free to migrate from one system to the other. We denote the chaotic PG system with the subscript 1 and the stable PG system with the subscript 2. The coupled system is governed by the following equations:

$$
\begin{aligned}
\frac{dx_1}{dt} &= x_1(1-x_1) - \frac{x_1 y_1}{1 + w_1 x_1} - \frac{x_1 z_1}{1 + w_2 x_1 + w_{31} y_1} + k_1(x_2 - x_1) \\
\frac{dy_1}{dt} &= -w_5 y_1 + \frac{w_4 x_1 y_1}{1 + w_1 x_1} - \frac{w_6 y_1 z_1}{1 + w_2 x_1 + w_{31} y_1} + k_2(y_2 - y_1) \\
\frac{dz_1}{dt} &= w_7 z_1{}^2 - \frac{w_8 z_1{}^2}{1 + w_9 x_1 + w_{10} y_1} + k_3(z_2 - z_1) \\
\frac{dx_2}{dt} &= x_2(1-x_2) - \frac{x_2 y_2}{1 + w_1 x_2} - \frac{x_2 z_2}{1 + w_2 x_2 + w_{32} y_2} + k_1(x_1 - x_2) \\
\frac{dy_2}{dt} &= -w_5 y_2 + \frac{w_4 x_2 y_2}{1 + w_1 x_2} - \frac{w_6 y_2 z_2}{1 + w_2 x_2 + w_{32} y_2} + k_2(y_1 - y_2) \\
\frac{dz_2}{dt} &= w_7 z_2{}^2 - \frac{w_8 z_2{}^2}{1 + w_9 x_2 + w_{10} y_2} + k_3(z_1 - z_2)
\end{aligned}
\tag{12}
$$

where $k_1, k_2,$ and k_3 are the migration coefficients of the prey, specialist predator and generalist predator populations, respectively. We assume that two systems differ only in the parameter w_3 in Equation (11); w_{31}, w_{32} are the parameters corresponding to systems 1 and 2, respectively.

In numerical simulations, we considered the following parameter values:

$$
w_1 = 1.4, w_2 = 1, w_4 = 1, w_5 = 0.16, w_6 = 0.1, w_7 = 0.1, w_8 = 0.5, w_9 = 8, w_{10} = 8,
\tag{13}
$$

which were taken from [9]. Choosing $w_{31} = w_{32} = 10$, then the coupled system given by Equation (11) remained chaotic for any coupling strength (migration rate). Choosing two different values of $w_3 (w_3 = 10$ and $w_3 = 1)$, then the PG model of Equation (11) showed chaotic dynamics and stable dynamics (Figure 10). For system 1, we set $w_{31} = 10$, so that the system showed chaotic dynamics, and for system 2, we set $w_{32} = 1$, so that the system showed stable dynamics. The initial condition for the simulation of the coupled system given by Equation (12) was $(x_1(0), y_1(0), z_1(0), x_2(0), y_2(0), z_2(0)) = (0.5, 0.2, 5, 0.6, 0.3, 7)$. We then investigated the effect of bi-directional migration on the two systems. For simplicity, we considered $k_1 = k_2 = k_3 = k$ and drew the bifurcation diagram of the coupled system given by Equation (12) with respect to the rate of migration k (Figure 11). We observed that the coupled system became stable through a Hopf bifurcation when the migration rate (k) crossed a threshold value, $(k_{PG}^* = 0.1)$ (Figure 11). We observed that weak migration destabilized the stable system, but if the strength of the migration was increased gradually, then the coupled system became stable. Further, we observed that for $k = 0.25$, the coupled system given by Equation (12) had a unique positive interior equilibrium $E_{PG}^* (0.3927, 0.2318, 1.2881, 0.2191, 0.1773, 1.1778)$. We also obtained the RH determinants $RH_1 = 1.9351 > 0, RH_2 = 2.2437 > 0, RH_3 = 0.7799 > 0, RH_4 = 0.0354 > 0,$ $RH_5 = 8.1643 \times 10^{-5} > 0,$ and $RH_6 = 4.8986 \times 10^{-8} > 0,$ which satisfied the RH stability criterion of order 6. The eigenvalues of the coupled system given by Equation (12) were $(-0.6763, -0.5218 +$

0.1623i, $-0.5218 - 0.1623i$, $-0.0180 + 0.1299i$, $-0.0180 - 0.1299i$, -0.1792). Hence, the system given by Equation (12) was stable around the positive interior equilibrium E^*_{PG} (Figure 12).

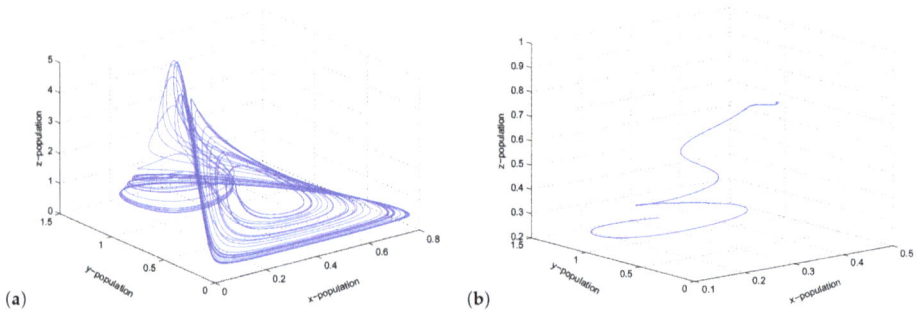

(a)   (b)

Figure 10. (**a**,**b**) Chaotic oscillations and stable focus of the system given by Equation (11) for $w_3 = 10$ and $w_3 = 1$, respectively.

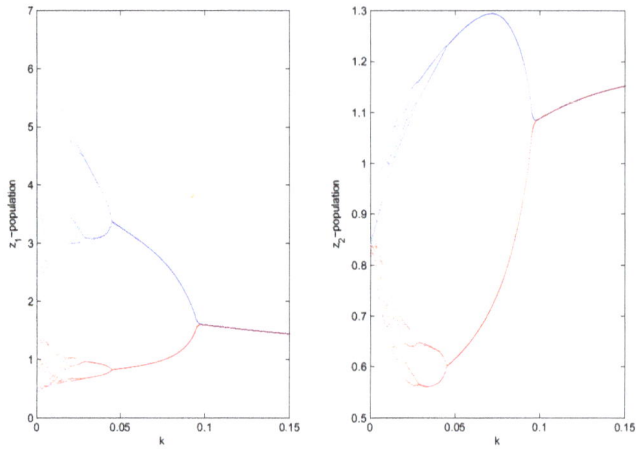

Figure 11. Bifurcation diagram for z_1 and z_2 populations of the coupled Priyadarshi–Gakkhar (PG)–PG system given by Equation (12) corresponding to the bifurcating parameter k, where $k \in [0, 0.15]$.

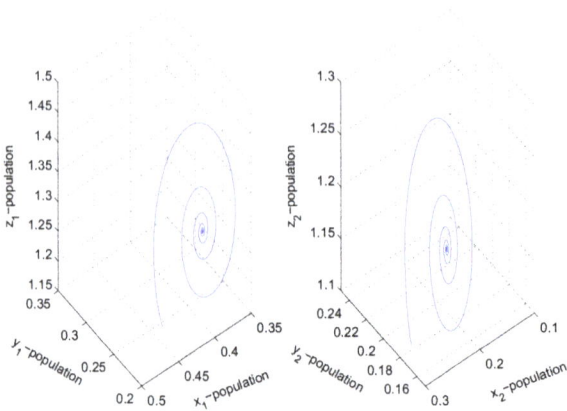

Figure 12. The figure shows stable dynamics of the coupled system given by Equation (12) for $k = 0.25$.

We calculated the maximum Lyapunov exponent (Figure 13) for the coupled systems given by Equations (4), (9) and (12) with respect to the coupling strength (k). We observed that if we increased the strength of migration, then the value of the maximum Lyapunov exponent became negative. The maximum Lyapunov exponent (Figure 13) confirmed that the coupled systems (HP–HP, UR–UR, and PG–PG) became stable from chaotic dynamics.

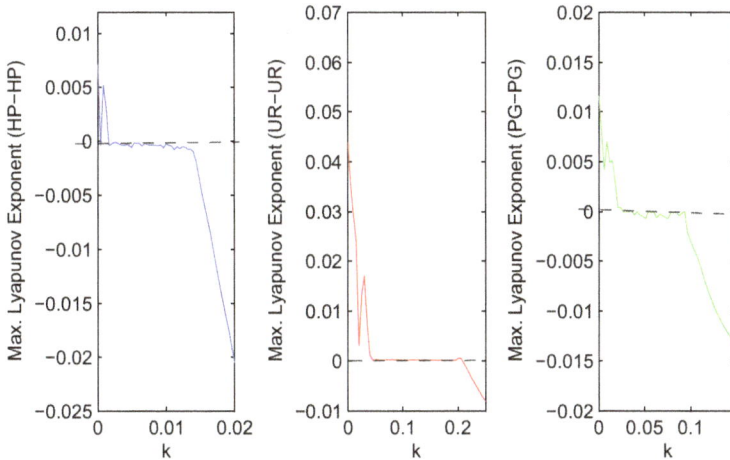

Figure 13. Largest Lyapunov exponent of the coupled systems given by Equations (4), (9) and (12) with respect to the parameter k.

4. Conclusions

The persistence of coupled unstable systems depends on the maintenance of the asynchronous behavior among populations. Several types of asynchronous behaviors, such as the existence of refuge [41], biased dispersal [42], fixed differences in parameters [43], and so on, can enhance the stability of predator–prey systems. In the present paper, we considered two ecological systems of the same type that were connected through migration. We also considered different sets of parameter values so that one system (HP-1/UR-1/PG-1) showed chaotic dynamics and the other system (HP-2/UR-2/PG-2) showed stable dynamics. The direction of migration was taken as bi-directional and depended on the density difference of the populations in the two patches. We studied the effect of bi-directional migration on the chaotic ecosystem and stable ecosystem by considering three different types of food webs. We observed that small migration destabilized the stable system, and the coupled system showed higher periodic and chaotic oscillations, but if the strength of the migration was increased gradually, then above a threshold value, all the coupled systems (HP/UR/PG) became stable. Bi-directional migration can replace chaotic oscillations by a stable steady state or stable limit cycle. Therefore, migration makes the system more regular. In the present work, migration was considered as the coupling force; the migration could be both ways depending on the density difference of each population in the two patches. If the migration strength was weak, then we observed that the chaotic system dominated the dynamic properties of the coupled system. For a low migration rate, the population density of each patch changed very slowly. Intuitively, migration has a stabilizing effect. However, if the change in the population densities due to trophic interactions is greater than the change due to migration, then the population dynamics are likely to be dominated by trophic interactions. Therefore, the population dynamics of a coupled system may be unstable. However, if the migration strength is high enough, then the population densities of each patch quickly converge to the average density of the two patches, which may stabilize the coupled system.

A possible explanation could be provided as follows: If the populations in a system show chaotic dynamics, the population densities oscillate in an unpredictable manner. The amplitude of the population oscillations may be high enough or may be very low. However, depending on the density difference of the populations in the two patches, all species start to migrate from one patch to the other (from higher to lower density). Bi-directional migration mediates the population densities, and none of the population densities increase or decrease drastically. Migration helps to balance the population densities in two patches. In terms of stability and extinction, populations in the coupled system will be less prone to extinction. Therefore, migration can prevent chaos in a coupled system and also enhance the stability and persistence of the system.

Author Contributions: N.P., S.S., M.M. and J.C. formulated the model. N.P. and S.S. performed mathematical analysis and numerical simulations. N.P., S.S., M.M. and J.C. wrote the paper.

Conflicts of Interest: The authors declare no conflicts of interest.

References

1. Malthus, T.R. *An Essay on the Principle of Population*; Joseph Johnson: London, UK, 1798.
2. Verhulst, P.F. Notice sur la loi que la population suit dans son accroissement. Correspondance Mathmatique et Physique Publie par A. *Quetelet* **1838**, *10*, 113–121.
3. Volterra, V. Fluctuations in the abundance of a species considered mathematically. *Nature* **1926**, *118*, 558–560.
4. Lotka, A.J. *Elements of Physical Biology*; Williams and Wilkins: Baltimore, MD, USA, 1925.
5. Gilpin, M.E. Spiral chaos in a predator-prey model. *Am. Nat.* **1979**, *113*, 306–308.
6. Hastings, A.; Powell, T. Chaos in three-species food chain. *Ecology* **1991**, *72*, 896–903.
7. Upadhyay, R.K.; Rai, V. Why chaos is rarely observed in natural populations. *Chaos Solitons Fractals* **1997**, *8*, 1933–1939.
8. Tanabe, K.; Namba, T. Omnivory creates chaos in simple food web models. *Ecology* **2005**, *86*, 3411–3414.
9. Priyadarshi, A.; Gakkhar, S. Dynamics of Leslie-Gower type generalist predator in a tri-trophic food web system. *Commun. Nonlinear Sci. Numer. Simul.* **2013**, *18*, 3202–3218.
10. McCann, K.; Hastings, A. Re-evaluating the omnivory-stability relationship in food webs. *Proc. R. Soc. Lond. B Biol. Sci.* **1997**, *264*, 1249–1254.
11. Chattopadhayay, J.; Sarkar, R. Chaos to order: Preliminary experiments with a population dynamics models of three tropic levels. *Ecol. Model.* **2003**, *163*, 45–50.
12. Pal, N.; Samanta, S.; Chattopadhyay, J. Revisited Hastings and Powell model with omnivory and predator switching. *Chaos Solitons Fractals* **2014**, *66*, 58–73.
13. Levins, R. Some demographic and genetic consequences of environmental heterogeneity for biological control. *Bull. Entomol.Soc. Am.* **1969**, *15*, 237–240.
14. Gilpin, M. (Ed.) *Metapopulation Dynamics: Empirical and Theoretical Investigations*; Academic Press: Cambridge, MA, USA, 1991.
15. Hanski, I.; Thomas, C.D. Metapopulation dynamics and conservation: A spatially explicit model applied to butterflies. *Biol. Conserv.* **1994**, *68*, 167–180.
16. Hanski, I. *Metapopulation Ecology*; Oxford University Press: Oxford, UK, 1999.
17. Dingle, H. Migration strategies of insects. *Science* **1972**, *175*, 1327–1335.
18. Dixon, A.F.G.; Horth, S.; Kindlmann, P. Migration in insects: Cost and strategies. *J. Anim. Ecol.* **1993**, *62*, 182–190.
19. Matthiopoulos, J.; Harwood, J.; Thomas, L.E.N. Metapopulation consequences of site fidelity for colonially breeding mammals and birds. *J. Anim. Ecol.* **2005**, *74*, 716–727.
20. Gyllenberg, M.; Soderbacka, G.; Ericsson, S. Does migration stabilize local population dynamics? Analysis of a discrete metapopulation model. *Mat. Biosci.* **1993**, *118*, 25–49.
21. McCallum, H.I. Effects of immigration on chaotic population dynamics. *J. Theor. Biol.* **1992**, *154*, 277–284.
22. Stone, L.; Hart, D. Effects of Immigration on the Dynamics of Simple Population Models. *Theor. Popul. Biol.* **1999**, *55*, 227–234.
23. Holt, R.D. Population dynamics in two-patch environments: Some anomalous consequences of an optimal habitat distribution. *Theor. Popul. Biol.* **1985**, *28*, 181–208.

24. Silva, A.L.J.; De Castro, L.M.; Justo, A.M.D. Synchronism in a metapopulation model. *Bull. Math. Biol.* **2000**, *62*, 337–349.

25. Reeve, J.D. Environmental variability, migration, and persistence in host-parasitoid systems. *Am. Nat.* **1988**, *132*, 810–836.

26. Reeve, J.D. Stability, variability, and persistence in host-parasitoid systems. *Ecology* **1990**, *71*, 422–426.

27. Taylor, A.D. Large-scale spatial structure and population dynamics in arthropod predator-prey systems. *Ann. Zool. Fenn.* **1988**, *25*, 63–74.

28. Ruxton, G.D. Low levels of immigration between chaotic populations can reduce system extinctions by inducing asynchronous regular cycles. *Proc. R. Soc. Lond. B Biol. Sci.* **1994**, *256*, 189–193.

29. Pal, N.; Samanta, S.; Chattopadhyay, J. The impact of diffusive migration on ecosystem stability. *Chaos Solitons Fractals* **2015**, *78*, 317–328.

30. Thieme, R. *Mathematics in Population Biology*; Princeton University Press: Princeton, NJ, USA, 2003.

31. Birkhoff, G.; Rota, G. *Ordinary Differential Equation*; John Wiley and Sons: Boston, MA, USA, 1989.

32. McCann, K.; Yodzis, P. Biological Conditions for Chaos in a Three-Species Food Chain. *Ecology* **1994**, *75*, 561–564.

33. Upadhyay, R.K.; Rai, V. Complex dynamics and synchronization in two non-identical chaotic ecological systems. *Chaos Solitons Fractals* **2009**, *40*, 2233–2241.

34. Parshad, R.D.; Kumari, N.; Kouachi, S. A remark on "Study of a Leslie–Gower-type tritrophic population model" [Chaos, Solitons & Fractals 14 (2002) 1275–1293]. *Chaos Solitons Fractals* **2015**, *71*, 22–28.

35. Straughan, B. *Explosive Instabilities in Mechanics*; Springer: Heidelberg, Germany, 1998.

36. Quittner, P.; Souplet, P. *Superlinear Parabolic Problems: Blow-Up, Global Existence and Steady States*; Birkhauser Verlag: Basel, Switzerland, 2007.

37. Parshad, R.D.; Quansah, E.; Black, K.; Beauregard, M. Biological control via "ecological" damping: An approach that attenuates non-target effects. *Math. Biosci.* **2016**, *273*, 23–44.

38. Parshad, R.D.; Upadhyay, R.K.; Mishra, S.; Tiwari, S.K.; Sharma, S. On the explosive instability in a three-species food chain model with modified Holling type IV functional response. *Math. Methods Appl. Sci.* **2016**, *40*, 5707–5726.

39. Parshad, R.D.; Quansah, E.; Beauregard, M.; Kouachi, S. On "small" data blow-up in a three species food chain model. *Comput. Math. Appl.* **2017**, *73*, 576–587.

40. Parshad, R.D.; Basheer, A.; Jana, D.; Tripathi, J.P. Do prey handling predators really matter: Subtle effects of a Crowley-Martin functional response. *Chaos Solitons Fractals* **2017**, *103*, 410–421.

41. Hassell, M.P. *The Dynamics of Arthropod Predator-Prey Systems*; Princeton University Press: Princeton, NJ, USA, 1978.

42. Bailey, V.A.; Nicholson, A.J.; Williams, E.J. Interaction between hosts and parasites when some host individuals are more difficult to find than others. *J. Theor. Biol.* **1962**, *3*, 1–18.

43. Chewning, W.C. Migratory effects in predator-prey models. *Math. Biosci.* **1975**, *23*, 253–262.

MDPI

Article

The Collapse of Ecosystem Engineer Populations

José F. Fontanari

Instituto de Física de São Carlos, Universidade de São Paulo, Caixa Postal 369, 13560-970 São Carlos SP, Brazil; fontanari@ifsc.usp.br

Received: 26 November 2017; Accepted: 8 January 2018; Published: 12 January 2018

Abstract: Humans are the ultimate ecosystem engineers who have profoundly transformed the world's landscapes in order to enhance their survival. Somewhat paradoxically, however, sometimes the unforeseen effect of this ecosystem engineering is the very collapse of the population it intended to protect. Here we use a spatial version of a standard population dynamics model of ecosystem engineers to study the colonization of unexplored virgin territories by a small settlement of engineers. We find that during the expansion phase the population density reaches values much higher than those the environment can support in the equilibrium situation. When the colonization front reaches the boundary of the available space, the population density plunges sharply and attains its equilibrium value. The collapse takes place without warning and happens just after the population reaches its peak number. We conclude that overpopulation and the consequent collapse of an expanding population of ecosystem engineers is a natural consequence of the nonlinear feedback between the population and environment variables.

Keywords: ecosystem engineering; colonization wavefront; cliodynamics; coupled map lattice

1. Introduction

There is hardly a landscape on Earth that has not been modified by past living beings as a result of the natural feedback between organisms and environment, whose study was initiated by Darwin in his last scientific book [1]. A recent alternative viewpoint: niche construction or ecosystem engineering, acknowledges a more active role of some species, the so-called ecosystem engineers, in modifying their environments to enhance their survival [2]. For instance, beavers—an oft-mentioned example of ecosystem engineer—cut trees, build dams, and create ponds, thus giving rise to new and safe landscapes [3]. In fact, since the areas flooded by dams increase the distance beavers can travel by water, which is safer than traveling by land, the modified landscape results in a net increase of the beavers' survival expectations [4]. Whereas the issue whether beavers and other nonhuman species qualify as ecosystem engineers is disputable [5], nobody contends that humans are the paramount ecosystem engineers, who have shaped the world into an (arguably) more hospitable place for themselves, most often with unlooked-for effects [6]. An extreme unforeseen effect of the engineering of landscapes, which is nonetheless ubiquitous in the history of civilizations, is the collapse of human societies caused by habitat destruction and overpopulation, among other factors [7].

The feature of the population dynamics of ecosystem engineers that makes it well suited to model human populations inhabiting isolated areas (e.g., islands and archipelagos) is that the growth of the population is determined by the availability of usable habitats, which in turn are created by the engineers through the modification, and consequent destruction, of virgin habitats. From a mathematical perspective, this feedback results in a density-dependent carrying capacity. This feature is the core of the continuous-time, space-independent mathematical model that Gurney and Lawton proposed to describe the population dynamics of ecosystem engineers [8]. The Gurney and Lawton model considers the quality of the habitats as dynamic variables, in addition to the density of engineers. There are three different types of habitats: virgin, usable (or modified) and degraded habitats. The transition from the virgin to the usable habitat is effected only in the presence of engineers. The modified habitats then degrade and eventually recover to become virgin habitats again. Virgin and degraded habitats are unsuitable for the growth of the engineer population. This ecosystem engineering approach seems a way more suitable to study the interplay between humans and their environment than the traditional predator-prey framework used in previous studies [9,10], although both approaches exhibit the characteristic population cycles that reflect the opposite interests of the interacting parts [11].

The Gurney and Lawton model becomes more effective (and instructive) to simulate the human-environment interaction if we use its recently proposed spatial formulation, where an initial small settlement of engineers is surrounded by vast areas (patches) of virgin habitats, and a fraction of the engineers are allowed to move between neighboring patches [12]. In time, a patch is an ecosystem, say, an island, that can potentially exhibit all three types of habitats as well as the engineer population, simultaneously. Hence the group of patches can be thought of as an archipelago. In this contribution we focus on the characterization of the speed of the colonization front and on global demographic quantities such as the total mean density of engineers.

Our main finding is that overpopulation is a natural outcome of the population dynamics of ecosystem engineers during the expansion phase to colonize the unexplored virgin patches. When all patches are explored, the population density plunges sharply towards its (local) equilibrium value. The collapse takes place just after the population reaches its peak number. This surprising outcome, which results from the nonlinear feedback between engineers and environment, could hardly be predicted without mathematics, thus lending credence to the tenets of the discipline Cliodynamics that advocates the mathematical modeling of historical processes [13,14].

The rest of the paper is organized as follows. In Section 2 we offer an overview of the discrete time version of Gurney and Lawton model of ecosystem engineers. In particular, we present the recursion equations that govern the local (single-patch) dynamics and summarize the relevant findings regarding the stability of the fixed-point solutions [12]. The coupled map lattice version of the discrete time model is then introduced in Section 3. The numerical solution of the coupled map lattice equations is presented and discussed in Section 4 for the case the patches are arranged in a chain with reflective boundary conditions, and the model parameters are set such that the local dynamics is attracted by a nontrivial fixed point. The focus is on the colonization scenario where an initial settlement of engineers placed in the central patch of the chain is allowed to disperse to neighboring patches. Finally, Section 5 is reserved to our concluding remarks.

2. The Discrete Time Version of the Gurney and Lawton Model

As pointed out, Gurney and Lawton have modeled the local population dynamics of ecosystem engineering using a continuous-time model [8]. Here we present a brief overview of a discrete time version of that model [12] that can be easily extended to incorporate the spatial dependence of the engineer population as well as of the habitat variables, following the seminal works on host-parasitoid systems [15,16] (see [17–19] for more recent contributions) as well as on Lotka-Volterra systems [20].

We begin by assuming that the population of engineers at generation t is composed of E_t individuals and that each engineer requires a unit of usable habitat to survive. Denoting by H_t the number of units of usable habitats available at generation t, so that in the equilibrium regime one has $\lim_{t\to\infty} E_t/H_t = 1$, we can use Ricker model [21] to write the expected number of engineers at generation $t + 1$ as

$$E_{t+1} = E_t \exp\left[r\left(1 - E_t/H_t\right)\right], \tag{1}$$

where r is the intrinsic growth rate of the population of engineers and H_t plays the role of a time-dependent carrying capacity for the population of engineers.

The essential ingredient of the Gurney and Lawton model, which sets it apart from the other population dynamics models [22], is the requirement that usable habitats be created by engineers working on virgin habitats. In particular, if we assume that there are V_t units of virgin habitats at generation t, then the fraction $C(E_t) V_t$ of them will be transformed in usable habitats at the next generation, $t+1$. Here $C(E_t)$ is any function that satisfies $0 \leq C(E_t) \leq 1$ for all E_t and $C(0) = 0$. Clearly, this function measures the efficiency of the engineer population to build usable habitats from the raw materials provided by the virgin habitats.

Usable habitats decay into degraded habitats that are useless to the engineers, in the sense they lack the raw materials needed to build usable habitats. Let δH_t denote the fraction of usable habitats that decay to degraded habitats in one generation, where $\delta \in [0,1]$ is the decay probability. Then the expected number of units of usable habitats at generation $t+1$ is simply

$$H_{t+1} = (1-\delta) H_t + C(E_t) V_t. \tag{2}$$

At first sight, one might think that the decay probability δ should be density dependent (i.e., $\delta = \delta(E_t)$), particularly in the case the habitat degradation resulted from the overexploitation of resources. However, in the Gurney and Lawton model the resources are represented by the virgin habitats, whose probability of change into usable habitats is in fact density dependent, $C = C(E_t)$. For example, in an island scenario, the virgin habitats can be thought of as the native forests whereas the usable habitats are the lands cleared for crops, whose degradation, due mainly to erosion and soil depletion of nutrients, is more suitably modeled by a constant decay probability δ, rather than by a density-dependent one.

Degraded habitats will eventually recover and become virgin habitats again. Denoting the fraction of degraded habitats that recover to virgin habitats in one generation by ρD_t we can write

$$D_{t+1} = (1-\rho) D_t + \delta H_t, \tag{3}$$

where $\rho \in [0,1]$ is the recovery probability. Finally, the recursion equation for the expected number of units of virgin habitats is simply

$$V_{t+1} = [1 - C(E_t)] V_t + \rho D_t. \tag{4}$$

As expected, $V_{t+1} + H_{t+1} + D_{t+1} = V_t + H_t + D_t = T$, where T is the (fixed) total store of habitats (e.g., the area of the island). Hence we can define the habitat fractions $v_t \equiv V_t/T$, $h_t \equiv H_t/T$ and $d_t \equiv D_t/T$ that satisfy $v_t + h_t + d_t = 1$ for all t. In addition, we define the density of engineers $e_t = E_t/T$ which, differently from the habitat fractions, may take on values greater than one. In terms of these intensive quantities, the above recursion equations are rewritten as

$$
\begin{align}
e_{t+1} &= e_t \exp[r(1 - e_t/h_t)] \tag{5}\\
h_{t+1} &= (1-\delta) h_t + c(e_t) v_t \tag{6}\\
v_{t+1} &= \rho(1 - v_t - h_t) + [1 - c(e_t)] v_t, \tag{7}
\end{align}
$$

where we have used $d_t = 1 - v_t - h_t$ and $c(e_t) \equiv C(Te_t)$.

To complete the model we must specify the density-dependent probability $c(e_t)$, which measures the engineers' efficiency to transform the virgin habitats into usable ones. The function $c(e_t)$ incorporates the collaboration and communication strategies that allowed the engineer ecosystems to build collective structures (e.g., termite mounds and anthills), which are their solutions to the external and internal threats to their survival [23–25]. For humans, this function incorporates the beneficial effects (from their perspective) of the technological advancements [26] that allowed a more efficient harvesting of natural resources. Alternatively, $c(e_t)$ can be viewed as a density-dependent resource depletion probability. Here we consider the function

$$c(e_t) = 1 - \exp(-\alpha e_t), \tag{8}$$

where $\alpha > 0$ is the productivity parameter, which measures the efficiency of the engineers in transforming natural resources into useful goods. For $\alpha \ll 1$ we have $c(e_t) \approx \alpha e_t$ and the term responsible for the depletion of virgin habitats in Equation (7) becomes $\alpha e_t v_t$, indicating a low-technological organization where, in a finite population scenario, it would be necessary the direct contact between one engineer and one unit of virgin habitat in order to transform it in one unit of usable habitat [27]. For $\alpha \gg 1$, however, a few engineers can transform all the available virgin habitats in just a single generation.

The discrete-time population dynamics of the ecosystem engineers, given by the system of recursion Equations (5)–(8), exhibits a complex dependence on the model parameters (r, δ, ρ and α) that was studied in great detail in [12]. For instance, Figure 1 illustrates the dependence on the growth rate r by showing the bifurcation diagram for the engineer density [28]. The period-doubling bifurcation cascade is expected since the source of nonlinearity of the population dynamics is Ricker's formula [21].

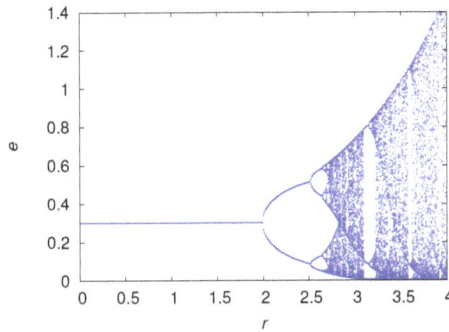

Figure 1. Bifurcation diagram for the local population dynamics (5)–(8) with parameters $\alpha = 0.1$, $\delta = 0.01$ and $\rho = 0.005$. The points on the y-axis show the values of the engineer density visited asymptotically from all initial conditions with $e_0 > 0$.

Here our interest is on the nontrivial fixed-point solutions only, which are obtained by setting $e_{t+1} = e_t = e^*$, $h_{t+1} = h_t = h^*$ and $v_{t+1} = v_t = v^*$ in Equations (5)–(8). Assuming $e^* > 0$ we have $h^* = e^*$ and $v^* = 1 - e^*(1 + \delta/\rho)$ with e^* given by the solution of the transcendental equation

$$\delta e^* = [1 - \exp(-\alpha e^*)][1 - e^*(1 + \delta/\rho)]. \tag{9}$$

In the limit of small density, i.e., $e^* \ll 1$, this equation reduces to

$$e^* \approx \frac{1 - \delta/\alpha}{1 + \delta/\rho}, \tag{10}$$

and $v^* = \delta/\alpha$, indicating that this fixed point is physical for $\delta < \alpha$ only. As expected, e^* increases with increasing α and ρ, and decreases with increasing δ. Although Equation (9) does not depend on the growth rate r, large values of this parameter lead to the instability of the fixed point e^* as illustrated in Figure 1. Finally, the trivial fixed point $e^* = 0$, $h^* = 0$ and $v^* = 1$ is stable for $\delta > \alpha$. We refer the reader to Ref. [12] for the detailed analysis of the local stability of the fixed points of the recursion Equations (5)–(8).

3. The Coupled Map Lattice Version of the Gurney and Lawton Model

The space-independent recursion Equations (5)–(8) govern the local or single-patch population dynamics, where, as already pointed out, a patch is a complete ecosystem (e.g., an isle) with the three types of habitats and a population of engineers. Those equations describe the growing phase of the population of engineers. Here we introduce another phase: the dispersal phase, which we assume takes place before the growing stage. In particular, we consider a system of N patches (e.g., an archipelago) and allow the engineers to circulate among neighboring patches, such that a fraction μ of the population in patch i is transferred to the K_i neighboring patches. Hence after the dispersal stage the population at patch i is

$$E'_{i,t} = (1 - \mu) \, E_{i,t} + \mu \sum_j E_{j,t}/K_j, \tag{11}$$

where the sum is over the K_i nearest neighbors of patch i. Note that since each term in the summation is divided by K_j, the fraction μ of the population lost by patch j is equally divided among the K_j neighboring patches. For simplicity, we assume that the total number of habitats T is the same for all patches, so that the effect of dispersal on the density of engineers is given by

$$e'_{i,t} = (1 - \mu) \, e_{i,t} + \mu \sum_j e_{j,t}/K_j. \tag{12}$$

for patch $i = 1, \ldots, N$. After dispersal of the engineers, the growing stage takes place within each patch according to the equations

$$
\begin{align}
e_{i,t+1} &= e'_{i,t} \exp\left[r \left(1 - e'_{i,t}/h_{i,t} \right) \right] \tag{13} \\
h_{i,t+1} &= (1 - \delta) \, h_{i,t} + c \left(e'_{i,t} \right) v_{i,t} \tag{14} \\
v_{i,t+1} &= \rho \left(1 - v_{i,t} - h_{i,t} \right) + \left[1 - c \left(e'_{i,t} \right) \right] v_{i,t}, \tag{15}
\end{align}
$$

for $i = 1, \ldots, N$. Together with Equation (12), these equations form a coupled map lattice (see, e.g., [29]) that describe the dynamics of the system of patches or metapopulation.

Since we are not interested on the formation of stationary spatial patterns, which for the Gurney and Lawton model appear only in the case the model parameters are such that the local dynamics (5)–(8) is chaotic [12], the only patch arrangements we will consider in this paper are chains with an odd number of patches and reflective boundary conditions (i.e., $K_1 = K_N = 1$ and $K_i = 2, \forall i \neq 1, N$). In addition, we will focus on a colonization or invasion scenario where at generation $t = 0$ only the central patch $i_c = (N + 1)/2$ of the chain is populated, whereas the other patches are composed entirely of virgin habitats (see, e.g., [16]). In particular, we set $e_{i_c,0} = h_{i_c,0} = v_{i_c,0} = 0.5$ and $e_{i,0} = h_{i,0} = 0$, $v_{i,0} = 1$ for all $i \neq i_c$.

In the next section we will study the time dependence of the mean density of engineers,

$$\langle e_t \rangle = \frac{1}{N} \sum_{i=1}^{N} e_{i,t}, \tag{16}$$

and the mean fraction of virgin habitats,

$$\langle v_t \rangle = \frac{1}{N} \sum_{i=1}^{N} v_{i,t}, \tag{17}$$

in the regime where the local dynamics is attracted to the nontrivial fixed point $e^* > 0$, so we do not need to worry about accuracy issues caused by the chaotic amplification of numerical noise [19].

4. Results

As our focus is on the time dependence of the global quantities $\langle e_t \rangle$ and $\langle v_t \rangle$, the results of this section are obtained solely through the numerical iteration of the coupled map lattice Equations (12)–(15). In addition, since we expect that the time to reach the borders of the chain scales linearly with the chain size N, the results are presented in terms of the rescaled time t/N.

Figure 2 shows the evolution of the mean density of engineers and the mean fraction of virgin habitats for several chain sizes N. It reveals the dramatic effect of the engineers' mobility, which allow the population to reach densities well above those the environment could support in a situation of equilibrium. The initial increase of the mean density $\langle e_t \rangle$ reflects the expansion phase of the engineers, which is accompanied by the monotone decreasing of the unexplored patches, as expected. This expansion halts only when the engineers reach the borders of the chain and the end of the availability of unexplored virgin habitats results in a sharp drop on their density, which then quickly converges to the stationary value $\langle e_\infty \rangle = e^*$. This scenario is corroborated by Figure 3 that shows the colonization wavefronts at three distinct times. The reason the size of the engineer density drop decreases with increasing N (see panel (a) of Figure 2) is simply because the contribution of the two high-density wavefronts is watered down by the equilibrium-density of the bulk of the chain. We note that the shape and height of the wavefronts are not affected by the chain size. The time evolution of the mean fraction of usable habitat is qualitatively similar to that shown in panel (a) for the density of engineers.

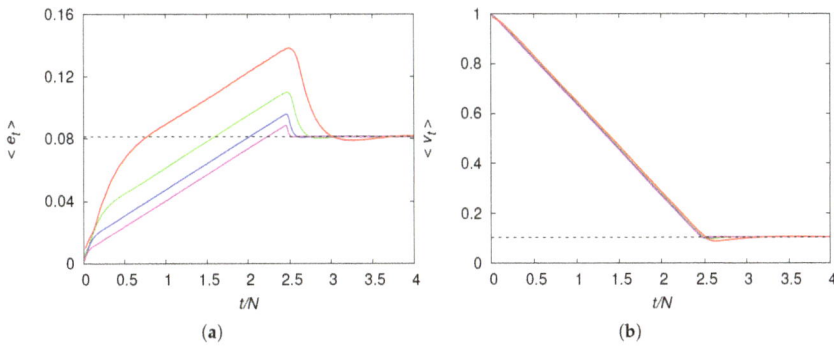

Figure 2. (**a**) Mean engineer density $\langle e_t \rangle$ as function of the rescaled time t/N; (**b**) Mean fraction of virgin habitats $\langle v_t \rangle$ as function of the rescaled time t/N. The chain sizes are $N = 51$ (red line), $N = 101$ (green line), $N = 201$ (blue line) and $N = 401$ (magenta line). The model parameters are $r = 1$, $\alpha = 1$, $\delta = 0.1$, $\rho = 0.01$ and $\mu = 0.1$. The fixed point solution $e^* \approx 0.0814$ and $v^* \approx 0.1041$ obtained from Equation (9) is shown by the dashed horizontal lines.

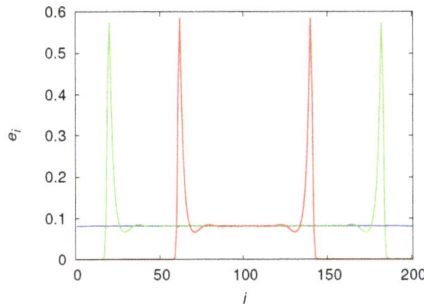

Figure 3. Density of engineers in patch $i = 1, \ldots, N$ at times $t/N = 1$ (red line), $t/N = 2$ (green line) and $t/N = 3$ (blue line). The chain size is $N = 201$ and the model parameters are $r = 1$, $\alpha = 1$, $\delta = 0.1$, $\rho = 0.01$ and $\mu = 0.1$.

Figure 2 suggests a direct way to calculate the mean speed ν of the colonization wavefronts, which is defined as the ratio between the distance from the center to the borders of the chain (i.e., $N/2$) and the time $\hat{t}$ to reach those borders, i.e.,

$$\nu = \frac{N}{2\hat{t}}. \tag{18}$$

In fact, $\hat{t}$ can be easily estimated by the time at which $\langle e_t \rangle$ is maximum. For instance, for all the chain sizes shown in Figure 2, we find $\hat{t}/N \approx 2.46$, so that $\nu \approx 0.20$. This means that it is necessary about five generations, on the average, for the wavefront peak to move between contiguous patches. Since the mean speed of the colonization wavefronts is very weakly influenced by the chain size, as illustrated in Figure 2, henceforth we will consider chains of size $N = 201$ only.

Figure 4 shows the dependence of the wavefront mean speed v on the growth rate r and on the dispersal probability μ. Although these two parameters have no effect whatsoever on the stationary solution e^* and v^*, they have a strong influence on the speed the population colonizes the unexplored patches. The speed v is a monotone increasing function of the dispersal probability μ, as expected, and the rate of increase of v decreases with increasing μ. The dependence of v on the growth rate r is more interesting since it exhibits a non-monotone behavior, which is best seen in the figure for large values of the dispersal probability but that actually happens for all values of μ. For instance, for $\mu = 0.1$ (cyan curve in Figure 4) the maximum of v occurs for $r \approx 0.9$ at which we find $v \approx 0.201$, whereas for $r = 1.5$ we find $v \approx 0.195$. This variation is not perceptible in the scale of Figure 4. Since only a fast growing population can guarantee a very large density at the borders of the expanding colony (see Figure 3) and hence take advantage of the neighboring unexplored patches, one should expect a steep increase of v with increasing r, despite the fact that r plays no role in the equilibrium situation. In fact, this is what one observes in Figure 4, provided that r is not too large. The smooth decrease of v for large r is probably due to the negative feedback of a large growth rate on the engineer population when the available fraction of useful habitats is not large enough. In fact, the maximum wavefront speed observed in the figure is a result of a fine tuning between the growth rate r and the potential to create usable habitats α from the virgin patches. Interestingly, although the first engineers to reach the virgin patches are doomed to extinction because there are no usable habitats in those patches (see Equation (13)), they build the usable habitats (see Equation (15)) for the next wave of migrants.

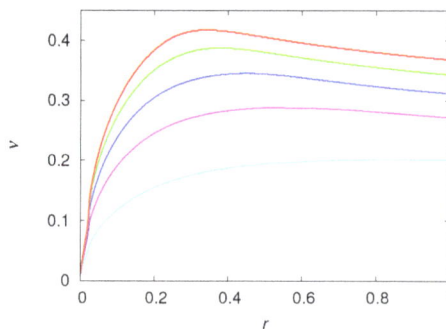

Figure 4. Mean speed of the colonization wavefronts v as function of the growth rate r for the dispersal probability $\mu = 0.9$ (red line), 0.7 (green line), 0.5 (blue line), 0.3 (magenta line) and 0.1 (cyan line). The chain size is $N = 201$ and the model parameters are $\alpha = 1, \delta = 0.1, \rho = 0.01$.

We have verified that the mean speed v of the colonization wavefronts is not influenced by the recovery probability ρ of the degraded habitats, as expected. In fact, the capacity of recovery of the degraded habitats that are left far behind the invasion front is completely irrelevant for the survival and growth of the pioneers in the colonization front. The dependence of v on the productivity rate α and on the decay probability δ is summarized in Figure 5. The results are restricted to the region $\alpha > \delta$ where we can guarantee the existence of a viable equilibrium population of engineers, i.e., $e^* > 0$. The mean speed v increases monotonically with α since the efficiency of the transformation of virgin habitats into usable habitats is crucial for the survival of the second wave of migrants in the colonization front, as pointed out before. In the same line of reasoning, if the recently created usable habitats decay too rapidly, then the colonization front will be delayed as shown in Figure 5. However, if the productivity α is large then the decay probability δ has only a negligible retarding effect on the mean speed of the wavefront.

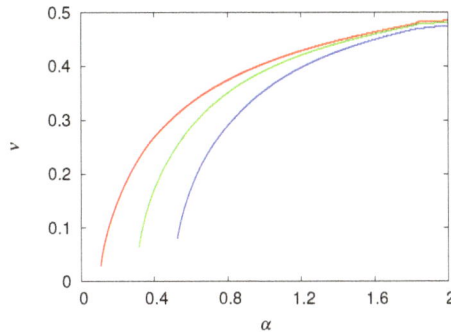

Figure 5. Mean speed of the colonization wavefronts v as function of the productivity rate α for the decay probability $\delta = 0.1$ (red line), 0.3 (green line), and 0.5 (blue line). The curves are shown for $\alpha > \delta$ so that the population is viable at equilibrium, i.e., $e^* > 0$. The chain size is $N = 201$ and the model parameters are $r = 0.5$, $a = 1$, $\rho = 0.01$ and $\mu = 0.9$.

The collapse of the population, which happens when there are no more virgin patches to be explored, can be made more spectacular by setting the model parameters such that the equilibrium population density is very small, i.e., $e^* \ll 1$. One way to achieve this is by setting $\rho \ll 1$ (see Equation (10)), resulting in the global measures $\langle e_t \rangle$ and $\langle v_t \rangle$ displayed in Figure 6. Let us consider first panel (b) of Figure 6 that shows the linear decrease of the fraction of virgin habitats with time, which ends when the colonization fronts reach the borders of the chain at time $t = \hat{t}$. At this moment we have $\langle v_{\hat{t}} \rangle \ll v^*$ and from then on the fraction of virgin habitats begins to increase very slowly following the time scale set by the recovery probability ρ (this last stage is barely seen on the scale of the figure).

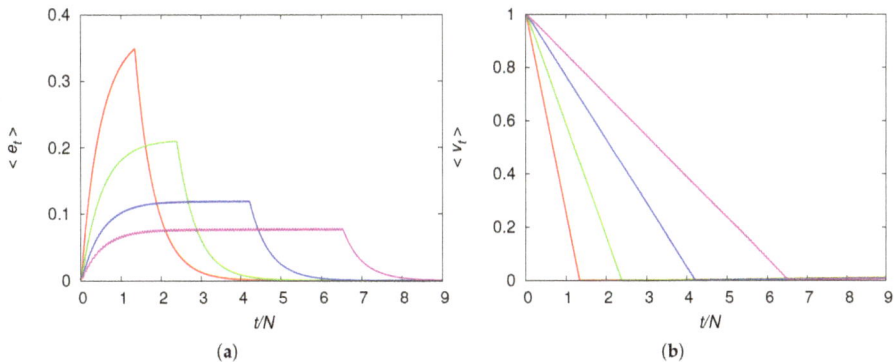

Figure 6. (a) Mean engineer density $\langle e_t \rangle$ as function of the rescaled time t/N; (b) Mean fraction of virgin habitats $\langle v_t \rangle$ as function of the rescaled time t/N. The dispersal probability is $\mu = 0.9$ (red line), $\mu = 0.1$ (green line), $\mu = 0.01$ (blue line) and $\mu = 0.001$ (magenta line) and the other model parameters are $r = 1$, $a = 1$, $\delta = 0.01$, $\rho = 10^{-5}$. The chain size is $N = 201$ and the fixed point solution is $e^* \approx 0.001$ and $v^* \approx 0.01$.

The time dependence of the density of engineers shown in panel (a) of Figure 6 is more instructive. The population with the highest mobility uses up the environmental resources quickly and reaches very high densities before plunging towards the low equilibrium density, whereas the population with the lowest mobility can maintain an average density value for a long time before exhausting the resources.

It is interesting that in this case the mean density of engineers exhibits a sort of metastable equilibrium, i.e., $\langle e_t \rangle$ is practically constant for a period of time significantly longer than the initial increase (see the blue and magenta curves in panel (a) of Figure 6), although the population is expanding through the unexplored patches.

The qualitative differences of the population density in these mobility extremes are easily understood with the aid of Figures 7 and 8 that show the population densities in each patch for $\mu = 0.9$ and $\mu = 0.001$, respectively. In fact, the reason the high mobility population (see Figure 7) attains such high densities before it collapses is simply that it reaches the borders before the usable habitats in the center of the chain can degrade appreciably. In the low mobility case (see Figure 8), the center of the chain is practically desert when the colonization front is moving towards the borders. The metastable equilibrium mentioned before is simply a consequence of the invariance of the shape (or, more pointedly, of the area) of the wavefronts, as the colonization fronts are the only places where usable habitats are found (the spatial distribution of usable habitats is indistinguishable from that of the engineers). We note that the density near the borders is higher than near the center of the chain because the usable habitats at the borders were created much later than those in the center and so had less time to decay into the degraded class.

Finally, we note that in the case of inviable patches, i.e., in the regime $\alpha < \delta$ where the nontrivial fixed point $e^* > 0$ is unphysical and $e^* = 0$ is stable, the colonization of the unexplored virgin patches fails and the engineers quickly go extinct. The reason is that the vast supply of unexplored patches is irrelevant if the production of usable habitats from them is not enough to balance the decay of the usable habitats into degraded habitats.

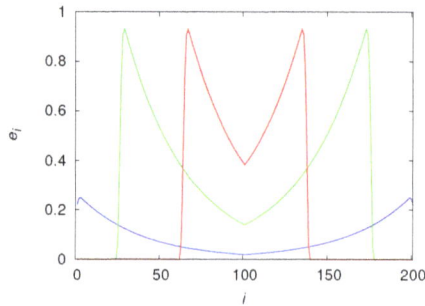

Figure 7. Density of engineers in patch $i = 1, \ldots, N$ at times $t/N = 0.5$ (red line), $t/N = 1$ (green line) and $t/N = 2$ (blue line). The chain size is $N = 201$ and the model parameters are $r = 1$, $\alpha = 1$, $\delta = 0.01$, $\rho = 10^{-5}$ and $\mu = 0.9$.

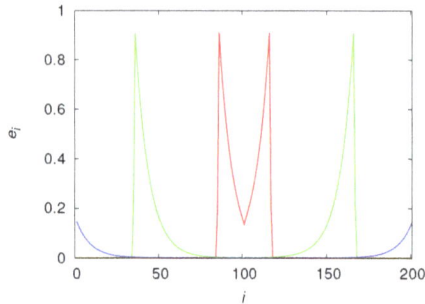

Figure 8. Density of engineers in patch $i = 1, \ldots, N$ at times $t/N = 1$ (red line), $t/N = 4$ (green line) and $t/N = 7$ (blue line). The chain size is $N = 201$ and the model parameters are $r = 1$, $\alpha = 1$, $\delta = 0.01$, $\rho = 10^{-5}$ and $\mu = 0.001$.

5. Discussion

The increase of the mean density of engineers $\langle e_t \rangle$ during the expansion phase of the colony is totally expected, of course, since as new patches are invaded by the engineers their overall density must increase. (We recall that in our scenario $\langle e_t \rangle$ varies from the arbitrarily set initial value $\langle e_0 \rangle = 0.5/N$ to the local equilibrium density $\langle e_\infty \rangle = e^*$, where N is the number of patches.) What is surprising is that the transient density reaches values much larger than the equilibrium density and plunges sharply when the colonization front hits the boundary of the available space (see, e.g., Figure 6). This phenomenon was observed en passant in Ref. [12] for the regime of chaotic local dynamics, and was somewhat concealed by the difficulty of controlling the numerical accuracy of hundreds of chaotically oscillating coupled patches. Here we have offered a detailed analysis of the colonization process for the numerically more manageable dynamic regime where the sole attractors are fixed points.

We note that if we set the recovery probability ρ of the degraded habitats to a small value, then the equilibrium population density $e^* \approx \rho\,(1/\delta - 1/\alpha)$ will be small too (see Equation (10)), and since the maximum mean density $\langle e_t \rangle$ does not depend on ρ, we get $e^*/\langle e_t \rangle \sim \rho$, which implies a catastrophic reduction of the population density around time $\hat{t}$. This is the reason we liken the sharp drop of the population density when the supply of unexplored resources is exhausted (see Figures 2 and 6) to the collapse of ancient human societies that overexploited their environment [7]. In that sense, we offer an alternative to the view that a population collapse involves the relatively quick decline of a population, which is established in the habitat considered. In our perspective, a collapse involves a population reaching physical or technological boundaries so that it can no longer sustain its over increased size.

In the context of the collapse of human societies, our population dynamics model produces two outcomes that are worth emphasizing. The first result is that for low mobility values the disaster comes without warning since the mean density $\langle e_t \rangle$ is practically constant before the collapse (see the blue and magenta curves in panel (a) of Figure 6), and the width and height of the wavefront are constant before the colonization front hits the chain border (see Figure 8). The disaster arrival is also silent for high mobility values since the mean density $\langle e_t \rangle$ and the width of the wavefront of colonization are still increasing when the collapse occurs (see red and green curves in panel (a) of Figures 6 and 7). This accords with Diamond's interpretation of the archaeological records of collapsed civilizations [7]: "In fact, one of the main lessons to be learned from the collapses of the Maya, Anasazi, Easter Islanders, and those other past societies (as well as from the recent collapse of the Soviet Union) is that a society's steep decline may begin only a decade or two after the society reaches its peak numbers, wealth, and power."

The second consequence of our model is that overpopulation is a natural outcome of the nonlinear dynamics of the ecosystem engineer population expanding over unexplored habitats. A rough global measure of the overpopulation at time t is given by the ratio $\langle e_t \rangle / e^*$ that equals one in the equilibrium situation. We note, however, that the local engineer density in the patches that are part of the colonization front are much higher than the overall mean density (see Figures 3, 7 and 8). This is so because the second wave of migrants finds empty patches composed mostly of usable habitats (meaning a large carrying capacity) that resulted from the work of the extinct first wave of migrants on the original virgin habitats.

As a cautionary note on the adequacy of our approach to describe human populations, we should mention that the discrete-time map (5)–(8) assumes non-overlapping generations, which is clearly not the case for humans. Our justification to use this approach is a pragmatic one: discrete-time maps can easily be extended to describe space-dependent problems, resulting in coupled map lattices that can be thoroughly studied numerically. In addition, since here we consider the regime where the only attractors are stable fixed points, we do not expect any significant differences between the discrete and the continuous time approaches. In this line, we note that replacing Ricker model, Equation (1), with a model that prevents large fluctuations on the population size, such as the Beverton-Holt model [30], yields qualitatively the same results as those reported here, supporting thus our conjecture that our findings are robust to model details.

We find it quite remarkable that the model proposed by Gurney and Lawton to study the population dynamics of ecosystem engineers [8], which seems to have been developed with an eye on the ecology of beavers [3], could provide such interesting insights on the collapse dynamics of past human societies, without incorporating specific traits of those societies [7]. For instance, one such a trait is existence of ruling elites that parasitize on the large mass of producers (commoners), using their workforce to produce luxury items and religious monuments [10]. This feature could easily be incorporated in our model by requiring that only the commoners modify the virgin habitats and that the elite members use a disproportionally large amount of usable habitats. Another interesting possibility for future research is the introduction of multiplicative noise affecting the availability of virgin habitats aiming at modeling the effects of droughts and bonanzas in our model ecosystem [31,32]. Nonetheless, our results show that the collapse of an expanding population of ecosystem engineers seems to be a robust, unavoidable consequence of the nonlinear feedback between the population and environment variables, so a more detailed modeling of human societies will probably have little effect on our findings.

Acknowledgments: This research was supported in part by grant 15/21689-2, Fundação de Amparo à Pesquisa do Estado de São Paulo (FAPESP) and by grant 303979/2013-5, Conselho Nacional de Desenvolvimento Científico e Tecnológico (CNPq).

Conflicts of Interest: The author declares no conflict of interest.

References

1. Darwin, C.R. *The Formation of Vegetable Mould through the Action of Worms, with Observations on their Habits*; John Murray: London, UK, 1881.
2. Odling-Smee, F.J.; Laland, K.N.; Feldman, M.W. *Niche Construction: The Neglected Process in Evolution*; Princeton University Press: Princeton, NJ, USA, 2003.
3. Wright, J.P.; Gurney, W.S.C.; Jones, C.G. Patch dynamics in a landscape modified by ecosystem engineers. *Oikos* **2004**, *105*, 336–348.
4. Dawkins, R. *The Extended Phenotype*; Oxford University Press: Oxford, UK, 1982.
5. Scott-Phillips, T.C.; Laland, K.N.; Shuker, D.M.; Dickins, T.E.; West, S.A. The Niche Construction Perspective: A Critical Appraisal. *Evolution* **2013**, *68*, 1231–1243.
6. Smith, B.D. The Ultimate Ecosystem Engineers. *Science* **2007**, *315*, 1797–1798.
7. Diamond, J. *Collapse: How Societies Choose to Fail or Succeed*; Penguin Books: New York, NY, USA, 2005.
8. Gurney, W.S.C.; Lawton, J.H. The population dynamics of ecosystem engineers. *Oikos* **1996**, *76*, 273–283.
9. Brander, J.A.; Taylor, M.S. The Simple Economics of Easter Island: A Ricardo-Malthus Model of Renewable Resource Use. *Am. Econ. Rev.* **1998**, *88*, 119–138.
10. Motesharrei, S.; Rivas, J.; Kalnay, E. Human and nature dynamics (HANDY): Modeling inequality and use of resources in the collapse or sustainability of societies. *Ecol. Econom.* **2014**, *101*, 90–102.
11. Turchin, P. Evolution in population dynamics. *Nature* **2003**, *424*, 257–258.
12. Franco, C.; Fontanari, J.F. The spatial dynamics of ecosystem engineers. *Math. Biosci.* **2017**, *292*, 76–85.
13. Turchin, P. *Historical Dynamics: Why States Rise and Fall*; Princeton University Press: Princeton, NJ, USA, 2003.
14. Turchin, P. Arise cliodynamics. *Nature* **2008**, *454*, 34–35.
15. Hassell, M.P.; Comins, H.N.; May, R.M. Spatial structure and chaos in insect population dynamics. *Nature* **1991**, *353*, 255–258.
16. Comins, H.N.; Hassell, M.P.; May, R.M. The spatial dynamics of host-parasitoid systems. *J. Anim. Ecol.* **1992**, *61*, 735–748.
17. Rodrigues, L.A.D.; Mistro, D.C.; Petrovskii, S. Pattern Formation, Long-Term Transients, and the Turing-Hopf Bifurcation in a Space- and Time-Discrete Predator-Prey System. *Bull. Math. Biol.* **2011**, *73*, 1812–1840.
18. Mistro, D.C.; Rodrigues, L.A.D.; Petrovskii, S. Spatiotemporal complexity of biological invasion in a space- and time-discrete predator-prey system with the strong Allee effect. *Ecol. Complex.* **2012**, *9*, 16–32.
19. Rodrigues, L.A.D.; Mistro, D.C.; Cara, E.R.; Petrovskaya, N.; Petrovskii, S. Patchy Invasion of Stage-Structured Alien Species with Short-Distance and Long-Distance Dispersal. *Bull. Math. Biol.* **2015**, *77*, 1583–1619.

20. Solé, R.V.; Bascompte, J.; Valls, J. Stability and complexity of spatially extended two-species competition. *J. Theor. Biol.* **1992**, *159*, 469–480.

21. Murray, J.D. *Mathematical Biology: I. An Introduction*; Springer: New York, NY, USA, 2003.

22. Turchin, P. *Complex Population Dynamics: A Theoretical/Empirical Synthesis*; Princeton University Press: Princeton, NJ, USA, 2003.

23. Fontanari, J.F. Imitative Learning as a Connector of Collective Brains. *PLoS ONE* **2014**, *9*, e110517.

24. Fontanari, J.F.; Rodrigues, F.A. Influence of network topology on cooperative problem-solving systems. *Theory Biosci.* **2016**, *135*, 101–110.

25. Reia, S.M.; Fontanari, J.F. Effect of group organization on the performance of cooperative processes. *Ecol. Complex.* **2017**, *30*, 47–56.

26. Basalla, G. *The Evolution of Technology*; Cambridge University Press: Cambridge, UK, 1989.

27. Gillespie, D.T. A General Method for Numerically Simulating the Stochastic Time Evolution of Coupled Chemical Reactions. *J. Comput. Phys.* **1976**, *22*, 403–434.

28. Sprott, J.C. *Chaos and Time-Series Analysis*; Oxford University Press: Oxford, UK, 2003.

29. Kaneko, K. Overview of Coupled Map Lattices. *Chaos* **1992**, *2*, 279–282.

30. Beverton, R.J.H.; Holt, S.J. *On the Dynamics of Exploited Fish Populations*; Her Majesty's Stationary Office: London, UK, 1957.

31. Fiasconaro, A.; Valenti, D.; Spagnolo, B. Nonmonotonic behavior of spatiotemporal pattern formation in a noisy Lotka-Volterra system. *Acta Phys. Pol. B* **2004**, *35*, 1491–1500.

32. Su, Y.-J.; Mei, D.-C. Effects of time delay on three interacting species system with noise. *Int. J. Theor. Phys.* **2008**, *47*, 2409–2414.

mathematics

MDPI

Article

Impact of Parameter Variability and Environmental Noise on the Klausmeier Model of Vegetation Pattern Formation

Merlin C. Köhnke * and Horst Malchow

Institute of Environmental Systems Research, School of Mathematics, Computer Science, Osnabrück University, Barbarastraße 12, 49076 Osnabrück, Germany; horst.malchow@uos.de
* Correspondence: merlin.koehnke@uni-osnabrueck.de; Tel.: +49-541-969-2573

Received: 5 October 2017; Accepted: 14 November 2017; Published: 23 November 2017

Abstract: Semi-arid ecosystems made up of patterned vegetation, for instance, are thought to be highly sensitive. This highlights the importance of understanding the dynamics of the formation of vegetation patterns. The most renowned mathematical model describing such pattern formation consists of two partial differential equations and is often referred to as the Klausmeier model. This paper provides analytical and numerical investigations regarding the influence of different parameters, including the so-far not contemplated evaporation, on the long-term model results. Another focus is set on the influence of different initial conditions and on environmental noise, which has been added to the model. It is shown that patterning is beneficial for semi-arid ecosystems, that is, vegetation is present for a broader parameter range. Both parameter variability and environmental noise have only minor impacts on the model results. Increasing mortality has a high, nonlinear impact underlining the importance of further studies in order to gain a sufficient understanding allowing for suitable management strategies of this natural phenomenon.

Keywords: Klausmeier model; pattern formation; self-organization; reaction–diffusion–advection model; environmental noise

1. Introduction

Vegetation patterns have been observed in many semi-arid regions in Africa, Australia, North America, South America, and Asia [1]. They were first described in 1941 [2]. In the 1950s, the occurrence and the spatial distribution of these patterns were investigated on the basis of air photographs (e.g., [3]). It has been reported that periodic patterns at the border between semi-arid and arid climates are even omnipresent [4]. The diversity in patterns makes pattern formation in arid and semi-arid areas an interesting case study for pattern formation in ecology [5]. Furthermore, semi-arid regions are highly dynamic (e.g., [6]), which makes them an intriguing research area in general.

Different soil types supporting pattern formation exist, such as clay, sand and silt [7]. Various shapes of vegetation patterns, such as circles (e.g., [8]), spots (e.g., [9]), and stripes, arcs or labyrinths (e.g., [10]) where bare ground and vegetated bands alternate have been reported. The patterns can be formed by plants such as grass, shrubs or trees [11]. Observations showing transitions from anisotropic vegetation spots that elongate [12] to isotropic banded patterns with increasing slope exist (e.g., [11]). With decreasing precipitation, a successive transition from homogeneous vegetation over gaps, labyrinths and spots to a bare desert state has been observed [13]. The last step from vegetation spots to the bare desert state is commonly abrupt and is referred to as a catastrophic shift [14], which can be theoretically explained by the existence of two alternative stable states [15]. Recent studies suggest that this transition can be investigated on satellite images and can be used to retrieve information about imminent regime shifts [16]. The wavelength of banded patterns tends to decrease with increasing slope and water input [7]. Some banded patterns show a slow uphill movement [17]. However, as a result of the slow speed, long-term data is necessary for reliable measurements. Usually, banded patterns form perpendicularly to the contour, but patterns parallel to the contour can also form under certain external impacts [18].

Early studies explain the formation of vegetation structures with the concept of islands of fertility (e.g., [19]). However, this does not give an appropriate explanation for the development [20]. More recent studies state that patterns emerge as a result of the coupling of reaction, diffusion and advection processes. Precipitation is not sufficient to ensure homogeneous vegetation in such areas. Thus, rainfall is the limiting factor of production [21], only allowing a bare desert state or patterned vegetation. The patterned vegetation facilitates the redistribution of the resource water [22] through diffusion and advection, yielding higher total plant biomass in comparison with homogeneous vegetation [23]. Furthermore, vegetation patterns can enhance accessibility of nutrients [24]. However, we note that different kinds of patterns can be observed within relatively small areas, which makes precipitation as the sole driving factor unlikely [20].

Different methods to model pattern formation and that describe this phenomenon by instabilities of uniform states exist [5]. Two common approaches are scale-dependent feedbacks (e.g., [25]) and competition for scarce resources (e.g., [26]). In the case of scale-dependent feedbacks, the range of facilitation must be shorter than the range of competition [11]. This is comparable to well-known systems with short-range activation and long-range inhibition in reaction–diffusion models [27,28]. In the case of vegetation pattern formation, the short-range activation can be given by increased soil permeability and shading of the plants, while the long-range activation can be given by growing roots competing for resources (e.g., [29]). However, we note that other approaches, such as stochastic models, explaining pattern formation as a consequence of noise-induced environmental fluctuations also exist (for a detailed review, see [1]). A large number of researchers have modelled the feedback between biomass growth and water use leading to pattern formation in semi-arid areas (e.g., [1,25,26,30,31]). More recent studies have also been focussing on the impact of heterogeneous environments [32] and heterogeneities in parameters [33] on pattern formation. They have pointed out that large patch sizes are a precondition for the widely considered self-organization [32].

Although many models describing pattern formation do already exist, the underlying mechanisms are still unclear. This article deals with the well-known reaction–diffusion–advection model by Klausmeier [30], which describes two-dimensional vegetation pattern formation as a function of the resource water. This model is commonly referred to as the Klausmeier model. It produces vegetation patterns that are banded if a gentle slope exists without any other environmental heterogeneities.

The aim of this study is to investigate whether variability or a change in different external conditions such as rainfall, evaporation or plant mortality has any influence on the system dynamics in a qualitative or in a quantitative manner. Here, the main focus is on the long-term dynamics of the model. To investigate the influence of variability in precipitation is particularly interesting, because constant rainfall is a strong assumption, as semi-arid areas are characterized by variable precipitation [34]. Periodicity can indeed have a strong influence on biomass dynamics [23]. Studies focussing on variability in precipitation already exist. However, the therein observed influence on the model behaviour is attributable either to model extensions of the Klausmeier model (e.g., [35]) or to a more detailed model structure (e.g., [36]). Therefore, it is interesting to investigate whether an influence on the classical Klausmeier model also exists. An investigation of the impact of changes in mean precipitation is particularly interesting because of its high influence on vegetation. However, it is still unclear how precipitation changes. Between 1955 and 2003, the total precipitation has been decreasing in western central Africa [37]. However, Leauthaud et al. [38] state that changes of −15% up to +8% in precipitation in semi-arid areas might be realistic depending on the scenario. Observations show that with high values of precipitation, a uniform plant distribution is reached, while at low values of precipitation, interactions between plants supporting self-organization are impossible because of the low biomass values [23]. Furthermore, the investigation of increasing mortality is justified as human pressure on the vegetation due to land use tends to increase, which augments the risk of desertification [39]. Evaporation might change as a result of its dependency on temperature, which is why the investigation of the three parameters, precipitation, evaporation, and mortality, is important. The detailed investigations of the impact of various model parameters can be further compared with observations and can help to validate the structure of the model. Furthermore, it can help in understanding and ranking the possible influence of climate change and human impacts on semi-arid ecosystems in different stages of development as a result of the extensive parameter studies, which might provide information for a better-suited resource management.

This paper is structured as follows: Section 2 describes the Klausmeier model and analyses it regarding the existence and stability of equilibria. The numerical results are described in Section 3 and are discussed and interpreted in Section 4. Section 5 closes with a summary of the main conclusions.

2. Model and Methods

2.1. Model Description

The classical Klausmeier model consists of two coupled nonlinear partial differential equations representing water and biomass dynamics. It is described by Equation (1) with $A(T) = A$:

$$\frac{\partial N}{\partial T} = JRWN^2 - MN + D\left(\frac{\partial^2}{\partial X^2} + \frac{\partial^2}{\partial Y^2}\right)N \qquad (1a)$$

$$\frac{\partial W}{\partial T} = A(T) - LW - RWN^2 + V\frac{\partial W}{\partial Y} \qquad (1b)$$

The plants can grow on a two-dimensional domain with $(X, Y) \in \mathbb{R}^2$. Plant biomass grows through water uptake with a yield J. Plant mortality is given by MN. Furthermore, plants can spread through diffusion $D\left(\frac{\partial^2}{\partial X^2} + \frac{\partial^2}{\partial Y^2}\right)N$. The precipitation parameter $A(T)$ increases the amount of water W homogeneously in each grid cell. In the original model, precipitation is assumed to be constant (see [30]), while in this study, it can be time-dependent, as this is a strong characteristic feature in semi-arid areas [34]. The loss of water is divided into evaporation LW with a rate L and water uptake by plant biomass RWN^2. Water advection $V\frac{\partial W}{\partial Y}$ only takes place in the downhill direction.

For numerical investigations, the model has been nondimensionalized in two different ways (see Appendix A). The nondimensionalized model used for the investigation of the influence of the precipitation and the mortality is given by

$$\frac{\partial n}{\partial t} = wn^2 - mn + \left(\frac{\partial^2}{\partial x^2} + \frac{\partial^2}{\partial y^2}\right)n \qquad (2a)$$

$$\frac{\partial w}{\partial t} = a(t) - w - wn^2 + v\frac{\partial w}{\partial y} \qquad (2b)$$

The nondimensionalized model for the investigation of the influence of the evaporation is given by

$$\frac{\partial n}{\partial t} = wn^2 - mn + \left(\frac{\partial^2}{\partial x^2} + \frac{\partial^2}{\partial y^2}\right)n \qquad (3a)$$

$$\frac{\partial w}{\partial t} = 1 - lw - wn^2 + v\frac{\partial w}{\partial y} \qquad (3b)$$

To obtain shorter computing times and to facilitate interpretations, the one-dimensional model has been used in the analyses throughout this paper:

$$\frac{\partial n}{\partial t} = wn^2 - mn + \frac{\partial^2 n}{\partial y^2} \tag{4a}$$

$$\frac{\partial w}{\partial t} = 1 - lw - wn^2 + v\frac{\partial w}{\partial y} \tag{4b}$$

This means that plant diffusion only takes place parallel to the advection. To use the one-dimensional model is justified because the long-term behavior is considered and after sufficiently long simulation times, there does not exist any difference in the x-direction. This is due to the balancing effect of diffusion in the absence of a Turing instability.

2.2. Nonspatial Equilibria

The conditions for the spatially homogeneous equilibria are given by

$$\frac{\partial w}{\partial t} = a - w - wn^2 = 0 \tag{5a}$$

$$\frac{\partial n}{\partial t} = wn^2 - mn = 0 \tag{5b}$$

The semi-trivial solution $n_1^* = 0$ and $w_1^* = a$ corresponds to a completely bare state in a biological sense. The non-trivial solutions are given by

$$w_{2,3}^* = \frac{a}{2} \pm \sqrt{\frac{a^2}{4} - m^2} \tag{6a}$$

$$n_{2,3}^* = \frac{2m}{a \pm \sqrt{a^2 - 4m^2}} \tag{6b}$$

Imaginary solutions do not make biological sense here wherefore the square roots have to be positive. Figure 1 shows a bifurcation diagram of the system for the control parameter a.

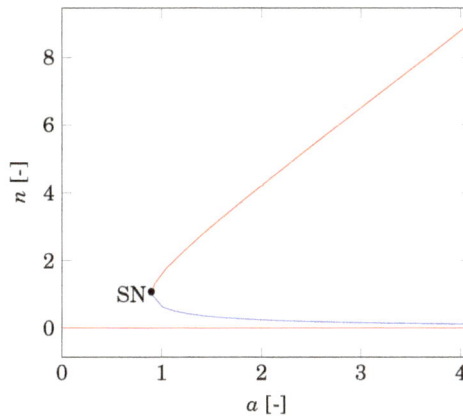

Figure 1. The bifurcation diagram for the control parameter a with $m = 0.45$ is shown. SN indicates the saddle-node bifurcation point. Red lines indicate stable equilibria, while the blue line indicates unstable equilibria.

The non-trivial equilibria only exist if $a \geq 2$ m, and $a = 2$ m corresponds to a saddle-node bifurcation point. In this case, only two equilibria exist. If $a < 2$ m, only the completely bare state exists. Graphically, the reason for the vanishing of the non-trivial equilibria at low precipitation values lies in the flatter curve shape of the water nullcline with decreasing a (see Figure A1).

To investigate whether the equilibria are stable, the Jacobian matrix J is considered. The Jacobian is given by

$$J = \begin{pmatrix} -n_i^{*2} - 1 & -2w_i^* n_i^* \\ n_i^{*2} & 2w_i^* n_i^* - m \end{pmatrix} \tag{7}$$

For the semi-trivial equilibrium, the Jacobian reads

$$J_t = \begin{pmatrix} -1 & 0 \\ 0 & -m \end{pmatrix} \tag{8}$$

The eigenvalues of J_t are thus $\lambda_{J_t,1} = -1$ and $\lambda_{J_t,2} = -m$. As $m = ML^{-1}$ and $M > 0 \wedge L > 0$, both eigenvalues are real and negative. This means that the desert state is a stable node.

The eigenvalues of the non-trivial equilibria can be calculated by

$$\det \begin{pmatrix} -n_i^{*2} - 1 - \lambda & -2w_i^* n_i^* \\ n_i^{*2} & 2w_i^* n_i^* - m - \lambda \end{pmatrix} = 0 \tag{9}$$

With $w^* = \frac{m}{n^*}$, Equation (9) simplifies to

$$\det \begin{pmatrix} -n_i^{*2} - 1 - \lambda & -2m \\ n_i^{*2} & m - \lambda \end{pmatrix} = 0 \tag{10}$$

This leads to the eigenvalues

$$\lambda_{\pm} = -\frac{1}{2}(1 + n_i^{*2} - m) \pm \sqrt{m(1 - n_i^{*2}) + \frac{1}{4}(1 + n_i^{*2} - m)^2} \tag{11}$$

Stability can further be investigated using the approach from Siteur et al. [40]. Equation (11) has the form

$$\lambda_{\pm} = \alpha \pm \sqrt{\beta + \alpha^2} \tag{12}$$

Table 1 shows the real parts of the eigenvalues of the Jacobian for the form given by Equation (12).

Table 1. Real parts of the eigenvalues of the Jacobian as a function of α and β. λ_+ refers to the addition of the square root, while λ_- refers to the subtraction of the square root in Equation (12).

	$\beta > 0$	$\beta < 0$
$\alpha > 0$	$\mathrm{Re}(\lambda_+) > 0$ $\mathrm{Re}(\lambda_-) < 0$	$\mathrm{Re}(\lambda_+) > 0$ $\mathrm{Re}(\lambda_-) > 0$
$\alpha < 0$	$\mathrm{Re}(\lambda_+) > 0$ $\mathrm{Re}(\lambda_-) < 0$	$\mathrm{Re}(\lambda_+) < 0$ $\mathrm{Re}(\lambda_-) < 0$

Positive β always yields a node or a saddle point. The lower non-trivial equilibrium is a saddle point, as positive values of β always refer to real parts of the eigenvalues with different signs. This holds because Equation (13) is true for $m > 0$ and $a > 2$ m, which is already a condition for the existence of the non-trivial equilibria:

$$\beta = m \left[1 - \left(\frac{2m}{a + \sqrt{a^2 - 4m^2}} \right)^2 \right] > 0 \tag{13}$$

If $\beta < 0$, the real parts of the eigenvalues will have the same signs. If they are negative, equilibria will be stable. Otherwise, equilibria will be unstable. Therefore, for stability of the larger equilibrium, Equations (14) and (15) must hold:

$$\beta = m \left[1 - \left(\frac{2m}{a - \sqrt{a^2 - 4m^2}} \right)^2 \right] < 0 \tag{14}$$

$$\alpha = -\frac{1}{2} \left[1 + \left(\frac{2m}{a - \sqrt{a^2 - 4m^2}} \right)^2 - m \right] < 0 \tag{15}$$

β is negative if $m > 0$ and $a > 2$ m; α is negative if $0 < m \leq 1$ and $a \geq 2$ m. Because different values of m do not make biological sense, and because $a < 2$ m does not yield the existence of the non-trivial equilibrium, the higher non-trivial equilibrium is stable.

2.3. Differential Flow-Induced Instability

If diffusion exists in a system and the diffusion constants differ, a Turing instability can occur under certain circumstances (for a detailed review, see, e.g., [41]). In this system, diffusion only exists in the differential equation describing biomass dynamics. Hence, a Turing instability is not possible.

Besides diffusion, a differential flow exists in the model. This means that advection of biomass and advection of water differ, which can result in instabilities. In the following, this is investigated on the basis of the method described in Rovinsky and Menzinger [42].

Assuming small perturbations:

$$n(y, t) = n^* + n'(y, t) \tag{16a}$$

$$w(y, t) = w^* + w'(y, t) \tag{16b}$$

yields the following expressions for the one-dimensional model:

$$\frac{\partial n'}{\partial t} = 2wnn' + wn'^2 + w'n^2 + 2nw'n' + w'n'^2 - mn' + \frac{\partial^2 n'}{\partial y^2} \tag{17}$$

$$\frac{\partial w'}{\partial t} = -w' - wn^2 - 2wnn' - wn'^2 - w'n^2 - 2w'nn' - w'n'^2 + wn^2 + v\frac{\partial w'}{\partial y} \tag{18}$$

We note that the consideration of the one-dimensional model might overestimate the stability [43]. As we consider small perturbations, we can ignore terms that are nonlinear in terms of n' and w'. This leads to Equation (19) for the evolution of the perturbations:

$$\begin{pmatrix} \frac{\partial n'}{\partial t} \\ \frac{\partial w'}{\partial t} \end{pmatrix} = A \begin{pmatrix} n' \\ w' \end{pmatrix} + \begin{pmatrix} D & 0 \\ 0 & 0 \end{pmatrix} \begin{pmatrix} \frac{\partial^2 n'}{\partial y^2} \\ \frac{\partial^2 w'}{\partial y^2} \end{pmatrix} + \begin{pmatrix} 0 \\ v \end{pmatrix} \begin{pmatrix} \frac{\partial n'}{\partial y} \\ \frac{\partial w'}{\partial y} \end{pmatrix} \tag{19}$$

where A is the Jacobian evaluated at the steady state with $a_{i,j}$ being the entries of the matrix. Non-uniform perturbations that are exponential in time given by Equation (20) are applied:

$$n'(y, t) = e^{\lambda t + iky} \tag{20a}$$

$$w'(y, t) = e^{\lambda t + iky} \tag{20b}$$

with λ being the perturbation growth rate, k being the wavenumber, and i being the imaginary unit. Equation (21) is the eigenvalue equation:

$$\det(M) = \begin{vmatrix} a_{11} - Dk^2 - \lambda & a_{12} \\ a_{21} & a_{22} + ikv - \lambda \end{vmatrix} = 0 \tag{21}$$

The eigenvalues $\lambda_\pm$ are given by

$$\lambda_\pm = \frac{1}{2}\left(\text{tr}(M) \pm \sqrt{\text{tr}(M)^2 - 4\det(M)}\right) \tag{22}$$

with

$$\text{tr}(M) = a_{11} + a_{22} - Dk^2 + ikv \tag{23}$$
$$\det(M) = \det(A) - a_{22}Dk^2 + ikv(a_{11} - Dk^2) \tag{24}$$

The saddle point is unstable against homogeneous perturbations and hence will not be further considered. To obtain instability in the non-trivial equilibrium, which is stable against homogeneous perturbations, the real part of one eigenvalue needs to be positive. This depends on the wavenumber k, on the slope v, on the mortality rate m and on the precipitation parameter a, which determine the stationary state. Figure 2 shows the dependence of the real part of the higher eigenvalue on these parameters.

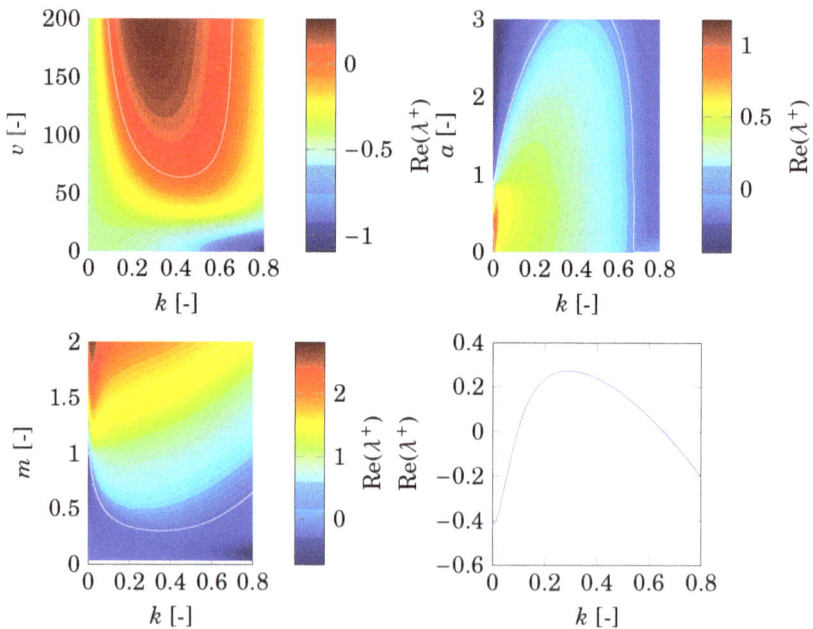

Figure 2. The dependence of the differential flow-induced instability on the wavenumber k, precipitation parameter a, slope v and mortality rate m is shown. Other parameters are given by the reference parameters for grass ($a = 2$, $m = 0.45$, and $v = 182.5$). The region of instability is restricted by the white line. The figure at the bottom right shows separately the dependence of the differential flow instability on the wavenumber for the reference values.

We note that the existence of the non-trivial equilibrium is restricted by $a \geq 2$ m. One can see that a differential flow-induced instability exists for the reference values. However, applying precipitation, slope and mortality values beyond a certain threshold leads to homogeneous vegetation.

The wavenumber corresponding to the maximum eigenvalue can be an indicator for the wavelength because it represents the fastest growing mode [44]. Hence, it can be seen that the wavelength tends to decrease with increasing precipitation and tends to increase with increasing mortality. The influence of the slope is weaker. However, a small increase in wavelength occurs for increasing slope.

We note that complex eigenvalues result in traveling waves for which the velocity can be calculated [41]. This corresponds to the uphill migration in this model. However, the analytical investigation lies beyond the scope of this study. The interested reader may find further information regarding the analytical treatment of wavetrains in the Klausmeier model in Sherratt [45].

2.4. Considered Parameters

The nondimensionalized parameters a, m and l are considered for precipitation, mortality, and evaporation, respectively. In fact, these parameters depend on various ecological parameters. Nonetheless, they are proportional to their dimensioned counterparts. Hence, it is justified to use these parameters for the numerical investigations. The default values shown in Appendix B were used for all other parameters.

2.5. Precipitation Model

The function modeling variability in precipitation in this study is given by

$$a(t) = \max \left\{ 6.2832 \cdot \sin \left(\frac{2\pi t}{12} + \frac{2\pi}{12} \left(12 - 4.5 \right) \right), 0 \right\} \tag{25}$$

This should represent both the dry character of semi-arid areas and the periodicity in precipitation observed in many semi-arid regions, that is, monsoonal and mediterranean rain [23]. The data on which the precipitation model is based is shown in Appendix C. The factor 6.2832 is used to obtain a mean value of $a = 2$ to ensure comparability between the results with variable and constant precipitation. The maximum function is used to consider only the positive part of the sine function. This represents the semi-aridity.

2.6. Stochasticity

Besides including variability in precipitation, the addition of environmental noise to reduce the dependence of the model on initial conditions is analyzed in this study. This can be done by the following:

$$\frac{\partial n_i^j}{\partial t} - \frac{\partial^2 n_i^j}{\partial y^2} = f \left(n_i^j, w_i^j \right) + g \left(n_i^j \right) \eta_i^j \tag{26}$$

Here, n_i^j represents the value of the plant biomass at time-step i and cell j. The term η_i^j represents white noise distributed as a normal distribution with $\mathcal{N}(0,1)$; $g(n_i^j)$ represents the density-dependent noise intensity. This shrinks with increasing biomass, which can be justified by the assumption that low biomass is more prone to environmental fluctuations, while larger population sizes have a stabilizing effect [46]. The function that describes the density-dependent noise intensity is given by

$$g \left(n_i^j \right) = \mu \cdot 0.8^{n_i^j} \tag{27}$$

The choice of the function $g(n_i^j)$ is arbitrary to a certain extent. It is a trade-off between a function that is not too high at high biomass values to destroy the shape of the biomass peaks, on the one hand, and a function that is still high enough at lower biomass values to support the formation of a new biomass peak, on the other hand.

2.7. Numerical Treatment

We consider a one-dimensional domain $[0, L]$ with $dx = 0.5$ and $L = 100/dx$ representing 50 m in dimensional terms. As initial values, every grid cell has an amount of water $w_{ini} = 5$ and a biomass that is given by a uniformly distributed random number in the interval [0.2, 0.4]. This is arbitrary to some degree. For the building up of biomass due to the positive biomass feedback loop, $n > 1$ is necessary in the first steps. This is achieved by the high initial water values but could also be achieved by random initial conditions with $n > 1$.

For the numerical solution of the model, the equations were discretized via the explicit Euler method with $h = 10^{-3}$ as the step size. The Forward-Time Central-Space (FTCS) explicit method has been used for diffusion and the Lax–Wendroff scheme has been used for advection. Periodic boundary conditions:

$$n(0) = n(L), \qquad\qquad n'(0) = n'(L) \qquad\qquad (28)$$
$$w(0) = w(L), \qquad\qquad w'(0) = w'(L) \qquad\qquad (29)$$

were used. This is biologically reasonable if one considers the grid as a particular section of the slope. For the numerical solution of the stochastic differential equation given by Eqaution (26), the Euler–Maruyama scheme has been applied (see [47]).

Unless otherwise stated, all data processing in this study was performed using Mat [48]. The data points for the uniform states in the bifurcation diagrams were calculated with the software XPPAUT [49]. Figures were drawn with Tikz [50].

3. Numerical Results

3.1. Precipitation

Figure 3 shows the impact of variations in variable and constant rainfall on existing vegetation patterns.

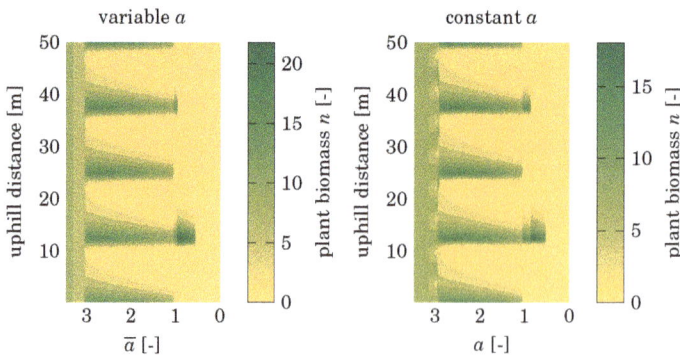

Figure 3. Results of the one-dimensional Klausmeier model after 1500 time-steps (to neglect transient dynamics) are shown for variable and constant precipitation. Resulting biomass peaks are shown for different values of mean precipitation. The maximum of plant biomass is located at 12.5 m uphill distance for each run to facilitate comparison and interpretation. As initial conditions, four peaks resulting from a model run with $\bar{a} = 2$ have been used.

In both cases, the system is in a state with four peaks (initial state) for a broad parameter range. In the case of variable precipitation, a critical precipitation value exists at which the four peaks vanish and homogeneous vegetation emerges. On the contrary, the system with constant precipitation shows a period doubling before homogeneous vegetation emerges. Considering lower parameter values of a, period halving in both cases can be observed. The critical values at which the period halving occurs are slightly different however. Furthermore, a critical value exists beyond which the bare desert state emerges.

Figure 4 shows the dependence of biomass patterns on the precipitation with constant random initial conditions.

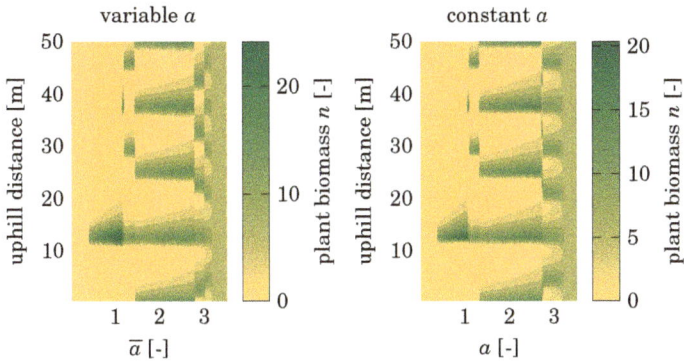

Figure 4. Results of the one-dimensional Klausmeier model after 1500 time-steps are shown for variable and constant precipitation. Resulting biomass peaks are shown for different values of mean precipitation. The maximum of plant biomass is located at 12.5 m uphill distance for each run. Random initial conditions have been used. Initial conditions have been the same for every value of $\bar{a}$.

In contrast to Figure 3, period doubling cannot be observed. Higher values of $\bar{a}$ lead to successive increases in the number of maxima in plant biomass. The parameter ranges in which certain numbers of peaks occur differ from those of Figure 3. However, four peaks are still the preferred state of the system. Variable and constant precipitation show the same general behavior. Nevertheless, the exact values at which certain numbers of maxima occur differ.

3.2. Stochasticity

To reduce the influence of the initial conditions on the wavenumber, stochasticity can be added to the model. We note that various numerical investigations have been made for different stochastic terms and for variable and constant precipitation. However, there have been no qualitative differences in the results of the model.

Figure 5 shows the long-term results of the one-dimensional model for different values of a and random initial conditions.

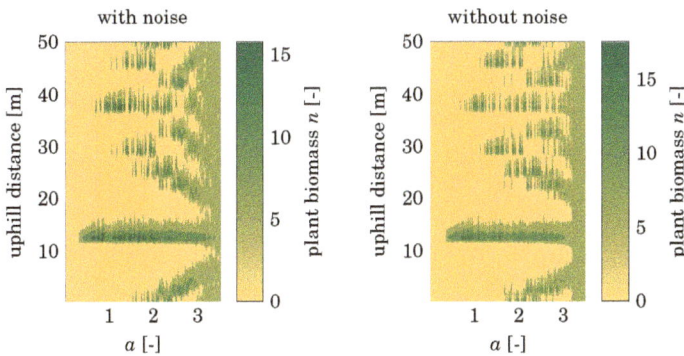

Figure 5. Results of the one-dimensional Klausmeier model after 1500 time-steps are shown for constant precipitation. Resulting biomass peaks are shown for different values of the parameter a. The maximum of plant biomass is located at 12.5 m uphill distance for each run. Varying random initial conditions have been used. For the generation of the left plot, the stochastic differential equation given by Equation (26) has been used for calculation of the plant biomass.

Clearly, there is no qualitative difference between the results of the model with noise and the results of the model without noise in terms of the wavenumber. Both cases show varying wavenumbers for different initial conditions. However, there are some parameter regions in which the model with noise leads to different wavenumbers in comparison with the model without noise.

3.3. Effect of Pattern Formation

As a result of the patterned vegetation, precipitation can be significantly lower while mortality and evaporation can be significantly higher before yielding the bare desert state. This is illustrated in the bifurcation diagrams in Figure 6.

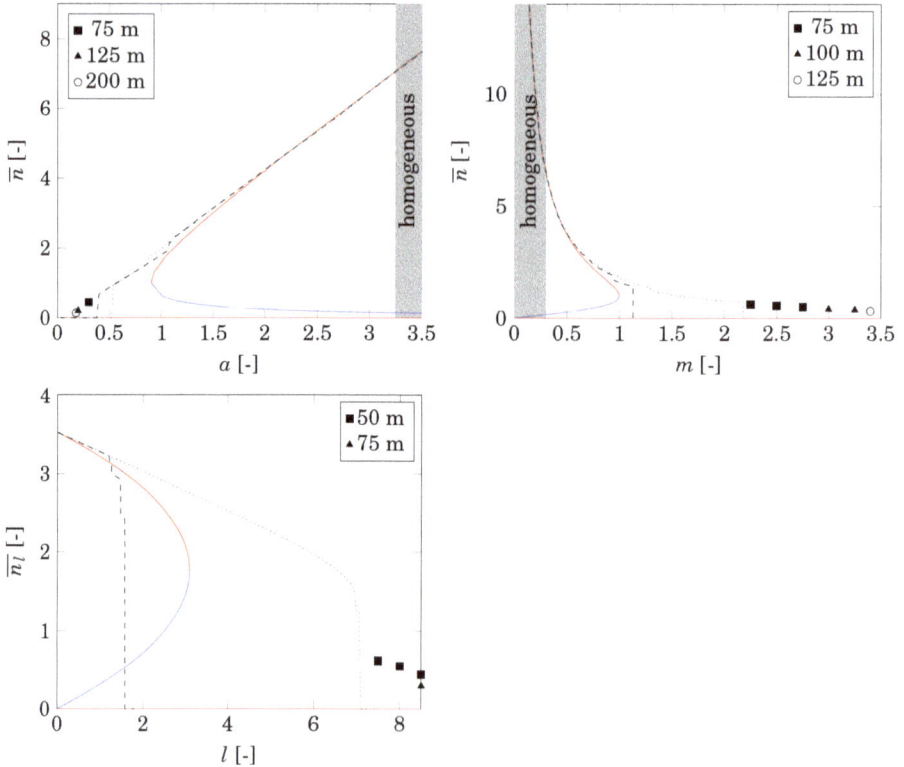

Figure 6. Bifurcation diagrams for different control parameters are shown. Red lines indicate stable equilibria, while the blue lines indicate unstable equilibria. The dashed line indicates the mean biomass with random initial conditions, while the dotted line indicates the mean biomass with four peaks as initial conditions. The square, the triangle, and the circle indicate the mean biomass for lower values of precipitation and higher values of mortality or evaporation with larger domain size and one biomass peak as initial conditions.

The biomass of the patterned state is in the same order of magnitude as the biomass of the uniform equilibrium in all cases. Furthermore, one can see that the wavenumber and hence the shift to the bare desert state depend on the size of the domain.

Below $a \approx 0.18$, the bare desert state is always reached, independent of the domain size and the initial conditions.

The transition to the bare desert state takes place at a significantly higher mortality in the case of existing patterns as initial conditions. The lowest value leading to pattern formation and the lowest value preserving an existing pattern differ by a factor of approximately 2 for a domain size of 50 m. Applying larger domain sizes enhances the parameter range yielding patterns for the case of one peak as initial conditions until approximately $m = 3.4$. Larger domain sizes have no effect in the case of random initial conditions yielding a factor of 3.1.

The lowest value of the evaporation rate leading to pattern formation differs from the lowest value preserving an existing pattern by a factor of 4.5. Applying only one peak as initial conditions yields a factor of 5.4. However, in the case of lower wavenumbers or larger domains, the mean biomass decreases sharply.

4. Discussion

4.1. Influence of Variable Precipitation

The formation of vegetation patterns can be achieved by variable and constant precipitation with some minor quantitative differences. Thus, parameter variability has only limited impacts on the model behavior in the case of precipitation.

The existing pattern as initial conditions persists over a broad parameter range. This supports the prevailing view of the necessity of improved environmental conditions for the reverse transition (e.g., [5]). The lowest precipitation value leading to pattern formation and the lowest precipitation value preserving existing patterns differ by a factor of 0.7 and are thus in accordance with the factor provided by Sherratt and Lord [51]. This can be explained as follows. In the case of random initial conditions, the strongest initial perturbation builds up to a biomass peak. In the case of existing patterns as initial conditions, the disturbances are very small. Thus, the strongest perturbation cannot build up quickly enough. If $n_{max} < 1$, the quadratic biomass term does not lead to fast biomass growth. Hence, the major perturbation has to be large enough to support a single biomass peak before the biomass drops under a critical value. This is not the case for random initial conditions.

Starting with random initial conditions, higher wavenumbers can occur in the case of constant precipitation. This can be explained as follows. Let a weak disturbance in plant biomass exist. In the case of variable precipitation, the year starts with a drought period. This can lead to biomass near to extinction before the rainy season begins. Conversely, in the case of constant precipitation, no drought period exists. Therefore, the biomass disturbance can lead to a biomass peak. This means that the initial biomass n must have a value that supports biomass growth with the given amount of water. Mathematically, this can be obtained by Equation (30) in the case of constant precipitation, keeping in mind that $n \geq 0$:

$$wn^2 - mn > 0 \longrightarrow wn > m \tag{30}$$

Furthermore, biomass needs to be sufficiently small so as not to build up enough in the rainy season. This means that Equation (31) must hold in the case of variable precipitation:

$$\left. wn - m \right|_{rainy} < \left. m - wn \right|_{drought} \tag{31}$$

As $w > m$ in the rainy season, this implies that the initial drought must lead to $n \ll 1$. The critical value depends on the length of the rainy season and precipitation in the rainy season.

We note that the influence of drought periods has also been investigated. However, the resulting figures are not of major significance; thus, they are not shown in this paper. Applying an already established pattern as initial conditions, the system remains in this state without changing the wavenumber unless the drought period leads to extinction.

4.2. Dependence on Initial Conditions

Applying varying random initial conditions for the same precipitation value, a strong dependence of the system behavior on the initial conditions became evident. An explanation for this dependence could be that two prerequisites for the formation of a biomass peak exist. First, a disturbance in biomass is needed so that biomass can build up. Second, a sufficiently large amount of water is required. Because in Figure 5 different initial conditions for each value of a have been applied, disturbances in biomass differ. If the amount of water in the system is large enough to support a specific wavenumber but no disturbance large enough and remote enough from the neighboring disturbance exist, biomass cannot build up. To overcome this shortcoming of the one-dimensional model, a stochastic term representing environmental noise has been added. This shows only a weak influence. On the one hand, the fact that multiplicative noise with zero mean has been used ensures that biomass cannot become negative and the noise does not add biomass to the system. On the other hand, this results in the fact that the effect is a maximum in regions with high biomass. Negative noise effects in those regions do not change the long-term behavior, because the diffusion of the neighbored cells is too high. Conversely, in the regions with low biomass, in which new biomass peaks should build up if precipitation is high enough to ensure a higher wavenumber, the effect of the noise is very small. Thus no peak can build up. Hence, the stochastic term is not suited to overcome the shortcoming of the dependence on the initial conditions.

4.3. Comparison with Nonspatial Equilibria

In comparison with the nonspatial equilibria, patterns are beneficial for the ecosystem. However, a lower bound of precipitation exists, beyond which the bare desert state is always reached. This is in accordance with Sherratt and Lord [51]. The reason for this lower bound is that, as a result of the evaporation, the upper bound on the amount of water on a grid cell is a, which is independent of the biomass.

In the case of random initial conditions and increasing mortality, larger domain sizes have no effect on the parameter range allowing for vegetation patterns (see Figure 6). This is opposite to the effect of precipitation and can be explained as follows. The precipitation parameter increases the amount of water in the system. For $n < 1$, the effect of a change in precipitation is low because of the multiplication with the square of the biomass. This is the case for the random initial conditions applied in this study. On the other hand, the mortality only depends on the biomass, which is not squared. For $n > 1$, the square of the biomass yields a higher effect of the change of precipitation in comparison to the mortality rate. This is the case for already formed patterns as initial conditions.

The application of higher evaporation rates is the only case in which a sharp decrease in plant biomass can be observed. The system remains in its initial state for a broad parameter range. This occurs because the influence of the evaporation rate depends on the amount of water, with $w_l \ll 1$ at the biomass peak. Furthermore, the change in the amount of water depends on the water uptake by plants, which in turn depends on the square of biomass. Changes in evaporation only have a small influence on the system, because $n_i^2 \gg l$. However, the peak cannot benefit from larger domain sizes because only water from imminent grid cells reaches the biomass peak. Water from remote cells is lost as a result of the evaporation.

The case of increasing evaporation and random initial conditions applied is the only case of all the analyses leading to extinction before the bifurcation point. This occurs because the evaporation rate at the bifurcation point is beyond 1. With random initial conditions applied, $n_l < 1$ holds for the first time-steps. Thus, biomass cannot build up effectively. Therefore, a large amount of water is needed to ensure the emergence of biomass peaks. If we apply $w_l = 5$ as initial values of water, the evaporation rate has a high impact. Hence, if $l \geq 1$, it makes the building up of the biomass unlikely. The exact value depends on the initial conditions, because these determine how fast a biomass peak can build up. If $n > 1$ and $w < 1$ would have been applied as initial conditions, the influence of the evaporation on the emergence of patterns would likely be significantly lower. We note that even at lower evaporation rates, the system would not reach the non-trivial equilibrium but would be attracted to the bare desert state. Hence, vegetation patterns are still beneficial in comparison with the uniform states.

In the case of decreasing precipitation or increasing mortality, a decrease in the wavenumber occurs via period halvings, which might explain regime shifts observed in reality. This will occur if the amount of water is not sufficient to ensure the present wavenumber. We note that small deviations are necessary. First, the amplitude of the four initial biomass peaks decreases as a result of the lack of water. Driven by the small deviation, one biomass peak can take up more water. This, in turn, leads to an even greater lack of water at the following peak, which consequently shrinks and gives an advantage to the next peak.

5. Conclusions

Variable precipitation has only minor impacts on the qualitative system behavior. It is doubtful whether the findings of the impact of variable precipitation make sense in a biological way. Because pattern formation in semi-arid areas is difficult to observe (e.g., caused by long time-scales), this cannot be supported by data. Moreover, we note that variability in precipitation has more complex influences on surface runoff (e.g., an increase due to saturated soils). These are not considered in the model.

A high influence of the mortality on the patterns also exists. Increasing the mortality rate (as is done, e.g., by wood cutting) can lead to severe changes in the vegetation pattern and in the total plant biomass in this model. The impact of the mortality rate on the total plant biomass is highly nonlinear. This underlines the importance of taking the mortality in analyses into account, as the biomass is crucial for populations depending on these resources.

To the authors' best knowledge, this study is the first taking the influence of the evaporation on the Klausmeier model or other models describing pattern formation into account. The reason might be that the prediction of changes in evaporation is very complex. Nevertheless, water as the limiting resource in semi-arid areas is closely linked to evaporation, justifying detailed investigations. However, as evaporation is proportional to the amount of water and $w_l \ll 1$ in the model, the effect of the evaporation rate is relatively low in the model.

In all investigations, a strong dependence on the initial conditions became evident. In particular, the preferred wavenumbers not only depended on initial conditions but also on the history (see also Sherratt and Lord [51]) in some cases. An attempt to reduce the dependence of the wavenumber on initial conditions has been presented. This did not succeed because of the density-dependence of the noise.

In all analyses, except in the case of increasing evaporation, patterns enhanced the parameter region for vegetation beyond the bifurcation point. This indicates the benefit of vegetation patterns for semi-arid ecosystems. It has been shown that the exact parameter regions can depend on the grid size.

Acknowledgments: The authors appreciated Matthew W. Adamson's critical reading of the manuscript. Furthermore, they are thankful for the comments of two anonymous referees.

Author Contributions: The results are part of Merlin C. Köhnke's Master's thesis under Horst Malchow's supervision.

Conflicts of Interest: The authors declare no conflict of interest.

Appendix A. Nondimensionalization

Appendix A.1. Nondimensionalization in Terms of Precipitation and Mortality

The classical Klausmeier model with physical parameters is given by Equation (A1) with $A(T) = A$:

$$\frac{\partial N}{\partial T} = JRWN^2 - MN + D\left(\frac{\partial^2}{\partial X^2} + \frac{\partial^2}{\partial Y^2}\right)N \tag{A1a}$$

$$\frac{\partial W}{\partial T} = A(T) - LW - RWN^2 + V\frac{\partial W}{\partial Y} \tag{A1b}$$

For nondimensionalization, constant values with the same physical units as the corresponding physical parameters are chosen. It follows that the ratio is dimensionless. Inserting the values

$$N = nN_0, \quad W = wW_0, \quad T = tT_0, \quad X = xX_0, \quad Y = yY_0$$

yields

$$\frac{\partial n}{\partial t} = JRwW_0n^2N_0T_0 - MnT_0 + D\left(\frac{\partial^2}{\partial x^2} + \frac{\partial^2}{\partial y^2}\right)n\frac{T_0}{X_0Y_0} \tag{A2a}$$

$$\frac{\partial w}{\partial t} = A(T)\frac{T_0}{W_0} - LwT_0 - RwT_0n^2N_0^2 + V\frac{T_0}{Y_0}\frac{\partial w}{\partial y} \tag{A2b}$$

In order to simplify the equations, the following are assumed:

$$LT_0 = 1 \rightarrow T_0 = L^{-1},$$

$$RT_0N_0^2 = 1 \rightarrow N_0 = L^{\frac{1}{2}}R^{-\frac{1}{2}}$$

$$JRW_0T_0N_0 = 1 \rightarrow W_0 = J^{-1}L^{\frac{1}{2}}R^{-\frac{1}{2}}$$

$$DT_0X_0^{-1}Y_0^{-1} = 1 \xrightarrow{X=Y} X_0 = Y_0 = L^{-\frac{1}{2}}D^{\frac{1}{2}}$$

$$a(t) = A(T)T_0W_0^{-1} \rightarrow a(t) = A(T)JL^{-\frac{3}{2}}R^{\frac{1}{2}}$$

As every constant value is defined, we obtain the following nondimensionalized parameters:

$$t = LT,$$

$$n = NL^{-\frac{1}{2}}R^{\frac{1}{2}}$$

$$w = WJL^{-\frac{1}{2}}R^{\frac{1}{2}}$$

$$x = XL^{\frac{1}{2}}D^{-\frac{1}{2}}$$

$$y = YL^{\frac{1}{2}}D^{-\frac{1}{2}}$$

$$v = VT_0Y_0^{-1} \rightarrow v = VL^{-\frac{1}{2}}D^{-\frac{1}{2}}$$

$$m = MT_0 \rightarrow m = ML^{-1}$$

Using these parameters yields the nondimensionalized form given by Equation (A3):

$$\frac{\partial n}{\partial t} = wn^2 - mn + \left(\frac{\partial^2}{\partial x^2} + \frac{\partial^2}{\partial y^2}\right)n \tag{A3a}$$

$$\frac{\partial w}{\partial t} = a(t) - w - wn^2 + v\frac{\partial w}{\partial y} \tag{A3b}$$

We note that this form only has the three parameters $a(t)$, v and m.
Figure A1 shows the nullclines of the system for different values of a.

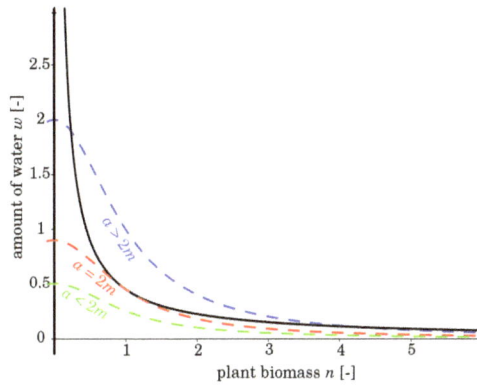

Figure A1. The nullclines of the nondimensionalized nonspatial system (see Equation (5)) are shown. Dashed lines represent the water nullclines, while the solid lines represent the biomass nullclines. Note that one solid line is represented by the ordinate.

Appendix A.2. Nondimensionalization in Terms of Evaporation

To investigate the influence of the evaporation on the system, another nondimensionalization has been taken. The first step is the same as in Appendix A.1, which again yields Equation (A2).
For simplification, the following are assumed:

$$RT_0 N_0^2 = 1$$
$$T_0 J R W_0 N_0 = 1$$
$$T_0 X_0^{-1} Y_0^{-1} D = 1$$
$$T_0 A W_0^{-1} = 1$$

Combining these assumptions, one obtains the following expressions for the constants:

$$T_0 = A^{-\frac{2}{3}} J^{-\frac{2}{3}} R^{-\frac{1}{3}}$$
$$N_0 = A^{\frac{1}{3}} J^{\frac{1}{3}} R^{-\frac{1}{3}}$$
$$W_0 = A^{\frac{1}{3}} J^{-\frac{2}{3}} R^{-\frac{1}{3}}$$
$$Y_0 = X_0 = D^{\frac{1}{2}} A^{-\frac{1}{3}} J^{-\frac{1}{3}} R^{-\frac{1}{6}}$$

This yields the following nondimensionalized parameters:

$$t = T A^{\frac{2}{3}} J^{\frac{2}{3}} R^{\frac{1}{3}}$$
$$n = N A^{-\frac{1}{3}} J^{-\frac{1}{3}} R^{\frac{1}{3}}$$
$$w = W A^{-\frac{1}{3}} J^{\frac{2}{3}} R^{\frac{1}{3}}$$
$$x = X D^{-\frac{1}{2}} A^{\frac{1}{3}} J^{\frac{1}{3}} R^{\frac{1}{6}}$$
$$y = Y D^{-\frac{1}{2}} A^{\frac{1}{3}} J^{\frac{1}{3}} R^{\frac{1}{6}}$$
$$l = L A^{-\frac{2}{3}} J^{-\frac{2}{3}} R^{-\frac{1}{3}}$$
$$v = V A^{-\frac{1}{3}} D^{-\frac{1}{2}} J^{-\frac{1}{3}} R^{-\frac{1}{6}}$$
$$m = M A^{-\frac{2}{3}} J^{-\frac{2}{3}} R^{-\frac{1}{3}}$$

Using these parameters, the nondimensionalized form to investigate the influence of the evaporation is given by Equation (A4):

$$\frac{\partial n}{\partial t} = wn^2 - mn + \left(\frac{\partial^2}{\partial x^2} + \frac{\partial^2}{\partial y^2} \right) n \qquad \text{(A4a)}$$

$$\frac{\partial w}{\partial t} = 1 - lw - wn^2 + v\frac{\partial w}{\partial y} \qquad \text{(A4b)}$$

Appendix B. Default Values

The default values for the physical parameters for grass given by Klausmeier [30] are shown in Table A1.

We note that $V \gg D$, because even on gentle slopes, the advection of water is assumed to be much faster than the plant dispersal.

Table A1. Physical parameters and their default values and units are shown for grass. The values are taken from Klausmeier [30].

Physical Parameter	Value	Unit
T		a
W		$kg\,m^{-2}$
N		$kg\,m^{-2}$
X		m
Y		m
A	533	$kg\,m^{-2}\,a^{-1}$
L	4	a^{-1}
R	100	$m^{-2}\,kg^{-2}\,a^{-1}$
J	0.003	—
V	365	$m\,a^{-1}$
M	1.8	a^{-1}
D	1	$m^2\,a^{-1}$

Appendix C. Precipitation Model

Figure A2 shows the average precipitation in Niamey, Niger for the period 1961–1990, with the corresponding function used to model variability in precipitation.

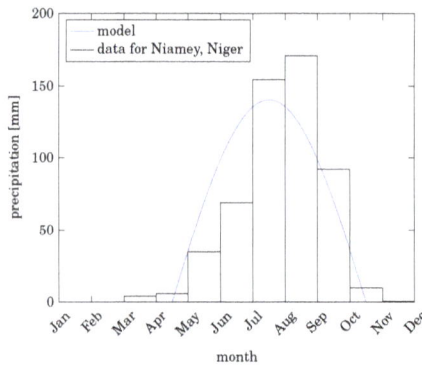

Figure A2. Average precipitation for Niamey, Niger (1961–1990) and the corresponding model are shown. Precipitation data retrieved from [52]. The geographic position of the station is N13°29′, E2°10′.

The region was chosen because it is a semi-arid region with good data coverage, where vegetation structures can be observed. The average annual precipitation in this region is 540.8 mm [52]. It can be seen that most rainfall occurs between May and September. In the remaining months, there are very low levels of rainfall. This is typical for semi-arid areas and justifies an investigation of the influence of such a variability on the results of the model.

References

1. Borgogno, F.; D'Odorico, P.; Laio, F.; Ridolfi, L. Mathematical models of vegetation pattern formation in ecohydrology. *Rev. Geophys.* **2009**, *47*, RG1005, doi:10.1029/2007RG000256.
2. Gillett, J. The plant formations of western British Somaliland and the Harar province of Abyssinia. *Misc. Bull. Inf.* **1941**, *1941*, 35–75.
3. Macfadyen, W.A. Vegetation patterns in the semi-desert plains of British Somaliland. *Geogr. J.* **1950**, *116*, 199–211.
4. Deblauwe, V.; Barbier, N.; Couteron, P.; Lejeune, O.; Bogaert, J. The global biogeography of semi-arid periodic vegetation patterns. *Glob. Ecol. Biogeogr.* **2008**, *17*, 715–723.
5. Meron, E. *Nonlinear Physics of Ecosystems*, 1st ed.; CRC Press: Boca Raton, FL, USA, 2015.
6. Gibbens, R.; McNeely, R.; Havstad, K.; Beck, R.; Nolen, B. Vegetation changes in the Jornada Basin from 1858 to 1998. *J. Arid Environ.* **2005**, *61*, 651–668.
7. Tongway, D.J.; Valentin, C.; Seghieri, J. (Eds.) *Banded Vegetation Patterning in Arid and Semiarid Environments*; Springer: New York, NY, USA, 2001.
8. Juergens, N. The biological underpinnings of Namib desert fairy circles. *Science* **2013**, *339*, 1618–1621.
9. Bromley, J.; Brouwer, J.; Barker, A.; Gaze, S.; Valentine, C. The role of surface water redistribution in an area of patterned vegetation in a semi-arid environment, south-west Niger. *J. Hydrol.* **1997**, *198*, 1–29.
10. White, L.P. Vegetation stripes on sheet wash surfaces. *J. Ecol.* **1971**, *59*, 615–622.
11. Barbier, N.; Couteron, P.; Lejoy, J.; Deblauwe, V.; Lejeune, O. Self-organized vegetation patterning as a fingerprint of climate and human impact on semi-arid ecosystems. *J. Ecol.* **2006**, *94*, 537–547.
12. Couteron, P.; Lejeune, O. Periodic spotted patterns in semi-arid vegetation explained by a propagation-inhibition model. *J. Ecol.* **2001**, *89*, 616–628.
13. Deblauwe, V.; Couteron, P.; Lejeune, O.; Bogaert, J.; Barbier, N. Environmental modulation of self-organized periodic vegetation patterns in Sudan. *Ecography* **2011**, *34*, 990–1001.
14. Scheffer, M.; Carpenter, S.; Foley, J.A.; Folke, C.; Walker, B. Catastrophic shifts in ecosystems. *Nature* **2001**, *413*, 591–596.
15. Scheffer, M.; Carpenter, S.R. Catastrophic regime shifts in ecosystems: Linking theory to observation. *Trends Ecol. Evol.* **2003**, *18*, 648–656.
16. Mander, L.; Dekker, S.C.; Li, M.; Mio, W.; Punyasena, S.W.; Lenton, T.M. A morphometric analysis of vegetation patterns in dryland ecosystems. *R. Soc. Open Sci.* **2017**, *4*, 160443.
17. Montana, C. The colonization of bare areas in two phase mosaics of an arid ecosystem. *J. Ecol.* **1992**, *80*, 315–327.
18. Maestre, F.T.; Reynolds, J.F.; Huber-Sannwald, E.; Herrick, J.; Smith, M.S. Understanding global desertification: Biophysical And socioeconomic dimensions of hydrology. In *Dryland Ecohydrology*; D'Odorico, P., Porporato, A., Eds.; Springer: Dordrecht, The Netherlands, 2006; pp. 315–332.
19. Charley, J.L.; West, N.E. Plant-induced soil chemical patterns in some shrub-dominated semi-desert ecosystems of Utah. *J. Ecol.* **1975**, *63*, 945–963.
20. Stewart, J.; Parsons, A.J.; Wainwright, J.; Okin, G.S.; Bestelmeyer, B.T.; Fredrickson, E.L.; Schlesinger, W.H. Modeling emergent patterns of dynamic desert ecosystems. *Ecol. Monogr.* **2014**, *84*, 373–410.
21. Dudney, J.; Hallett, L.M.; Larios, L.; Farrer, E.C.; Spotswood, E.N.; Stein, C.; Suding, K.N. Lagging behind: Have we overlooked previous-year rainfall effects in annual grasslands? *J. Ecol.* **2016**, *105*, 484–495.
22. Bergkamp, G.; Cammeraat, L.H.; Martinez-Fernandez, J. Water movement and vegetation patterns on shrubland and an abandoned field in two desertification-threatened areas in Spain. *Earth Surf. Proc. Land.* **1996**, *21*, 1073–1090.
23. Noy-Meir, I. Desert ecosystems: Environment and producers. *Annu. Rev. Ecol. Syst.* **1973**, *4*, 25–51.
24. Schlesinger, W.H.; Raikes, J.A.; Hartley, A.E.; Cross, A.F. On the spatial pattern of soil nutrients in desert ecosystems. *Ecology* **1995**, *77*, 364–374.
25. Lefever, R.; Lejeune, O. On the origin of tiger bush. *Bull. Math. Biol.* **1997**, *59*, 263–294.
26. von Hardenberg, J.; Meron, E.; Shachak, M.; Zarmi, Y. Diversity of vegetation patterns and desertification. *Phys. Rev. Lett.* **2001**, *87*, 198101.
27. Nicolis, G.; Prigogine, I. *Self-Organization in Nonequilibrium Systems: From Dissipative Structures to Order Through Fluctuations*; Wiley: New York, NY, USA, 1977.
28. Gierer, A.; Meinhardt, H. A theory of biological pattern formation. *Kybernetik* **1972**, *12*, 30–39.

29. Lefever, R.; Barbier, N.; Couteron, P.; Lejeune, O. Deeply gapped vegetation patterns: On crown/root allometry, criticality and desertification. *J. Theor. Biol.* **2009**, *261*, 194–209.

30. Klausmeier, C.A. Regular and Irregular Patterns in Semiarid Vegetation. *Science* **1999**, *284*, 1826–1828.

31. Rietkerk, M.; Boerlijst, M.C.; van Langevelde, F.; HilleRisLambers, R.; van de Koppel, J.; Kumar, L.; Prins, H.H.T.; de Roos, A.M. Self-organization of vegetation in arid ecosystems. *Am. Nat.* **2002**, *160*, 524–530.

32. Sheffer, E.; von Hardenberg, J.; Yizhaq, H.; Shachak, M.; Meron, E. Emerged or imposed: a theory on the role of physical templates and self-organisation for vegetation patchiness. *Ecol. Lett.* **2012**, *16*, 127–139.

33. Yizhaq, H.; Sela, S.; Svoray, T.; Assouline, S.; Bel, G. Effects of heterogeneous soil-water diffusivity on vegetation pattern formation. *Water Resour. Res.* **2014**, *50*, 5743–5758.

34. Chesson, P.; Gebauer, R.L.E.; Schwinning, S.; Huntly, N.; Wiegand, K.; Ernest, M.S.K.; Sher, A.; Novoplansky, A.; Weltzin, J.F. Resource pulses, species interactions, and diversity maintenance in arid and semi-arid environments. *Oecologia* **2004**, *141*, 236–253.

35. Ursino, N.; Contarini, S. Stability of banded vegetation patterns under seasonal rainfall and limited soil moisture storage capacity. *Adv. Water Resour.* **2006**, *29*, 1556–1564.

36. Guttal, V.; Jayaprakash, C. Self-organization and productivity in semi-arid ecosystems: Implications of seasonality in rainfall. *J. Theor. Biol.* **2007**, *248*, 490–500.

37. Aguilar, E.; Barry, A.A.; Brunet, M.; Ekang, L.; Fernandes, A.; Massoukina, M.; Mbah, J.; Mhanda, A.; do Nascimento, D.J.; Peterson, T.C.; et al. Changes in temperature and precipitation extremes in western central Africa, Guinea Conakry, and Zimbabwe, 1955–2006. *J. Geophys. Res.* **2009**, *114*, D02115.

38. Leauthaud, C.; Demarty, J.; Cappelaere, B.; Grippa, M.; Kergoat, L.; Velluet, C.; Guichard, F.; Mougin, E.; Chelbi, S.; Sultan, B. Revisiting historical climatic signals to better explore the future: Prospects of water cycle changes in Central Sahel. *Proc. Int. Assoc. Hydrol. Sci.* **2015**, *371*, 195–201.

39. Millennium Ecosystems Assessment. *Ecosystems and Human Well-Being: Desertification Synthesis*; World Resources Institute: Washington, DC, USA, 2005.

40. Siteur, K.; Siero, E.; Eppinga, M.B.; Rademacher, J.D.; Doelman, A.; Rietkerk, M. Beyond Turing: The response of patterned ecosystems to environmental change. *Ecol. Complex.* **2014**, *20*, 81–96.

41. Malchow, H.; Petrovskii, S.V.; Venturino, E. *Spatiotemporal Patterns in Ecology and Epidemiology: Theory, Models, and Simulation*; Mathematical and Computational Biology, Chapman and Hall/CRC: Boca Raton, FL, USA, 2008.

42. Rovinsky, A.B.; Menzinger, M. Chemical instability induced by a differential flow. *Phys. Rev. Lett.* **1992**, *69*, 1193–1196.

43. Siero, E.; Doelman, A.; Eppinga, M.B.; Rademacher, J.D.M.; Rietkerk, M.; Siteur, K. Striped pattern selection by advective reaction-diffusion systems: Resilience of banded vegetation on slopes. *Chaos* **2015**, *25*, 036411.

44. Sherratt, J.A. An analysis of vegetation stripe formation in semi-arid landscapes. *J. Math. Biol.* **2005**, *51*, 183–197.

45. Sherratt, J.A. Pattern solutions of the Klausmeier model for banded vegetation in semi-arid environments II: Patterns with the largest possible propagation speeds. *R. Soc. A* **2011**, *467*, 3272–3294.

46. Siekmann, I.; Malchow, H. Fighting enemies and noise: Competition of residents and invaders in a stochastically fluctuating environment. *Math. Model. Nat. Phenom.* **2016**, *11*, 120–140.

47. Maruyama, G. Continuous Markov processes and stochastic equations. *Rend. Circ. Mat. Palermo* **1955**, *4*, 48–90.

48. MATLAB: Version 8.1 (R2013a); The MathWorks Inc.: Natick, MA, USA, 2013.

49. Ermentrout, B. *Simulating, Analyzing, and Animating Dynamical Systems: A Guide to Xppaut for Researchers and Students*; Society for Industrial and Applied Mathematics: Philadelphia, PA, USA, 2002.

50. Tantau, T. The TikZ and PGF Packages. Manual for Version 3.0.1. Available online: http://sourceforge.net/projects/pgf/ (accessed on 6 July 2017).

51. Sherratt, J.A.; Lord, G.J. Nonlinear dynamics and pattern bifurcations in a model for vegetation stripes in semi-arid environments. *Theor. Popul. Biol.* **2007**, *71*, 1–11.

52. Deutscher Wetterdienst. Klimatafel von Niamey (Aéro)/Niger (In German). Available online: http://www.dwd.de/DWD/klima/beratung/ak/ak_610520_kt.pdf (accessed on 15 April 2017).

Σ *mathematics*

MDPI

Review

Ecological Diversity: Measuring the Unmeasurable

Aisling J. Daly *, Jan M. Baetens and Bernard De Baets

KERMIT, Department of Data Analysis and Mathematical Modelling, Ghent University, Coupure links 653,
B-9000 Ghent, Belgium; jan.baetens@ugent.be (J.M.B.); bernard.debaets@ugent.be (B.D.B.)
* Correspondence: aisling.daly@ugent.be

Received: 1 May 2018; Accepted: 5 July 2018; Published: 10 July 2018

Abstract: Diversity is a concept central to ecology, and its measurement is essential for any study of ecosystem health. But summarizing this complex and multidimensional concept in a single measure is problematic. Dozens of mathematical indices have been proposed for this purpose, but these can provide contradictory results leading to misleading or incorrect conclusions about a community's diversity. In this review, we summarize the key conceptual issues underlying the measurement of ecological diversity, survey the indices most commonly used in ecology, and discuss their relative suitability. We advocate for indices that: (i) satisfy key mathematical axioms; (ii) can be expressed as so-called effective numbers; (iii) can be extended to account for disparity between types; (iv) can be parameterized to obtain diversity profiles; and (v) for which an estimator (preferably unbiased) can be found so that the index is useful for practical applications.

Keywords: diversity; richness; evenness; effective numbers; estimators; mathematical ecology

1. Introduction

A central aim of ecology is to understand the processes that sustain biodiversity, which is critically important for the viability of ecosystems [1]. Biodiversity loss degrades ecosystem functionality, and in drastic cases can even lead to mass extinctions and total ecosystem collapse [2]. In recent years, this vulnerability has worsened to critical levels as a significant portion of the Earth's species are driven to extinction largely due to human actions [3]. Unprecedentedly high extinction rates have caused extensive and damaging changes in ecosystem structure and functionality and have brought particular urgency to the issue of biodiversity loss [4].

Biodiversity is a rich concept that admits a wide range of possible definitions. In its broadest sense, it has been defined as the variety of life forms at all organizational levels of an ecosystem, ranging from molecules over individuals to species [5]. Examples of different definitions of biodiversity thus include species (taxonomic), functional, genetic, phylogenetic, and chemical diversity, among very many others.

Any meaningful study of biodiversity, no matter which aspect is in question, must involve its quantitative measurement. This is a complex task both conceptually and practically. Biodiversity is quantified by constructing mathematical functions generally known as *diversity indices*. The use of such indices permits comparisons between different spatial regions, temporal periods, taxa (species), functional groups, or trophic levels. These measures are therefore essential tools for ecological monitoring and conservation, as well as for any efforts to study and address the biodiversity crisis [6].

However, there is no consensus on which indices are more suitable and informative than others. The available indices are so numerous and disparate in their ecological interpretation and mathematical behaviour that researchers must start by asking themselves the most basic question of all: what is even meant by 'diversity'?

This review aims to provide guidance to those seeking to use diversity indices in ecological settings. Due to the multitude of available indices, such reviews arise each year in an attempt to bring some order and insight to this overwhelming body of work. This review focuses on the mathematical foundations of diversity indices, which have consequences for their use and, perhaps more importantly, their misuse.

What is driving this proliferation of diversity indices? In recent years, the era of "Big Data" has spread to ecology, in particular microbial ecology. This has driven a reassessment of which diversity indices to use to summarize and analyze these data [7]. This has led to increased focus on the mathematical behaviour of different measures, since there has been an increasing recognition that ecologists (often non-experts in mathematics) are using measures of diversity without proper consideration of their behaviour and limitations. This has led to difficulties in interpreting studies, and especially in comparing them. As we shall see, the choice of index is highly dependent on the particular study context. But there are a number of important points to consider.

We have settled on five recommendations, each of which we motivate in detail in subsequent sections. The choice of a diversity index should ensure that:

(1)   it satisfies key mathematical axioms;
(2)   it can be converted to an effective number;
(3)   it can be extended to account for disparity between types;
(4)   it can be parameterized to obtain diversity profiles; and
(5)   an estimator (preferably unbiased) can be obtained to allow the index to be used in practical applications.

We first address in Section 2 the fundamental question of how to conceptualize diversity, and the difficulties in settling on a single definition. We then outline briefly the three components of diversity in Section 3. In Section 4, we survey the axioms that characterize diversity indices, and discuss which are necessary for indices to have meaningful mathematical properties and ecological interpretations. The diversity indices most used in ecological applications are described in Section 5, along with their mathematical properties—in particular, their conversion to effective numbers (which endows a common set of mathematical properties and ecological behaviours) and the calculation of diversity profiles. In Section 6, we discuss the partitioning of diversity, which allows diversity indices to be used across different landscape scales, and is a mathematical and conceptual problem with both theoretical and practical implications. We then address in Section 7 the different techniques that can be used to estimate diversity indices in practical settings. Finally, in Section 8 we summarize the conclusions and recommendations that may be drawn from this review of diversity indices and their use in ecology.

2. Defining Diversity

Dozens upon dozens of different diversity indices can be found in scientific literature. Such an abundance of indices and their often discrepant behaviour has led to such confusion that some authors have concluded that the very concept of diversity is meaningless. Even as far back as 1971, Hurlbert was moved to declare that *"the term 'species diversity' [...] now conveys no information other than 'something to do with community structure'; species diversity has become a nonconcept"* [8], a reproval often repeated in the intervening years. Although there is an important nuance in this criticism, as we shall shortly discuss, it is also true that since then the picture has only become busier, with the plethora of diversity indices already to be found in the literature being joined by new indices proposed each and every year. We may identify four broad reasons why this is so.

First and most importantly, 'biodiversity' is such a broad concept that it can and has been defined in many different ways, depending on researchers' specific needs and interests. These definitions can range from species or morphological diversity to functional or chemical diversity, and any number of others in between. One review—now already more than 20 years old—unearthed no fewer than 85 different definitions [5]. Thus biodiversity, as a general ecological concept, has been described (rather mildly) as "extremely confusing" [9]. It is also clear that different diversity indices can therefore measure patently different aspects of diversity. But even once researchers have decided which form of diversity they wish to measure, quantifying biodiversity is still problematic because there is no single index that can provide a suitable summary.

A second problem is the fact that the concept of diversity is often confounded with the indices that measure it. Jost illustrates this issue with the example of a sphere: its radius can be used as an index of its volume, but these two quantities are obviously not equivalent [10]. Analogously, the most commonly used diversity measure, the Shannon index, is actually a measure of entropy. Entropy refers to uncertainty in information: it is more difficult to predict the identity of an individual (in terms of its species, functional group, or whichever biodiversity aspect is in question) in a very diverse system, whereas this prediction is less uncertain in a system with only a few types. Hence the former system has a higher entropy than the latter. Entropy therefore shares important conceptual similarities with diversity, and hence entropy and diversity measures also share many (but not all) of the characterizing axioms that we will examine in Section 4. Unsurprisingly, entropic and other information-theoretic indices have a long history of use in ecology [11,12], and have therefore contributed significantly to the explosion in number of ecological diversity indices. But although entropy measures are reasonable and frequently used indices of diversity, this of course does not imply that entropy is equivalent to diversity. Similar arguments can be made regarding many other diversity indices.

Third, indices typically condense all relevant information about an ecosystem's diversity into a single real number, so that there are unfathomably many ways of calculating such an index from the complex and extensive data on the ecosystem in question. Different indices can weigh different components of these data more strongly than others, and can even entirely neglect some [13]. Needless to say, this freedom of formulation has permitted the emergence of dozens of different indices, which often provide wildly different estimates of ostensibly the same quantity.

A final confounding issue is the relatedness but non-equivalence of the concept of diversity across different scientific disciplines. The obvious similarities between diversity measurement in, for example, sociology or economics compared to ecology, have encouraged the adoption of indices by one field from another, thus swelling their number without proper consideration of their conceptual differences. As an example, numerous indices have been developed in economics to measure the diversity or inequality of income, corporate productivity, or racial representation [14–16]. But a key conceptual difference often underlies economic and ecological studies of diversity: the latter take into account the actual abundances of the different species present in the ecosystem, while the former are instead often concerned with an abstract allocation of commodities (species), and so take no account of actual abundances [17].

Before we delve further into ecological diversity and its mathematical representations, it is important to emphasize that none of the numerous diversity indices are wrong per se. On the contrary, each index has its own unique properties that are useful for specific applications. The key point of Hurlbert's criticism relates to diversity as a unified concept: since raw diversity indices exhibit such a wide variety of mathematical behaviours, they cannot all give reasonable results when directly inserted into any general equation or formula of diversity [8]. Thus it is critically important to consider the purpose for which an index will be used, when addressing how it is constructed mathematically and how it should be interpreted ecologically.

The purpose of an index's use will of course depend strongly on the context of the study. Examples might include: classifying shifts in diversity after a disturbance; ranking ecosystems in terms of their biodiversity (to determine which are more in need of protection); detecting the effects of external (typically anthropogenic) factors on diversity; understanding interactions between diversity at different trophic levels (i.e., how changes in one can affect another); or searching for mechanistic explanations of diversity changes [6]. For each of these goals and many others, some indices will be more suitable than others, and a poor choice may lead to misleading or even false conclusions.

In this review, we will focus on species diversity (hereafter referred to merely as 'diversity' for ease of reference). But while species diversity is by far the most commonly considered form of diversity in ecology, it should be noted that other forms of diversity may in some contexts be more important. In particular, intra-species diversity (for example in individuals' age or body size) has been recognized as often playing a more significant role in ecosystem functioning and dynamics than species diversity [18]. We focus this review on species diversity since it is more often the focus of study, and can be more daunting to newcomers to the field, due to the large and disparate amount of theory and practice that has already been established.

3. Components of Diversity

It is generally understood that diversity can be divided into three components: richness, evenness, and disparity [10]. Most indices (including almost all the classical and best established indices) do not account for disparity, and hence this component of diversity is often neglected in reviews of the topic. In this section, we provide definitions of these three components (illustrated in Figure 1), as well as a brief summary of their ecological significance.

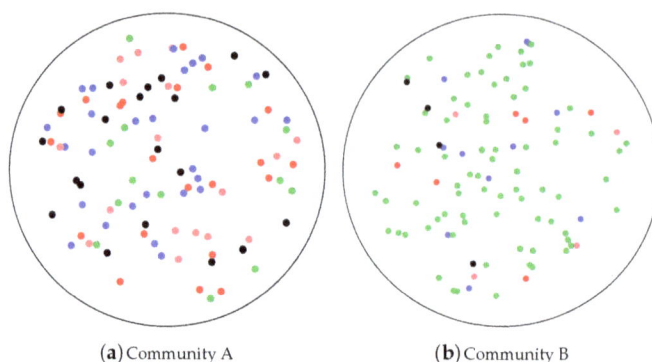

(**a**) Community A　　　　　　　(**b**) Community B

Figure 1. Representation of two communities of 100 individuals of different species, indicated by their colour. Richness is the absolute number of species in a community: in both communities it is equal to five. Evenness is the equitability of the species abundance distribution: in Community A all species are present in equal abundances and so it is perfectly even, while Community B is very uneven since it is dominated by the green species. The disparity is the level of similarity between species: for example, the red and the pink species are more similar in colour than the red and the black species.

3.1. Richness

The absolute number of species present in the population of interest is referred to as its *richness*. Two assumptions underlie the definition of richness [19]. First, that a classification of types exists and is known. If such a classification would not exist, any richness calculation would become difficult since it might not be clear to which class or taxon any particular individual belongs. This issue is especially relevant to microbial ecology, where the definition of species is particularly shaky due to distinct life history features like frequent genetic exchanges and the lack of sexual reproduction [20]. The second assumption is that each class is equally distinct, so that no two classes are more or less similar than any others.

In ecological terms, increased richness has been shown to both enhance community functionality (i.e., increase productivity) and to stabilize it in the face of disturbances [21,22]. Two mechanisms are believed to drive this effect [23]. The complementarity mechanism assumes that different species use different resources, even if only slightly. Then communities with higher richness will use more of the available resources, thus increasing their productivity. On the other hand, the selection mechanism assumes that different species contribute differently to the community's overall functionality, namely that some species are more productive than others. Then richer communities are more likely to contain more productive species, enhancing the community productivity.

From a mathematical and quantitative perspective, richness is the most straightforward component of diversity, being a simple enumeration of the different types present in a community. Of course, in practice ecologists can only count those species that they can find: rarely do they encounter communities that can be fully censused, for obvious practical reasons. Instead, the richness of a community must often be estimated from a sample or subset of the community, a topic that will be discussed in Section 7.

3.2. Evenness

Aside from the total number of species in a community, the distribution of their abundances is also an important component of diversity. If a species is represented by only a few individuals, it should be clear that it contributes less to the community's diversity than a species represented by several thousand individuals. The equitability of a community's species abundance distribution is referred to as its *evenness*. A community is perfectly even if every species is present in equal proportions, and uneven if one species dominates the abundance distribution.

The important role of species evenness in ecological diversity has received less attention than that of species richness, but is also inarguable [24]. Evenness has been shown to be a key factor in preserving the functional stability of ecosystems [25], as well as in enhancing productivity by increasing the representation of each species' functional traits [26]. In contrast, uneven communities are often more susceptible to invasion and less resilient to stresses and disturbances [27].

Unlike richness, there is no consensus on how to measure evenness. The literature relating to the formulation and choice of an evenness index is a self-contained and well-developed sub-field of the diversity index literature. This topic is more narrowly focused, since it only considers one component of diversity, and thus we restrict ourselves here to a short summary of the most important considerations relating to evenness, and mention several extensive reviews of this topic for the interested reader.

Many researchers have attempted to address the problem of choosing one evenness index from the available legion by listing desirable criteria for an index, and then assessing how well these are satisfied by candidate indices. In their influential review paper on this subject, Smith and Wilson [28] list 14 criteria for evenness indices, and subdivide these into four essential requirements and 10 desirable features. The latter have not been as widely accepted as the former, in part because some of the desirable features conflict with each other. The list has also been modified and/or expanded by others, see for example [29–32].

Thus the question of an index's desirable or appropriate mathematical behaviour is much less settled. Most reviews on the topic conclude by stating that there is no universal way to measure evenness, and thus researchers must choose the index most suited for their particular needs [28,30,33–35]. While this seriously limits the comparability of different studies, it also reflects the reality that studies are generally interested in different aspects of evenness, which consequently are optimally described by different indices.

3.3. Disparity

As we will shortly see, most classical diversity indices account for two components of diversity—richness and evenness—and thus implicitly assume that distinct species have nothing in common. That is to say, they do not account for any disparity between species. This conception of diversity is referred to as *species-neutral diversity* [19]. It implies, for instance, that a community of six dramatically different species is considered to be no more diverse than a community of six species of butterflies.

Hence there are relatively fewer diversity indices that reflect the more realistic and nuanced situation of varying dissimilarities between species. Such indices are termed *similarity-sensitive*. The reason for the dearth of similarity-sensitive indices may well be the fact that "theoretical ecologists have been hesitant to introduce new diversity indices when the profusion of similarity *in*sensitive indices is already perceived to form an impenetrable jungle" [36].

Although wading into this "jungle" may be a forbidding prospect for many users of diversity indices, it is worthy of the effort. Indices of neutral diversity depend too much on the notion of species, a concept based upon the division of organisms into different classes that is known to be problematic [37]. Classical indices (such as species richness) depend entirely on this division, and thus behave badly in the face of taxonomic reclassification [36].

To account for species disparity, similarity-sensitive indices incorporate in their formulation some measure of similarity. Many scientific fields make use of similarity measures, for example in biology for taxonomic classification, in economics for analyzing investment policy, in information theory for document filtering, or in genetics for sequence analysis [38]. A similarity measure is related in a complementary way to the mathematical concept of a distance metric, since *dis*similarity can often be represented as a distance. However, many ecological dissimilarity measures do not fulfill all the properties necessary to be considered a distance metric [19], and hence cannot make use of the many theorems and results developed by mathematicians relating to distance metrics.

Within ecology, various methods for measuring similarity between species have already been developed. Some are genetic [39], others are functional [40], taxonomic [41], morphological [42], or phylogenetic [43]. They generally associate with each focal species some data concerning the characteristics deemed to be important, such as a list of functional traits, a DNA sequence, a location on a phylogenetic tree, and so on. The similarity values are then computed in terms of some notion of difference between the associated data, depending on its particular characteristics.

4. Axioms Characterizing a Diversity Index

Due to the fluid definition of diversity, the mathematical formulations of diversity indices are also extremely varied. This is due to the high-dimensional nature of ecological data, which a diversity index must synthesize and summarize, typically by mapping it to a scalar. The number of ways in which this can be achieved is virtually unlimited.

While this might provide a pithy summary statistic of diversity for easy comprehension and comparison, it can also result in a significant loss of information. Thus the precise mathematical formulation of a diversity index will affect how a community's diversity can be connected to its structure, functionality and other significant features, and therefore have important implications for the index's use in drawing ecological conclusions.

For these reasons, it is essential to understand the mathematics underlying a particular index. In an effort to bring rigour to this endeavour, researchers turned to the question of whether an axiomatic basis could be identified for any of the numerous proposed diversity indices [9]. Such a set of axioms would identify the most important properties of diversity indices, as well as allow for their differentiation based on which axioms they satisfy or fail to satisfy. The axiomatic foundations of entropy measures were addressed in this way quite some time ago [44], while the axioms for non-entropic families of indices were outlined relatively later [45].

Let us now consider a community of S species, where p_i represents the proportional abundance of species i, so that $0 \leq p_i \leq 1$ for $i = 1, ..., S$ and $\sum_{i=1}^{S} p_i = 1$. If the abundances are measured by counting individuals, then the relative abundance p_i represents the probability of randomly selecting an individual of species i from amongst all individuals in the community. If the abundances are measured in terms of biomass, then the relative abundance p_i represents the relative share of the community's biomass that is stored in individuals of species i.

The richness of the community is simply S, while its evenness is calculated from the vector of relative abundances $\mathbf{p} = (p_1, ..., p_S)$. Calculating the disparity additionally requires a similarity measure, which we discuss later in Section 5.

The most commonly assumed axioms that a diversity index $H(\mathbf{p})$ should satisfy are also common to entropy measures, and are as follows [46].

(1) **Symmetry.** $H(p_1, ..., p_S)$ must be a symmetric function.

This ensures that species are interchangeable and their order in the abundance vector has no effect on diversity.

(2) **Continuity.** $H(p_1, ..., p_S)$ must be a continuous function.

(3) **Evenness.** The diversity measure is maximal for a fixed number of species S when all species abundances are equal, i.e., $H(p_1, ..., p_S) \leq H(1/S, ..., 1/S)$, for any $(p_1, ..., p_S)$.

Additional axioms that are also satisfied by most indices of diversity were adopted later, and require additional definitions [19]. First, we define a *transfer of abundance* following the terminology of Patil and Taillie [47]. This involves taking two species i and j with relative abundances $p_i > p_j > 0$, and modifying these to increase p_j by ϵ and decrease p_i by ϵ, where $p_i - p_j > \epsilon > 0$. Hence we have made the more common species a bit less common, and the rarer species a bit less rare.

More precisely, if we have two communities C and C' with respective relative abundance vectors **p** and **q** with $m := \dim(\mathbf{p}) = \dim(\mathbf{q})$, where $\dim(\cdot)$ denotes the dimension of a vector, we say that C leads to C' by a transfer of abundance if $\exists i, j \in \{1, ..., m\}$ such that

$$q_k = \begin{cases} p_k & \text{if } k \notin \{i, j\} \\ p_i - \epsilon & \text{if } k = i \\ p_j + \epsilon & \text{if } k = j \end{cases}, \tag{1}$$

where $p_i - p_j > \epsilon > 0$.

Then we may state the following axiom.

(4) **Principle of transfers.** A transfer of abundance must increase diversity.

This axiom is clearly a stronger version of Axiom 3: if every transfer of abundance must increase diversity, then the index must be maximized when no more transfers between common and rare species are possible. This occurs when all species have the same abundance, i.e., for $(p_1, ..., p_S) = (1/S, ..., 1/S)$.

Indices that satisfy Axiom 4 must have a property known as *Schur-concavity*, which earned the right to a specific name due to its importance in econometrics for measuring income inequality [48]. The property is defined as follows [49]. For indices of the form $H(\mathbf{p}) = \sum_i V(p_i)$, it holds that H is Schur-concave if

$$V(p_j + \epsilon) - V(p_j) \geq V(p_i) - V(p_i - \epsilon), \tag{2}$$

where $p_i > p_i - \epsilon \geq p_j + \epsilon > p_j \geq 0$. If V is differentiable, then this condition can be replaced by

$$V'(p_i) \leq V'(p_j), \tag{3}$$

where $p_i > p_j \geq 0$ and $p_i + p_j \leq 1$. The proof that Schur-concave indices satisfy Axiom 4 is considered standard and is thus generally omitted in reviews of the topic (for details, see e.g., Solomon [50]).

For ecological purposes, Axiom 4 is important because it permits a partial ordering of communities based on their diversities (discussed later in Section 5.5).

For the next axiom, we first rigorously define the term *introducing a species*.

For two communities C and C' with respective relative abundance vectors **p** and **q** with $\dim(\mathbf{p}) = m$ and $\dim(\mathbf{q}) = m + 1$, we say that C leads to C' by introducing a new species if $\exists i \in \{1, ...m\}$ such that

$$q_k = \begin{cases} p_k & \text{if } k \neq i \\ p_i - \epsilon & \text{if } k = i \\ \epsilon & \text{if } k = m + 1 \end{cases}, \tag{4}$$

where $0 < \epsilon < p_j$. Note that this definition differs only slightly from that of transferring abundance given in Equation (1). In Equation (4), $p_j = 0$ which implies in this notation that we have transferred abundance from one species to another that previously had zero abundance.

Now we may state the following axiom:

(5)   **Monotonicity in number of species.** The introduction of a new species must increase diversity.

The monotonicity axiom holds for indices of the form $H(\mathbf{p}) = \sum_i p_i I(p_i)$, where $I(p_i)$ is a decreasing function of p_i over $[0,1]$ (this proof is straightforward).

A final axiom was proposed still later [51] as a necessary condition for any ecological diversity index.

(6)   **Replication principle.** The diversity of a pooled sample of n maximally distinct (i.e., no shared species) and equally diverse sub-communities is n times the diversity of a single sub-community. That is, if for $k = 1, ..., n$ the communities $\mathbf{p}^k = (p_1^k, ..., p_S^k)$ do not have any common species and $H(\mathbf{p}^k) = \delta$ for every k, then $H(p_1^1, ..., p_S^1, p_1^2, ..., p_S^2, ..., p_1^n, ..., p_S^n) = n\delta$.

This axiom is important because, as we shall see in later sections, ecologists often measure diversity by considering the ratio of within-group diversity ('alpha' diversity) to the total pooled diversity ('gamma' diversity). But using this ratio, which reflects compositional similarity, to make inferences about diversity is not always justified: some indices, when used in this way, do not accurately reflect changes in diversity. This issue will be discussed in detail in Section 6, but the key point is that for these problematic indices, the ratio of within-group diversity to total pooled diversity will always approach unity when within-group diversity is high, often leading to nonsensical results. Such misbehaviour is prevented by the replication principle, which ensures that this ratio does indeed reflect the compositional similarity of the different communities. More specifically, this principle ensures that when a region is partitioned into distinct sub-communities of equal diversity, then the region's total diversity is equal to the diversity of any subset of sub-communities, plus the diversity of the sub-communities not contained in the subset. Though this axiom would seem intuitive, we shall see that it is not satisfied by all indices. Indices that do not satisfy Axiom 6 can hence provide misleading results.

The axioms in this section also allow us to distinguish between indices of diversity and entropy, since the latter have a more restrictive definition [19]. Entropy measures are concave functions of the relative abundances (typically referred to as probabilities in that setting), whereas this is not necessarily the case for diversity indices [10]. Additionally, while entropy measures are linear functions of these abundances/probabilities, diversity indices are often non-linear, and so the average diversity of a set of communities is generally not the average of their diversities.

In general, the axioms listed in this section are agreed to be desirable for ecological diversity indices, since they ensure consistent and meaningful behaviour. However, if an index fails one of these axioms, this should be considered as a warning sign rather than a fatal flaw, since the index may still be useful in contexts where that axiom is less important. For example, if an index fails Axiom 4 it should not be used to rank communities by their diversities, which is frequently but not always the goal of ecological studies.

In any case, enumerating the axioms as we have done here, and ascertaining which of them are satisfied by a particular index (as we will address in the subsequent section) at least ensures that the researcher is aware of the underlying mathematical behaviour of their chosen index, and can thus take into account the effects this may have on subsequent ecological interpretations—forewarned is forearmed.

5. Diversity Indices

In this section we provide an overview of the diversity indices that are most widely used in ecology. We classify these in several groups: classical indices, effective number indices, similarity-sensitive indices, and finally parametric families of indices. This is not a strict classification since overlap between these groups is possible. We begin with the oldest, simplest, and therefore most commonly used indices: the classical indices of ecological diversity.

5.1. Classical Indices

5.1.1. Definitions

The most widely used diversity indices can account for richness and evenness, but do not account for species disparity. Instead they take as input the vector of relative abundances $\mathbf{p} = (p_1, ..., p_S)$, where S is the total number of species in the community. Thus these indices measure species-neutral diversity (see Section 3.3) and are known as classical diversity indices.

The simplest index, which is still used surprisingly often by ecologists as a measure of diversity despite its obvious shortcomings for this purpose [52], is simply species richness itself:

$$H_{SR}(\mathbf{p}) = S. \tag{5}$$

The reason underlying species richness' poor performance as an estimate of diversity should be clear: it takes no account of the species abundance distribution (i.e., evenness). In ignoring completely one key component of diversity, species richness gives exceptionally rare species equal weight as exceptionally common species, an entirely unintuitive estimate of diversity. Hence of the fundamental axioms in Section 4, species richness satisfies all but Axiom 4 since it ignores the relative abundances of species.

At the opposite extreme, the *Berger–Parker diversity index* entirely ignores rare species in the community [53]. It is defined as the reciprocal of the relative abundance of the most common species:

$$H_{BP}(\mathbf{p}) = \frac{1}{\max_i p_i}, \tag{6}$$

and thus estimates the relative dominance of this species as a proxy for the entire community's diversity. This index does not satisfy Axioms 4, 5, and 6 since it ignores all species abundances other than that of the most common species.

A much more balanced estimate of diversity is provided by the *Shannon diversity index*, also known as the Shannon–Wiener index, the Shannon–Weaver index and the Shannon entropy. It measures the uncertainty in the outcome of a sampling process [54], and is given by:

$$H_{Sh}(\mathbf{p}) = -\sum_{i=1}^{S} p_i \ln(p_i). \tag{7}$$

In Shannon's original information-theoretic formulation, the logarithm was given in base 2, so that the Shannon entropy represents the number of yes/no questions necessary to determine an object's classification. Since then, it has become more common to use either the natural logarithm (particularly in ecology) or the base 10 logarithm [55]. This index satisfies all the axioms in Section 4 aside from the replication principle (Axiom 6).

The Shannon index is also the basis of Pielou's evenness index, which is given by $J = H_{Sh}/H_{Sh}^*$, where H_{Sh}^* is the maximum value of H_{Sh} (a function of S). However, this index is in fact a very poor estimate of evenness, since it depends strongly on species richness [8]. Although this weakness is widely known, Pielou's evenness is still the most widely used evenness index in the ecological literature [55].

The *Simpson diversity index* represents the probability that two individuals taken at random from the community of interest (with replacement) represent the same species, and thus takes values in the unit interval [56]. Like the Shannon index, it satisfies all axioms in Section 4 aside from the replication principle (Axiom 6). As originally proposed by Simpson [57], it is given by:

$$H_{Si}(\mathbf{p}) = \sum_{i=1}^{S} p_i^2. \tag{8}$$

However, this index is not a very intuitive measure of diversity since higher values indicate lower diversity (note that it still satisfies Axiom 3 since diversity is high for *low* values of the index, which occurs when the species proportions are equal). For this reason, two other formulations of this index are more often used. Most common is the *Gini-Simpson diversity index*, also called the probability of interspecific encounter (PIE), which represents the probability that the two individuals represent different species, and is thus the complement of Simpson's original formulation [58]:

$$H_{GS}(\mathbf{p}) = 1 - H_{Si}(\mathbf{p}). \tag{9}$$

In contrast, the *Simpson dominance index* is the reciprocal of Simpson's original formulation. The term 'dominance' has been attached to this formulation since it gives more weight to common species than to rare species. It is given by:

$$H_{SD}(\mathbf{p}) = \frac{1}{H_{Si}(\mathbf{p})}. \tag{10}$$

The name 'Simpson index' has been used interchangeably for all three of the formulations shown here. Hence, when encountering this index, researchers should take care to note which form of the measure has been employed.

The Simpson index is also occasionally used as a measure of evenness, but this approach is not appropriate since the index also varies with richness. To be safely used as an index of evenness, the richness effect should first be eliminated by dividing the index by its maximum value, which depends on S.

5.1.2. Issues with Classical Indices

The most widely used measures of diversity in ecology are species richness, the Shannon index and the Simpson index [59]. Together, they provide a perfect example of the problems discussed in Sections 2 and 3: each index may be used to compare different communities to each other (although we shall discuss shortly why even this can be problematic), but diversity values of the different indices cannot be directly compared to each other. This is because each index has a different fundamental meaning in terms of the diversity it measures.

Richness is simply the number of species in a community, the Simpson index represents the probability that two randomly selected individuals belong to the same species, and the Shannon index is a measure of the entropy or disorder of the community. These three indices measure conceptually different features of a community, and hence have different units: richness is in units of species, the Simpson index is a probability, and the Shannon index is an entropy measure, and therefore has units of bits of information.

Thus, diversity measurements calculated using different classical indices are not immediately comparable, which is a serious drawback for any scientific study. But even measurements calculated using the *same* index can present important issues of comparison.

To illustrate this problem, let us consider the simplest possible case: a community composed of S equally-common species. In virtually any ecological context, it seems reasonable to say that a community C_1 with ten equally-common species is twice as diverse as a community C_2 with five equally-common species. But calculating for example the Shannon entropy, we find $H_{Sh}(\mathbf{p}_1) = 2.30$ for the first community and $H_{Sh}(\mathbf{p}_2) = 1.61$ for the second. How should we understand the difference in diversity between these two communities? The diversity of the first community is not twice that of the second, although our intuition tells us otherwise. Furthermore, it is also unclear what these values mean in absolute terms: should we consider a diversity of 2.3 to be high or low? Without intuitive units, such judgements are not immediately clear.

Diversity analysis using classical indices also suffers from their occasionally extreme non-linearity. As an example, if a perfectly even community of one million species is confronted with some disaster that wipes out all but 100 species, the Gini-Simpson index of this community will drop from 0.999999 to 0.99. So despite the fact that more than 99% of the species of the pre-catastrophe community have disappeared, the Gini-Simpson diversity index only drops by 1%. Cursory study of the catastrophe using this index would probably conclude that the community's diversity was not greatly affected, while it is clear that the opposite is true. The Shannon index demonstrates the same non-linearity problem, but to a lesser degree.

Unfortunately, in practice many ecologists have not been too concerned that classical diversity indices may give results that are misleading or difficult to interpret [59]. In their view, the actual values of the indices are unimportant, as long as they can be used to calculate the statistical significance of the change in diversity following a disturbance [46]. But in many cases this is not a reasonable basis for study conclusions, since the statistical significance of a change in the diversity index often has little to do with the actual magnitude or ecological significance of the change [52]. Furthermore, one index may indicate a statistically significant change in diversity while another index does not.

Other researchers were not content with this state of affairs, and proposed a solution: the use of so-called effective numbers.

5.2. Effective Numbers

Converting classical indices to effective numbers of species endows them with a set of common mathematical properties and ecological behaviours. After conversion, diversity is always measured in units of number of species, allowing for easy comparison and interpretation. This also means that the serious misinterpretations spawned by the non-linearity of most diversity indices can be avoided. By unifying various indices in this way, the use of effective numbers enhances our understanding and interpretation of diversity measurement.

Converting the diversity of a community to its effective number equivalent reduces to finding an equivalent community (having the same value of the index as the initial community in question) that is perfectly even [10]. For example, if a community has a diversity of 18.2 effective species, this means that it is slightly more diverse than a community of 18 totally dissimilar and equally abundant species: there are "effectively" 18.2 species.

This concept is hence related to the better known "effective population size", since both involve a measure that summarizes the nature of a community through a comparison to an abstract 'ideal' community. Effective diversity makes a comparison to a perfectly even community, while effective population size makes a comparison to a community with certain key properties such as constant size, an even sex ratio, and consisting only of breeding adults [60].

Conversion to effective numbers of species is straightforward: after calculating the diversity index for D equally-common species (each species therefore having a relative abundance of $1/D$), this is set equal to the community's 'raw' diversity value, and the equation is solved for D. MacArthur named D as the community's *effective number of species* [11].

Other fields have recognized the importance of effective number indices many years ago, though the concept goes by different names depending on the discipline. In physics it is known as the number of states associated with a given entropy, and in economics it is called the numbers equivalent of a diversity measure [47].

As an example of the conversion procedure, consider a community whose relative species abundance distribution is given by $\mathbf{p} = (0.41, 0.21, 0.08, 0.25, 0.04, 0.01)$. The Simpson diversity of this community is $H_{Si}(\mathbf{p}) = 0.2828$. To convert this diversity to its effective number equivalent, we need to find a community of D equally abundant species that also has a Simpson diversity of 0.2828. We therefore have that $p_i = \frac{1}{D}$ for $i = 1,...,6$ and that $H_{Si}(\mathbf{p}) = 0.2828 = \sum_{i=1}^{6} p_i^2 = \sum_{i=1}^{6} \frac{1}{D^2} = \frac{6}{D^2}$. It only remains to solve for D. We obtain $D = 4.61$, implying that our six-species community is effectively as diverse as a community of 4.61 equally abundant species.

Thus any diversity index can easily be converted into an effective number index. The use of effective number indices to assess changes in community diversity has become more established in ecological literature in recent years [61–65], as researchers recognize the advantages they bring compared to classical indices, most particularly the well-founded comparisons that they permit between different studies and even different indices.

5.3. Similarity-Sensitive Indices

Conversion to effective numbers allows researchers to compare indices whose mathematical formulations imply that they would otherwise account differently for richness and evenness. To address the third component of diversity, disparity (see Section 3.3), a further extension of classical indices is needed: the inclusion of a similarity measure.

Similarity-sensitive indices incorporate a similarity matrix encoding pairwise species similarities, which becomes an additional input of the index along with the species abundance vector. For a community of S species, an $S \times S$ matrix $Z = (Z_{ij})$ is constructed, where Z_{ij} is a measure of the similarity between species i and j. We assume that $0 \leq Z_{ij} \leq 1$, with 0 indicating total dissimilarity and 1 indicating identical species; therefore $Z_{ii} = 1$. Although we might assume that similarity matrices are always symmetric (i.e., $Z_{ij} = Z_{ji}$), this is not necessarily the case. The definition of a similarity matrix does not require symmetry, and indeed useful non-symmetric similarity matrices do exist. These are useful in cases where the direction of comparison is important.

Similarity matrices are often constructed based on distance metrics, which encode the distances between every pair of species in a multi-dimensional trait space. This is a well-established approach since it ensures that the distance matrix is Euclidean and the similarity matrix is positive definite [66]. This allows for the use of various techniques from linear algebra such as matrix decomposition or transformation.

While it is generally assumed that a similarity measure takes values in the unit interval, other formulations are possible. Genetic measures of similarity, frequently used in microbial ecology and microbiology, are often expressed as percentages so that their values lie in the interval $[0, 100]$ [67]. Other typical measures of inter-species distance d_{ij} range instead between zero and infinity, but these can be scaled to the unit interval through various transformations. One of the simplest uses the formula $Z_{ij} = e^{-ud_{ij}}$, where u is a constant. Although in comparison to other transformations this brings an additional degree of freedom to the diversity index (through the parameter u), it has nevertheless been recommended as a good choice of transformation [36], since varying u allows the user to control the distribution of similarity values. When $u = 0$, the similarity is equal to one irrespective of the value of d_{ij}. As $u \to \infty$, the similarity matrix tends to the identity matrix and thus the index tends to species-neutral diversity.

A seminal example of a similarity-sensitive diversity index is *Rao's quadratic diversity index*, which is often called Rao's quadratic entropy although it is in fact not an entropy measure. It is defined as the expected dissimilarity between two individuals selected at random from the community (with replacement) [68], and is given by:

$$H_R(\mathbf{p}) = \sum_{i,j=1}^{S} d_{ij} p_i p_j,$$ (11)

where d_{ij} is the dissimilarity between species i and j. Here, d_{ij} is not necessarily obtained from a distance metric. Note that H_R reduces to the Gini-Simpson diversity index in the case where $d_{ij} = 1$ for all $i \neq j$, and $d_{ii} = 0$ for all i.

However, H_R (along with other similarity-sensitive indices) violates Axioms 1 and 3 from Section 4, which imply respectively that diversity is invariant to permutation of the species abundance vector, and that diversity is maximal for a perfectly even community.

5.4. Parametric Families of Indices

Conversion to effective numbers and accounting for species disparity are two important improvements to classical indices, which address several of their weaknesses with respect to well-reasoned diversity measurement. Yet there is another important issue with classical indices: they make *a priori* assumptions about how the species present in the community contribute to its diversity. Thus, for example, species richness assumes that all species contribute equally to diversity. However, it is very arguable that this leads to a good estimate of diversity. Intuitively, we would not wish for a species representing less than 0.1% of a community to contribute just as much to its diversity as a species representing 99% of the community.

While other classical indices do weigh some species more than others, this is often done in an implicit and inflexible way as a consequence of their mathematical formulation. This key fact is not always accounted for, or even recognized. For example, the Shannon index is biased more towards evenness than richness. Since richness weighs rare species just the same as abundant species, this implies that the Shannon index gives more significance to common species. In some study settings, this may not be a problem and can even be an asset, for example if the focus of the study is dominance in a community. But in other settings this is far from ideal, and there is no way to account for it if only the Shannon index is used without any independent analyses of richness and evenness.

These deficiencies can be addressed by using *parametric families of indices*, also referred to in the literature as *multivariate* or *compound indices*, which include an additional parameter that tunes the relative contribution of different species in the community to its diversity [6,69]. Each member of an index family is then defined by a specific value of the parameter.

If we wish to use such indices to avoid the implicit bias inherent in other diversity indices, how should they be formulated? Two schools of thought have developed [13]. The first, due largely to economists, proposes that different species should be given different weights in the index based on their characteristic features. The second approach, found mainly in ecology, weighs different species according to their relative abundances. This has been justified by observing that species' functional contributions typically vary with their abundance in the community [70].

Ideally, we would like to vary this balance or weighing of species. Then we are not wedded to a single *a priori* judgement of which species contribute more to the community's diversity, but can investigate various scenarios. This would mean introducing an additional parameter that tunes this weighing. Thus the index becomes a function of both the species abundance vector and the sensitivity parameter.

Parametric families of diversity indices introduce exactly such a parameter, which controls the index's sensitivity to species abundances and therefore avoids the issue of bias towards one component of diversity compared to another. This parameter can be varied according to the user's interests and needs. In this way, additional insights can be gained into the community's composition under different assumptions of whether rare or common species are thought to play more important roles.

Although the use of these indices for ecological diversity studies is only recently becoming more widespread, this type of index is not new. Parametric families of diversity indices first emerged decades ago, initially as information-theoretic measures that were first introduced to ecology by MacArthur [71].

The *Rényi entropy* was formulated to generalize several other entropy measures, notably the Shannon entropy. The Rényi entropy of order $\alpha \in [0, \infty] \setminus \{1\}$ is given by:

$$H_\alpha(\mathbf{p}) = \frac{1}{1-\alpha} \ln \sum_{i=1}^{S} p_i^\alpha, \tag{12}$$

where α is a parameter that modulates the index's sensitivity to species abundances; the case $\alpha = 1$ is excluded since the index is not defined for this value. Rather, the Shannon entropy is the limiting case of this entropy as $\alpha \to 1$. Note that the Rényi entropy was originally defined using the base 2 logarithm (so as to measure information content in bits), but in ecology the natural logarithm is much more frequently used.

A similar index was proposed by Tsallis [72] and is known as the *HCDT entropy* or the *Tsallis entropy*. It is given by:

$$^q H(\mathbf{p}) = \frac{1}{1-q} \left(1 - \sum_{i=1}^{S} p_i^q \right), \tag{13}$$

where q is the index's sensitivity parameter. For $q = 0$, the index is equivalent to species richness minus one, for $q = 1$ it is equivalent to the Shannon entropy, and for $q = 2$ it is equivalent to the Gini-Simpson index.

Most of the diversity indices, including all generalized entropy measures used in ecology and those mentioned above, are monotonic functions of the sum $\sum_{i=1}^{S} p_i^q$, or limits of such functions as q approaches unity [73]. Such indices include: species richness, Shannon entropy, all Simpson indices, all Rényi entropies, and many others. All of these measures result in the same expression for diversity when converted to their effective number equivalent, which are known as the *Hill numbers* after their originator [51]. This family of indices is given by:

$$^q D(\mathbf{p}) = \left(\sum_{i=1}^{S} p_i^q \right)^{1/(1-q)}, \tag{14}$$

where $q \in [0, \infty] \setminus \{1\}$ is known as the *order* of the diversity. Again, the case $q = 1$ is excluded since the index is undefined for this value; however, the limit exists and will be discussed shortly. The Hill numbers are the only family of ecological diversity indices that are known to satisfy all six axioms in Section 4 [46].

For all indices that are a function of $\sum_{i=1}^{S} p_i^q$, their effective number diversity depends only on the value of q and the relative species abundances, and not on the index's mathematical formulation. For example, the Simpson diversity index, the Simpson dominance index, and the Gini-Simpson index give the same effective number diversity, namely the Hill number of order 2:

$$^2 D(\mathbf{p}) = 1/ \left(\sum_{i=1}^{S} p_i^2 \right). \tag{15}$$

We also note the link to another family of diversity indices that were proposed by Patil and Taillie [47], and which reintroduced the Hill numbers from a different mathematical viewpoint, namely the use of a sensitivity parameter β that is linked to Hill's q by $q = \beta + 1$ [74].

The order q of the diversity index indicates its sensitivity to common and rare species. The diversity index of order zero ($q = 0$) is species richness: it is completely insensitive to relative species abundances. All values of q less than one result in diversity indices that disproportionately favour rare species, while all values of q greater than one lead to diversity indices that disproportionately favour the most common species.

The critical point, which weighs all species by their frequency without favouring any, occurs when $q = 1$. Note that $^q D$ is undefined at $q = 1$, but the limit exists and is equal to:

$$^1 D(\mathbf{p}) = \exp \left(-\sum_{i=1}^{S} p_i \ln p_i \right) = \exp \left(H_{\text{Sh}}(\mathbf{p}) \right). \tag{16}$$

This is the exponential of the Shannon entropy, a quantity that has deep connections to biology, information theory, physics, and mathematics, since it is a measure that is uniquely able to weigh elements of a system exactly by their frequency [75]. These elements might be species in a community, spin configurations in a particle system, or DNA sequences in a genome [76,77]. While the Shannon entropy (Equation (7)) provides a measure of how disordered these systems are (and hence is an index for how diverse they are), it is biased towards common elements rather than rare ones. The transformation to its exponential ^1D (Equation (16)) resolves this problem by not favouring any elements above others, and thus avoiding bias in its measurement. It is for these reasons that ^1D finds a role in various scientific disciplines.

Leinster and Cobbold [36] proposed a similarity-sensitive extension of the Hill numbers: as well as the relative abundance vector $\mathbf{p}$ and a sensitivity parameter $q \in [0, \infty]$, their index includes a similarity matrix Z. Then for $q \in [0, \infty] \setminus \{1\}$, the *Leinster-Cobbold diversity* of order q is given by:

$$^qD^Z(\mathbf{p}) = \left(\sum_{i=1}^{S} p_i \left((Z\mathbf{p})_i \right)^{q-1} \right)^{1/(1-q)}. \tag{17}$$

The Leinster-Cobbold family includes—either directly or upon simple transformation—species richness (Equation (5)), Rao's quadratic index (Equation (11)), the Shannon index (Equation (7)), the Gini-Simpson index (Equation (9)), the Berger–Parker index (Equation (6)), the Hill numbers (Equation (14)), and the Tsallis entropies (Equation (13)). Thus almost all of the measures discussed in Section 5, and others still, can be subsumed in one family of measures that are both effective numbers and similarity-sensitive.

Most of these indices are retrieved by setting the similarity matrix Z equal to the identity matrix, so that $Z_{ij} = 0$ (total dissimilarity) if $i \neq j$, and $Z_{ij} = 1$ (total similarity) if $i = j$. This is referred to as the 'naive' Leinster-Cobbold diversity, in the sense that it ignores species disparity. Then for example at $q = 0$ the naive Leinster-Cobbold index is equivalent to species richness. At the other extreme, for $q = \infty$ the naive index corresponds to the Berger–Parker index (Equation (6)), which depends only on the most abundant species and ignores all others.

The cases $q = 1$ and $q = \infty$ are excluded from the definition in Equation (17) because $^qD^Z(\mathbf{p})$ is not a valid expression for these values. However, the index does converge at these values, to

$$^1D^Z(\mathbf{p}) = \frac{1}{((Z\mathbf{p})_1)^{p_1} ((Z\mathbf{p})_2)^{p_2} \cdots ((Z\mathbf{p})_S)^{p_S}}, \tag{18}$$

as $q \to 1$, and to

$$^\infty D^Z(\mathbf{p}) = \frac{1}{\max_{i \in \{1,\dots,S | p_i \neq 0\}} (Z\mathbf{p})_i}, \tag{19}$$

as $q \to \infty$.

To close this section, we note that all of these parametric families are deeply linked. Most obviously, the naive Leinster-Cobbold indices (Equation (17)) are by definition equivalent to the Hill numbers (Equation (14)): $^qD = {}^qD^I$. Deeper links exist between the other indices due to the previously noted fact that all of them are monotonic functions of the sum $\sum_{i=1}^{S} p_i^q$, or limits of such functions as $q \to 1$ [73]. Thus, the Tsallis entropy (Equation (13)) is the base q logarithm of the q-th Hill number (Equation (14)),

$$^qH = \log_q {}^qD, \tag{20}$$

while the Rényi entropy (Equation (12)) is the natural logarithm of the Hill number of the same order,

$$H_\alpha = \ln {}^\alpha D. \tag{21}$$

These transformations are useful for the practical estimation of the Hill numbers (Section 7), since conversion to an entropy measure provides an easier approach to bias correction. Conversely, the Hill numbers can permit a more intuitive conception of diversity (in terms of effective number of species) which may be more useful for theoretical works.

5.5. Diversity Profiles

Different indices give different estimates of diversity, so it may happen that when calculating how a community's diversity has shifted after a disturbance, one index indicates that the diversity has increased, whereas another indicates that it has decreased [74]. Therefore, restricting the analysis to a single index can lead to a biased or distorted estimate of diversity, and consequently conclusions based on this measurement may be misleading or incorrect.

This can be avoided through the use of parametric families of diversity indices: as discussed in Section 5.4, these indices include a sensitivity parameter that controls the weighing of common and rare species in the community. Varying this parameter allows for a wider view of a community's diversity and provides more information than any single statistic.

A *diversity profile* is a graph that does exactly this: it shows at once multiple values of a community's diversity, so that they may easily be compared. For any parametric family $H(\mathbf{p}, q)$ where q is the index's sensitivity parameter, its diversity profile is a graph of $H(\mathbf{p}, q)$ against q. Changes in community diversity can then be studied by comparing profiles. These may be two different communities, or the same community at different points in time, typically before and after some ecological disturbance.

Communities' diversity profiles define a partial order on their diversities [78]. If the two communities are comparable, then one profile will lie pointwisely above the other, and the former community can be said to be more diverse than the latter. However, the profiles of the communities may intersect, in which case the communities are not directly comparable (and hence the ordering is only partial). This is again an unavoidable consequence of reducing the complex concept of diversity to a relatively simple numerical analysis, which diversity profiles are not able to entirely overcome. However, they are much more meaningful than univariate indices, and even when two communities are not directly comparable, studying where their profiles intersect can reveal which sections of the community (common or rare species) have shifted [74].

The region of a diversity profile where q is small gives information about species richness and rare species, since here $H(\mathbf{p}, q)$ is affected almost as much by rare species as common ones. The tail where q is large gives information about dominance and common species, since here $H(\mathbf{p}, q)$ is almost entirely unaffected by rare species. As the sensitivity parameter q increases, the diversity $H(\mathbf{p}, q)$ drops. More precisely, the diversity profile is always decreasing and continuous.

As an illustration, we plot in Figure 2 three classical (univariate) indices, and two parametric families of indices. For the same species abundance vector, we compute the value for univariate indices and the diversity profile for the parametric families of indices. Since univariate indices do not account for differential sensitivity to rare or common species, they do not vary with q. One of the parametric families is similarity-sensitive (the Leinster-Cobbold index), so that some pairs of species are more similar than others, while the other parametric family is insensitive to disparity (the Hill numbers). For the Leinster-Cobbold index, we use a random sensitivity matrix satisfying the conditions outlined in Section 5.4, i.e., it is symmetric and all its elements lie in the unit interval.

We first note that the univariate indices give different estimates of diversity, since they are not effective number indices. Hence they do not have a common unit, and the diversity values they give for the same community are significantly different (as discussed in Section 5.2).

For the parametric families, we notice the effect of including a similarity measure to account for disparity. For all values of q, the similarity-sensitive index gives a lower value of diversity than the naive index. This is to be expected since the naive index treats all species as equally different, whereas the similarity-sensitive index considers some species to be less distinct than others.

The steepness of the left-hand side of the profiles, where q is small, gives us information about the rare species in the community. As q increases, these rare species are given less weight by the index, and therefore the steeper the drop of the profile, the more rare species there are in the community. Again we notice that the naive index considers there to be more rare species than the similarity-sensitive index, since the slope of the former is steeper. In fact, the slope of the similarity-sensitive profile is so low that we can surmise that the similarity measure considers the rare species to be very similar.

Figure 2 also illustrates graphically the mathematical relationships between the indices that were discussed in Section 5.4: the Hill number of order 0 is equal to species richness, the Hill number of order 1 is equal to the exponential of the Shannon index, and the Hill number of order 2 is equal to the Gini-Simpson index.

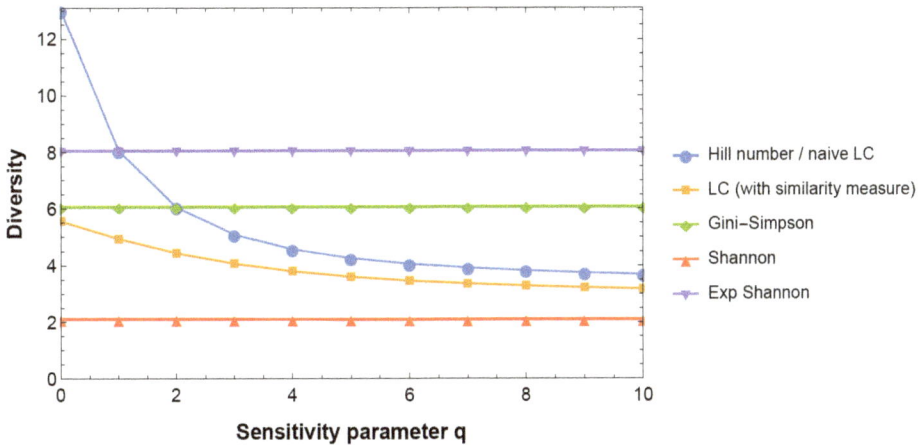

Figure 2. Comparison of diversity indices for a community given by the species abundance vector $(0.0407, 0.179, 0.0144, 0.00403, 0.31, 0.00858, 0.109, 0.0643, 0.0999, 0.021, 0.0638, 0.0184, 0.0677)$. Parametric families of indices (Hill numbers and Leinster-Cobbold index family) result in profiles, while univariate indices are not functions of q.

To illustrate the useful application of diversity profiles in ecological studies, we provide several examples. First, we show in Figure 3 the diversity profiles calculated by Leinster and Cobbold [36] relating to an experiment comparing the microbial communities in the guts of lean and overweight humans [79].

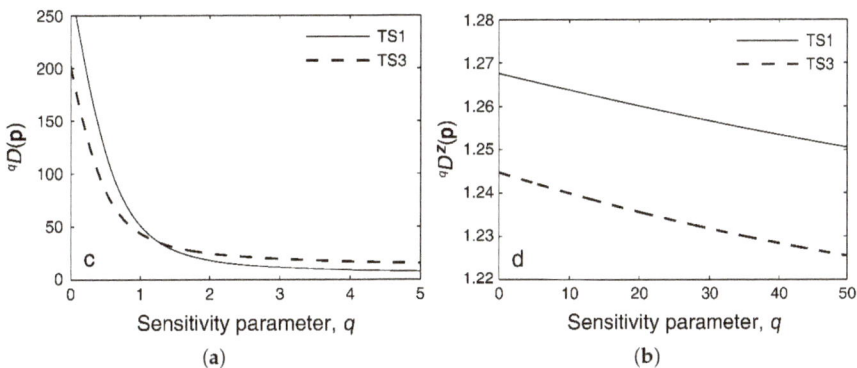

Figure 3. Comparison of diversity profiles using (**a**) a naive similarity measure and (**b**) a genetic similarity measure [36] for datasets representing two human gut microbiomes (TS1 and TS3) [79].

The diversity profiles for two particular test subjects from that study were compared, one subject being overweight and the other not. Since only a fraction of microbial species has been isolated and given taxonomic classifications, it was not possible for the authors to partition the microbes into species (this issue was discussed in Section 3.3). Instead, the authors turned to DNA sequencing data, which they used to construct a similarity matrix. Using the naive similarity matrix (Figure 3a), the diversity profiles cross at $q \approx 1$. This suggests that the community in the lean subject's gut is richer but less evenly distributed than that of the overweight subject. However, using a similarity matrix based on genetic measures (Figure 3b), the diversity in the lean subject is seen to be greater for all values of q, a conclusion supported by other analyses conducted by the authors [79].

Another study, again using the Leinster-Cobbold index, conducted a meta-analysis of microbial diversity studies by calculating the diversity profiles of the various microbial communities, in order to investigate whether the use of these profiles altered the interpretation or conclusions of the initial studies [61]. Aside from insights into community structure that were not detectable with classical (univariate) diversity measures, the authors also found that similarity-based and naive diversity profiles only agreed on which community was the most diverse in approximately 50% of cases, another strong argument for incorporating similarity information into diversity quantification [61].

In a conservation study of reptiles in Madagascar, Nopper et al. [80] used the Hill numbers to calculate diversity profiles. Their goal was to detect changes in the reptile communities' composition in order to assess the success of different forest management strategies. Their results indeed demonstrate that such differences existed in the Madagascar forests: significant differences were detected in community compositions between used and less-used forests when management rules were respected, whereas when these rules were ignored, the diversity profiles showed no significant differences between the reptile communities in used and less-used forests, implying homogenization of reptile community composition under this scenario [80]. Diversity profiles allow for the identification of such patterns in community composition, which are stronger conclusions than merely calculating similar values of diversity, as would be the only possibility with a classical index.

Genetic diversity profiles were used in another study to assess the effect of sample size for a genetic analysis—the authors' goal was to study how genetic diversity calculations were affected by minimum sample size and the effect of relatives (i.e., high genetic similarity) in the sample [81]. To do so, they compared the results of two classical (univariate) indices with diversity profiles calculated using the Hill numbers, for sampled populations of amphibians. The comparison indicated good agreement between the classical indices and the profiles at the relevant q values, but differences were found in profiles of rare alleles, which require higher sample sizes to calculate univariate measures. The authors took these results as evidence for the usefulness of their method for calculating minimum sample sizes, and for pinpointing possible underestimation of diversity due to rare alleles, all in service of accurate quantification of genetic diversity.

Other practical applications, particularly conservation studies, are often more interested in how diversity varies temporally rather than spatially. As an example, Iacchei et al. [82] undertook a study of temporal variations in the genetic composition of populations of two planktonic copepod species. Conventional population genetic analyses were able to characterize the structures of the populations, but found no significant differences in these over time. In contrast, the use of diversity profiles allowed for the identification of seasonal turnover patterns, particularly in rare species, adding important nuance to the researchers' understanding of these communities. The authors therefore concluded that their results "highlight the complementary insights" that can be obtained from these two techniques, which are especially useful for examining subtle temporal shifts in community composition.

Such practical applications of diversity indices often involve considering diversity at different scales across an ecosystem or landscape. Parametric families of indices are well suited for this purpose, since they can be meaningfully decomposed into components representing these different scales, a technique known as the *partitioning* of diversity.

6. Partitioning of Diversity

In Section 3, the three key components of diversity were discussed: richness, evenness, and disparity. Diversity can be partitioned according to these components. For example, the Simpson evenness index is derived from the Simpson diversity index by removing the effect of richness [83] (the Simpson index already neglects disparity). However, if we still wish to consider diversity as a unified concept of at least richness and evenness (and possibly also disparity), then diversity can also be partitioned according to the scales over which it is considered.

If we think of an ecological landscape, we will typically find one large community of species that aggregates smaller sub-communities, in terms of their spatial organization or other features. So when we consider the diversity of this landscape, to what are we referring?

There is the total diversity of the landscape: this is known as its *gamma diversity*. There is also the diversity of the sub-communities, which is known as the landscape's *alpha diversity*, or within-community diversity. More specifically, alpha diversity refers to the diversity of a uniform habitat of a fixed size [19]. The link between alpha and gamma is the diversity representing the differences between the sub-communities: its *beta diversity*, or between-community diversity. These concepts are illustrated in Figure 4.

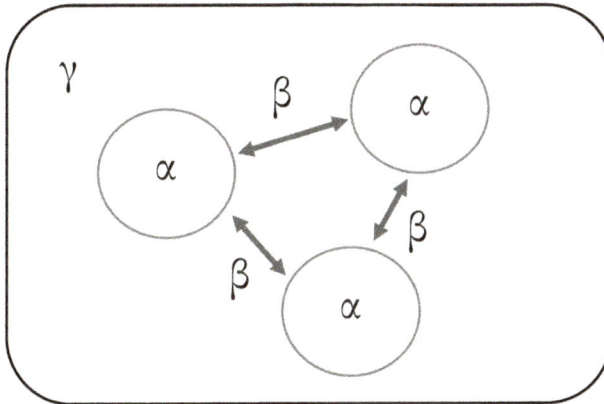

Figure 4. Partitioning diversity according to landscape scale. Gamma diversity is the total diversity of the landscape. Alpha diversity is the diversity within sub-communities. Beta diversity is the diversity between the sub-communities.

Alpha and gamma diversity are straightforward to conceive and measure, while beta diversity is trickier. It is agreed that beta diversity links alpha and gamma diversity, but how exactly should the former be derived from the latter?

The first partitioning to be proposed was multiplicative [84], so that $D_\alpha D_\beta = D_\gamma$. Later, others insisted that the independence of alpha and beta diversity is not necessary, since their dependence can be accounted for in other ways, and thus proposed an additive formulation [67], so that $D_\alpha + D_\beta = D_\gamma$.

In practice, ecologists generally measure alpha and gamma diversity, and then derive beta diversity from these. This has led to the emergence and use of a wide variety of measures of beta diversity, similar to the proliferation of biodiversity measures, despite beta diversity at least benefiting from an agreed definition, unlike biodiversity: beta diversity refers to the average dissimilarity in composition among sub-communities. The multitude of beta diversity indices can be attributed to similar causes as those underlying the vastness of the set of biodiversity indices. Different ecological studies wish to interrogate different research questions relating to beta diversity, and thus focus on different aspects of it, while the various beta diversity measures can have quite different statistical properties [85]. This has led to the further issue of devising measures that capture diversity at these different scales while also respecting the axioms in Section 4.

Why, then, are ecologists interested in wading into this complex topic? The multiplicative and additive partitions of diversity are appreciated and widely used because they provide a single set of values of alpha and beta diversity for a given landscape (sampling) scale, and are thus very suitable techniques for analyzing patterns or changes in diversity across multiple scales. Hence, they are able to provide ecologists with an intuitive and accessible measure of species diversity, and its spatial and temporal variations [85]. They are thus particularly appreciated by conservation ecologists, who are interested in large-scale ecological processes affecting biodiversity.

The disadvantage of the simplicity of additive or multiplicative diversity partitioning is that these methods also result in a significant loss of information, as the multidimensional features of a community are projected to a much lower-dimensional space. For example, the beta diversity obtained by diversity partitioning does not directly measure the differences in composition among the sub-communities, but is instead an average of this diversity, which is not found in any particular sub-community [86]. Unsurprisingly, as with diversity in the broader sense, there is no single index that provides a perfect measurement of beta diversity (for an in-depth discussion of beta diversity measurement, we direct the interested reader to the review in [87]).

Yet, just as with biodiversity indices, users of diversity partitions have in recent years fought their way through an analogous 'jungle' of indices, measures and metrics to arrive at the beginnings of a consensus. There is increasing agreement that the Hill numbers (Equation (14)) should be partitioned to measure beta diversity [88–90]. This is because the Hill numbers, as effective numbers and in contrast to most other indices, obey the replication principle (Axiom 6, discussed in Section 4).

The replication principle was first proposed by Hill [51] merely as a desirable characteristic of his eponymous family of indices, but it was later shown by Jost [91] that indices satisfying this principle permit the partitioning of gamma diversity into independent alpha and beta components, through a multiplicative partitioning. Hence any change in the magnitude of their partitioned components has a simple interpretation [86].

To illustrate the importance of the replication principle to a sensible partitioning of diversity, we use a simple example following Jost [46]. We consider a landscape made up of twenty island communities of equal size and with the same diversity, that do not share any species; hence the islands have identical relative abundance distributions, but no common species. Take as an example the relative abundance distribution given by $\mathbf{p}$ = (0.057, 0.012, 0.052, 0.064, 0.026, 0.060, 0.034, 0.012, 0.043, 0.034, 0.052, 0.049, 0.054, 0.060, 0.037, 0.052, 0.089, 0.040, 0.078, 0.044, 0.030, 0.021). If this landscape was now under threat, for example from rising sea levels, and finite resources dictated that only some of its diversity could be protected, how would a suitable fraction be identified?

Using the Gini-Simpson index (Equation (9)), we would find that the alpha diversity of each island is 0.95, while the gamma diversity of the total landscape is 0.998 (note that to calculate D_γ we must rescale $\mathbf{p}$ since the pooled community has twenty times as many species as each island). So if in the worst case only one island can be protected, this would imply that we have saved $0.95/0.998 = 95\%$ of the landscape's diversity—a fantastic result for relatively little effort. But then how much of the total diversity has been lost? Together, the nineteen doomed islands have a diversity of 0.997, so we have lost $0.997/0.998 = 99.9\%$ of the total diversity. Somehow, we have simultaneously saved 95% of the landscape's diversity, and lost 99.9% of it.

This nonsensical result is due to the Gini-Simpson index failing the replication principle, and hence its alpha and beta components do not scale sensibly when communities are pooled. Similar (though less egregious) results would be obtained with other indices that fail Axiom 6. In contrast, if we used an index that satisfies Axiom 6 (like the effective number equivalent of the Gini-Simpson index, or the Hill numbers), this would sensibly tell us that each of the twenty islands represents $1/20 = 5\%$ of the total landscape diversity, while the other nineteen collectively represent $19/20 = 95\%$ of it. Intuitive and consistent behaviour of this kind underlies the consensus forming around the choice of the Hill numbers for diversity partitioning.

A recent review of the topic provided the following guidelines [85]. First, that beta diversity should be measured using species richness (i.e., using $q = 0$), but the authors also acknowledged that depending on the research context, it may be more enlightening to also vary q so as to measure other Hill numbers as well, thereby obtaining a diversity profile for each community (see Section 5.5). Second, not to use entropy measures in diversity partitioning. Entropy measures are often not effective number indices, and also generally do not satisfy the replication principle (Axiom 6 in Section 4), leading to misleading results (a deeper discussion of this point can be found in [46]). Third, to make the necessary statistical adjustments to remove alpha/beta dependence when using an additive partitioning (see e.g., [92]).

As a consequence of this last point, an even more recent consensus is emerging around the use of the multiplicative Hill number partitioning for measuring beta diversity, and thus avoiding the additive partitioning [93]. This is due to the recognition that the independence of alpha and beta diversity is a desirable feature of diversity partitioning; among other reasons, it simplifies the statistical approaches used to detect patterns in beta diversity changes [87].

Finally, we note that the Hill family is not unique in its decomposition into independent alpha and beta components. This is also possible with other families of indices, but the analyses are more complicated than those required for the Hill numbers, and these families generally do not share the advantageous characteristics of the Hill family indices, most notably that they are effective numbers and they satisfy the replication principle, the usefulness of which we have justified in previous sections [91].

7. Practical Estimation of Diversity Indices

7.1. Background

When it comes to assessing the diversity of a community in practice, the species proportions $\mathbf{p} = (p_1, ..., p_S)$ are generally unknown. Without a full census of the community, calculation of its diversity using an index is not possible. Instead, this must be estimated from sampled data.

For the purpose of this discussion, we assume that we have a sample of n individuals from the community in question. Most estimators of diversity assume that this sample is random, namely that its observations are independent and identically distributed [94]. But this requires either an infinite population, or sampling with replacement. Generally, neither of these requirements are met: populations are finite and ecological studies typically sample populations without replacement [95]. This issue is avoided by considering a sample size much smaller than the population size, so that the effect of sampling without replacement from a finite population can be neglected [96]. Thus the problem reduces to estimating $\mathbf{p}$ using an appropriate sampling design, and then estimating the index as a function of the estimated species abundance vector.

A good estimator of a diversity index is one that is unbiased and with minimal variance [19]. Therefore, an appropriate measure of an estimator's error is the mean squared error (MSE), which is the sum of squares of the bias and the variance [97]. Bias generally arises for two reasons: unobserved species (those present in the community but not in the sample), and the non-linearity of indices with respect to the species abundances. The former source of bias is exacerbated the more strongly the index depends on S, the richness of the community. Hence, for example, the bias of the Simpson index (Equation (8)) is less than that of the Shannon entropy (Equation (7)) since the former assigns less weight to rare species, which are less likely to be sampled.

As with the indices themselves, their estimators should be selected carefully to ensure that their formulation is appropriate for the particular application. Interrogation of this point is key, particularly since non-experts often resort to dedicated statistical software or packages whose underlying and disparate estimation techniques can result in different calculations of the same quantity [7].

For most diversity indices, numerous estimators exist. The statistical details of their derivation and bias reduction are beyond the scope of this review, so in this section we do not aim for an in-depth discussion of these various estimators, but we instead focus on whether an unbiased estimator exists for a particular index, and point the interested reader towards more exhaustive reviews. Unbiased estimators are of course preferable to biased estimators, but in practice many techniques exist for reducing bias, often to minimal levels, so that good estimates of diversity can still be obtained with a biased index estimator.

7.2. Estimators of Diversity Indices

7.2.1. Richness

Estimates of richness are obviously strongly correlated with the size of the sample used to obtain them: the total number of species in a sampling unit increases with its area and the number of detected individuals, and therefore comparing two different sampling units will be subject to bias. To address this, methods have been proposed that reduce samples of different sizes to a single standard size, so that they may be comparable in terms of their richness, a procedure known as *rarefaction* [55].

The two most frequently used richness estimators are the Chao estimators (Chao1 and Chao2) [98,99], and the jackknife method [95]. The Chao estimators estimate the number of unobserved species based on the number of species that are observed either once or twice. The jackknife method is based on resampling, and aims to reduce the estimator's bias by considering subsets of the sample, which are obtained by removing a certain number of observations (this number is known as the order of the jackknife method).

Brose et al. [100] provide a decision tree for choosing between these estimators, based either on the estimated range of sample coverage or community evenness. Both these estimators can be efficiently implemented and calculated, hence their popularity. For an overview of other richness estimation techniques, see for example [101].

7.2.2. Simpson Index

Simpson [57] provided an unbiased estimator of his own index (Equation (8)), which is based on the number of pairs of individuals drawn without replacement:

$$\hat{H}_{\text{Si}} = \frac{1}{n(n-1)} \sum_{i=1}^{S} n_i(n_i - 1) \tag{22}$$

where n_i is the number of sampled individuals of species i.

7.2.3. Shannon Index

Due to its importance in numerous scientific fields, the Shannon entropy (Equation (7)) has seen the development of many different estimators (Marcon [19] provides an overview). The details of these estimators are beyond the scope of this review, but for our purposes we note that the estimation of the Shannon index is subject to more bias than the estimation of Simpson's index, since the former is more sensitive to richness, i.e., to the number of species that are sampled.

An estimator frequently used in ecology is due to Chao and Shen [102], and is based on the steepness of the rarefaction curve—recall that the rarefaction procedure reduces bias in estimating richness, and thus reduces the same bias in the Shannon estimator.

7.2.4. Hill Numbers

Chao and Jost [103] formulated a low-bias estimator of the Hill numbers (Equation (14)), and is based on the discovery rates of new species with respect to the sample size n. This estimator therefore also allows for the estimation of a community's diversity profile, which can significantly improve the analysis of diversity shifts in sampled communities.

Haegeman et al. [104] used a different approach, by proposing two different estimators of the Hill family through the generalization of Chao's richness estimators. These estimators provide a lower and an upper bound to delimit the range in which the true Hill diversity values are expected to lie. For larger q, including the indices of Simpson and Shannon, this range is relatively narrow and thus these estimators yield a good estimate of these Hill numbers. But as q decreases to approach zero, this unsurprisingly diverges as the Hill diversities approach species richness, whose estimate is inherently biased. Thus in general, Hill numbers of larger order can be estimated with less uncertainty. This is again unsurprising since the larger the order, the less weight is given to rare species, which are less likely to be sampled.

7.2.5. Leinster-Cobbold index

The Leinster-Cobbold indices provide a similarity-sensitive extension of the Hill numbers, and so their estimation is related but more complex. An additional complication in this similarity-sensitive case is that we have no information about the similarity between unobserved and observed species. Either the estimator can be corrected for these unobserved species, or their contribution to diversity can be explicitly estimated by assuming that the average similarity between unobserved and observed species is equal to the average similarity between observed species [105].

Using the latter approach, Marcon et al. [105] provide low-bias estimators of the Leinster-Cobbold index family (Equation (17)), along with a multiplicative partitioning that can be used to estimate alpha and beta diversity (see Section 6). They make use of a bias-corrected estimator originally developed for the Tsallis entropy (Equation (13)), through the transformation in Equation (20). The bias-corrected estimator is first applied to a similarity-sensitive extension of the Tsallis entropy, which is then transformed into the Leinster-Cobbold index.

8. Conclusions

In this review, we have provided an overview of the key conceptual issues with which researchers are confronted when they seek to measure ecological diversity, and we have surveyed and compared the most common diversity indices used in ecology. In general, we recommend that a diversity index is selected that:

(1) satisfies the key axioms in Section 4;
(2) can be expressed as an effective number (Section 5.2);
(3) can be extended to account for species disparity if necessary (Section 5.3);
(4) can be parameterized to obtain diversity profiles (Section 5.5); and
(5) an estimator (preferably unbiased) can be obtained so that the index can be used in practical applications (Sections 6 and 7).

In the relevant sections of this paper, we have elaborated on each recommendation by motivating its importance for reasonable and realistic diversity measurement.

Overall, the Hill numbers have been recommended for quantifying species diversity for several decades [106], and despite the significant progress in measuring and estimating diversity that has been achieved in the intervening years, this consensus has not changed [86] and we do not deviate from it. This family of indices has recently been extended to measure similarity-sensitive diversity [36], phylogenetic diversity [107] and functional diversity [108]. However, these types of diversity do not yet dispose of a knowledge base as well-developed as that of species diversity, due to their more complex nature (for a review of these issues in functional diversity measurement, see e.g., [109]; for phylogenetic diversity, see e.g., [63]). The ability of the Hill numbers to be extended to measure these different types of diversity hints at their power as a unifying framework for measuring biodiversity.

However, this general recommendation is of course not always the best choice. Depending on the particular context of an ecological study, other indices may be more appropriate. As discussed in Section 2, the vast number and variety of available diversity indices allows researchers to be flexible in their choice of index, with the key stipulation that the underlying definition of the index should first be considered carefully to ensure that it is appropriate for the particular application, and will not lead to misinterpretations.

Thus more specific study contexts may benefit from other recommendations. For example, in conservation ecology, the goal of a study is oftentimes to rank different communities (or sites) by their diversities, since only certain sites can be selected for protection or management [6]. In these cases, parametric families of indices are generally preferred since they provide information about both richness and evenness, and so outperform classical indices at discriminating between such sites. On the other hand, a ranking of sites by their diversity cannot always be obtained using a parametric family of indices, since their diversity profiles impose only a partial order on diversity. Hence a well-selected univariate index may be more appropriate in this particular setting.

In sum, when one is faced with the task of selecting a diversity index, it is enlightening to consider Baumgärtner's observation that all diversity indices require, to a greater or lesser extent, that the user makes *a priori* value judgements about which aspects of biodiversity are more important [13]. This includes, for example, the choice of the sensitivity parameter q in parametric families of indices. In choosing the value of q, the user specifies whether rare or common species are more important for the type of diversity they wish to study. Hints about the most appropriate value(s) to study might be gleaned from other insights into the community, but q cannot itself be measured or inferred from the community. Hence, there is no correct value of q in any application, and it must instead be chosen by the user based on the goals of the study.

This advice applies to all indices, not just those involving a sensitivity parameter. Diversity indices merely provide a summary of an inherently complex and multidimensional concept: a community's structure. Indices achieve this summarization in different ways by emphasizing different aspects of diversity. To avoid confusion and misinterpretation, users should first define their objectives and then choose the appropriate measure for the specific problem.

Author Contributions: A.J.D., J.M.B. and B.D.B. wrote the paper.
Funding: This research was funded by a UGent–BOF GOA project "Assessing the biological capacity of ecosystem resilience", grant number BOFGOA2017000601'.

Conflicts of Interest: The authors declare no conflict of interest.

References

1. Tilman, D.; Isbell, F.; Cowles, J. Biodiversity and ecosystem functioning. *Annu. Rev. Ecol. Evol. Syst.* **2014**, *45*, 471–493. [CrossRef]
2. Dunne, J.; Williams, R. Cascading extinctions and community collapse in model food webs. *Philos. Trans. R. Soc. Lond. B Biol. Sci.* **2009**, *364*, 1711–1723. [CrossRef] [PubMed]
3. Hart, S.; Usinowicz, J.; Levine, J. The spatial scales of species coexistence. *Nat. Ecol. Evol.* **2017**, *1*, 1066. [CrossRef] [PubMed]
4. Mendes, R.; Evangelista, L.; Thomaz, S.; Agostinho, A.; Gomes, L. A unified index to measure ecological diversity and species rarity. *Ecography* **2008**, *31*, 450–456. [CrossRef]
5. DeLong, D. Defining biodiversity. *Wildl. Soc. Bull. (1973–2006)* **1996**, *24*, 738–749.
6. Morris, E.; Caruso, T.; Buscot, F.; Fischer, M.; Hancock, C.; Maier, T.; Meiners, T.; Müller, C.; Obermaier, E.; Prati, D.; et al. Choosing and using diversity indices: Insights for ecological applications from the German Biodiversity Exploratories. *Ecol. Evol.* **2014**, *4*, 3514–3524. [CrossRef] [PubMed]

7.   Lucas, R.; Groeneveld, J.; Harms, H.; Johst, K.; Frank, K.; Kleinsteuber, S. A critical evaluation of ecological indices for the comparative analysis of microbial communities based on molecular datasets. *FEMS Microbiol. Ecol.* **2017**, *93*, fiw209. [CrossRef] [PubMed]

8.   Hurlbert, S. The nonconcept of species diversity: A critique and alternative parameters. *Ecology* **1971**, *52*, 577–586. [CrossRef] [PubMed]

9.   Ricotta, C. Through the jungle of biological diversity. *Acta Biotheor.* **2005**, *53*, 29–38. [CrossRef] [PubMed]

10. Jost, L. Entropy and diversity. *Oikos* **2006**, *113*, 363–375. [CrossRef]

11. MacArthur, R. Patterns of species diversity. *Biol. Rev.* **1965**, *40*, 510–533. [CrossRef]

12. Pielou, E. The measurement of diversity in different types of biological collections. *J. Theor. Biol.* **1966**, *13*, 131–144. [CrossRef]

13. Baumgärtner, S. *Measuring the dIversity of What? And for What Purpose? A Conceptual cOmparison of Ecological and Economic Measures of Biodiversity*; Interdisciplinary Institute for Environmental Economics: Heidelberg, Germnay, 2004.

14. Lambert, P.; Aronson, J. Inequality decomposition analysis and the Gini coefficient revisited. *Econ. J.* **1993**, *103*, 1221–1227. [CrossRef]

15. Amroabady, B.; Renani, M.; Tayebi, S. Analysis of diversity in companies using entropy index. *Int. J. Econ. Perspect.* **2017**, *11*, 1133–1144.

16. Mora Villarrubia, R.; Ruiz-Castillo, J. *Entropy-Based Segregation Indices*; Technical Report; Universidad Carlos III, Departamento de Economia: Madrid, Spain, 2010.

17. Maignan, C.; Ottaviano, G.; Pinelli, D.; Rullani, F. *Bio-Ecological Diversity vs. Socio-Economic Diversity: A Comparison of Existing Measures*; Working Papers; Fondazione Eni Enrico Mattei: Milan, Italy, 2003; Volume 13.

18. Ferrer, J.; Prats, C.; López, D. Individual-based modelling: An essential tool for microbiology. *J. Biol. Phys.* **2008**, *34*, 19–37. [CrossRef] [PubMed]

19. Marcon, E. Mésures de la Biodiversité. Ph.D. Thesis, AgroParisTech, Paris, France, 2015.

20. Ogunseitan, O. *Microbial Diversity*; Blackwell Science Ltd.: Oxford, UK, 2005.

21. Bell, T.; Newman, J.; Silverman, B.; Turner, S.; Lilley, A. The contribution of species richness and composition to bacterial services. *Nature* **2005**, *436*, 1157–1160. [CrossRef] [PubMed]

22. Grman, E.; Lau, J.; Schoolmaster, D.; Gross, K. Mechanisms contributing to stability in ecosystem function depend on the environmental context. *Ecol. Lett.* **2010**, *13*, 1400–1410. [CrossRef] [PubMed]

23. Hooper, D.; Dukes, J. Overyielding among plant functional groups in a long-term experiment. *Ecol. Lett.* **2004**, *7*, 95–105. [CrossRef]

24. Hillebrand, H.; Bennett, D.; Cadotte, M. Consequences of dominance: A review of evenness effects on local and regional ecosystem processes. *Ecology* **2008**, *89*, 1510–1520. [CrossRef] [PubMed]

25. Wilsey, B.; Potvin, C. Biodiversity and ecosystem functioning: Importance of species evenness in an old field. *Ecology* **2000**, *81*, 887–892. [CrossRef]

26. Lemieux, J.; Cusson, M. Effects of habitat-forming species richness, evenness, identity, and abundance on benthic intertidal community establishment and productivity. *PLoS ONE* **2014**, *9*, e109261. [CrossRef] [PubMed]

27. Wittebolle, L.; Marzorati, M.; Clement, L.; Balloi, A.; Daffonchio, D.; Heylen, K.; De Vos, P.; Verstraete, W.; Boon, N. Initial community evenness favours functionality under selective stress. *Nature* **2009**, *458*, 623–626. [CrossRef] [PubMed]

28. Smith, B.; Wilson, J. A consumer's guide to evenness indices. *Oikos* **1996**, *76*, 70–82. [CrossRef]

29. Eliazar, I.; Sokolov, I. Measuring statistical evenness: A panoramic overview. *Phys. A Stat. Mech. Appl.* **2012**, *391*, 1323–1353. [CrossRef]

30. Ricotta, C. A recipe for unconventional evenness measures. *Acta Biotheor.* **2004**, *52*, 95–104. [CrossRef] [PubMed]

31. Ginebra, J.; Puig, X. On the measure and the estimation of evenness and diversity. *Comput. Stat. Data Anal.* **2010**, *54*, 2187–2201. [CrossRef]

32. Jost, L. The relation between evenness and diversity. *Diversity* **2010**, *2*, 207–232. [CrossRef]

33. Alatalo, R. Problems in the measurement of evenness in ecology. *Oikos* **1981**, *37*, 199–204. [CrossRef]

34. Tuomisto, H. An updated consumer's guide to evenness and related indices. *Oikos* **2012**, *121*, 1203–1218. [CrossRef]

35. Kvalseth, T. Evenness indices once again: Critical analysis of properties. *SpringerPlus* **2015**, *4*, 232. [CrossRef] [PubMed]

36. Leinster, T.; Cobbold, C. Measuring diversity: The importance of species similarity. *Ecology* **2012**, *93*, 477–489. [CrossRef] [PubMed]

37. Hey, J. The mind of the species problem. *Trends Ecol. Evol.* **2001**, *16*, 326–329. [CrossRef]

38. Stirling, A. A general framework for analysing diversity in science, technology and society. *J. R. Soc. Interface* **2007**, *4*, 707–719. [CrossRef] [PubMed]

39. Hughes, A.; Inouye, B.; Johnson, M.; Underwood, N.; Vellend, M. Ecological consequences of genetic diversity. *Ecol. Lett.* **2008**, *11*, 609–623. [CrossRef] [PubMed]

40. Botta-Dukát, Z. Rao's quadratic entropy as a measure of functional diversity based on multiple traits. *J. Veg. Sci.* **2005**, *16*, 533–540. [CrossRef]

41. Shimatani, K. On the measurement of species diversity incorporating species differences. *Oikos* **2001**, *93*, 135–147. [CrossRef]

42. Pavoine, S.; Ollier, S.; Pontier, D. Measuring diversity from dissimilarities with Rao's quadratic entropy: Are any dissimilarities suitable? *Theor. Popul. Biol.* **2005**, *67*, 231–239. [CrossRef] [PubMed]

43. Hardy, O.; Senterre, B. Characterizing the phylogenetic structure of communities by an additive partitioning of phylogenetic diversity. *J. Ecol.* **2007**, *95*, 493–506. [CrossRef]

44. Rényi, A. On Measures of Entropy and Information. *Proc. Fourth Berkeley Symp. Math. Stat. Prob.* **1961**, *1*, 547–561.

45. Davydov, D.; Weber, S. A simple characterization of the family of diversity indices. *Econ. Lett.* **2016**, *147*, 121–123. [CrossRef]

46. Jost, L. Mismeasuring biological diversity: Response to Hoffmann and Hoffmann (2008). *Ecol. Econ.* **2009**, *68*, 925–928. [CrossRef]

47. Patil, G.; Taillie, C. Diversity as a concept and its measurement. *J. Am. Stat. Assoc.* **1982**, *77*, 548–561. [CrossRef]

48. Reardon, S.; Firebaugh, G. Measures of multigroup segregation. *Sociol. Methodol.* **2002**, *32*, 33–67. [CrossRef]

49. Hoffmann, S. *Generalized Distribution Based Diversity Measurement: Survey and Unification*; Technical Report; Otto-von-Guericke University Magdeburg, Faculty of Economics and Management: Magdeburg, Germany, 2008.

50. Solomon, D. *Ecological Diversity in Theory*; Chapter A Comparative Approach to Species Diversity; International Co-Operative Publishing House: Fairland, MA, USA, 1979; pp. 29–35.

51. Hill, M. Diversity and evenness: A unifying notation and its consequences. *Ecology* **1973**, *54*, 427–432. [CrossRef]

52. Tuomisto, H. A consistent terminology for quantifying species diversity? Yes, it does exist. *Oecologia* **2010**, *164*, 853–860. [CrossRef] [PubMed]

53. Berger, W.; Parker, F. Diversity of planktonic foraminifera in deep-sea sediments. *Science* **1970**, *168*, 1345–1347. [CrossRef] [PubMed]

54. Shannon, C. A mathematical theory of communication. *Bell Syst. Tech. J.* **1948**, *27*, 379–423. [CrossRef]

55. Borcard, D.; Gillet, F.; Legendre, P. *Numerical Ecology with R*; Springer: New York, NY, USA, 2018.

56. Keylock, C. Simpson diversity and the Shannon–Wiener index as special cases of a generalized entropy. *Oikos* **2005**, *109*, 203–207. [CrossRef]

57. Simpson, E. Measurement of diversity. *Nature* **1949**, *163*, 688. [CrossRef]

58. Jost, L.; Chao, A. *Diversity Analysis*; Taylor & Francis: Milton Keynes, UK, 2008.

59. Pallmann, P.; Schaarschmidt, F.; Hothorn, L.; Fischer, C.; Nacke, H.; Priesnitz, K.; Schork, N. Assessing group differences in biodiversity by simultaneously testing a user-defined selection of diversity indices. *Mol. Ecol. Resour.* **2012**, *12*, 1068–1078. [CrossRef] [PubMed]

60. Hare, M.; Nunney, L.; Schwartz, M.; Ruzzante, D.; Burford, M.; Waples, R.; Ruegg, K.; Palstra, F. Understanding and estimating effective population size for practical application in marine species management. *Conserv. Biol.* **2011**, *25*, 438–449. [CrossRef] [PubMed]

61. Doll, H.; Armitage, D.; Daly, R.; Emerson, J.; Goltsman, D.; Yelton, A.; Kerekes, J.; Firestone, M.; Potts, M. Utilizing novel diversity estimators to quantify multiple dimensions of microbial biodiversity across domains. *BMC Microbiol.* **2013**, *13*, 259. [CrossRef] [PubMed]

62. Armitage, D.; Gallagher, K.; Youngblut, N.; Buckley, D.; Zinder, S. Millimeter-scale patterns of phylogenetic and trait diversity in a salt marsh microbial mat. *Front. Microbiol.* **2012**, *3*, 293. [CrossRef] [PubMed]

63. Tucker, C.; Cadotte, M.; Carvalho, S.; Davies, T.; Ferrier, S.; Fritz, S.; Grenyer, R.; Helmus, M.; Jin, L.; Mooers, A.; Pavoine, S.; et al. A guide to phylogenetic metrics for conservation, community ecology and macroecology. *Biol. Rev.* **2016**, *92*, 698–715. [CrossRef] [PubMed]

64. Wang, S.; Loreau, M. Biodiversity and ecosystem stability across scales in metacommunities. *Ecol. Lett.* **2016**, *19*, 510–518. [CrossRef] [PubMed]

65. Kang, S.; Rodrigues, J.; Ng, J.; Gentry, T. Hill number as a bacterial diversity measure framework with high-throughput sequence data. *Sci. Rep.* **2016**, *6*, 38263. [CrossRef] [PubMed]

66. Buckland, S.; Yuan, Y.; Marcon, E. Measuring temporal trends in biodiversity. *AStA Adv. Stat. Anal.* **2017**, *101*, 461–474. [CrossRef]

67. Lande, R. Statistics and partitioning of species diversity, and similarity among multiple communities. *Oikos* **1996**, *76*, 5–13. [CrossRef]

68. Rao, C. Diversity: Its measurement, decomposition, apportionment and analysis. *Sankhya Indian J. Stat. Ser. A* **1982**, *44*, 1–22.

69. Lamb, E.; Bayne, E.; Holloway, G.; Schieck, J.; Boutin, S.; Herbers, J.; Haughland, D. Indices for monitoring biodiversity change: Are some more effective than others? *Ecol. Indic.* **2009**, *9*, 432–444. [CrossRef]

70. Magurran, A. *Measuring Biological Diversity*; Blackwell Science Ltd.: Oxford, UK, 2004.

71. MacArthur, R. Fluctuations of animal populations and a measure of community stability. *Ecology* **1955**, *36*, 533–536. [CrossRef]

72. Tsallis, C. Possible generalization of Boltzmann-Gibbs statistics. *J. Stat. Phys.* **1988**, *52*, 479–487. [CrossRef]

73. Ricotta, C. On parametric evenness measures. *J. Theor. Biol.* **2003**, *222*, 189–197. [CrossRef]

74. Tóthmérész, B. Comparison of different methods for diversity ordering. *J. Veg. Sci.* **1995**, *6*, 283–290. [CrossRef]

75. Tuomisto, H. Commentary: Do we have a consistent terminology for species diversity? Yes, if we choose to use it. *Oecologia* **2011**, *167*, 903–911. [CrossRef]

76. Suyari, H. On the most concise set of axioms and the uniqueness theorem for Tsallis entropy. *J. Phys. A Math. Gen.* **2002**, *35*, 10731. [CrossRef]

77. Schmidt, T.; Matias Rodrigues, J.; Mering, C. Limits to robustness and reproducibility in the demarcation of operational taxonomic units—Supplementary Information. *Environ. Microbiol.* **2014**, *17*, 1689–1706. [CrossRef] [PubMed]

78. Rousseau, R.; Van Hecke, P.; Nijssen, D.; Bogaert, J. The relationship between diversity profiles, evenness and species richness based on partial ordering. *Environ. Ecol. Stat.* **1999**, *6*, 211–223. [CrossRef]

79. Turnbaugh, P.; Hamady, M.; Yatsunenko, T.; Cantarel, B.; Duncan, A.; Ley, R.; Sogin, M.; Jones, W.; Roe, B.; Affourtit, J.; et al. A core gut microbiome in obese and lean twins. *Nature* **2009**, *457*, 480–484. [CrossRef] [PubMed]

80. Nopper, J.; Ranaivojaona, A.; Riemann, J.; Rödel, M.O.; Ganzhorn, J. One forest is not like another: The contribution of community-based natural resource management to reptile conservation in Madagascar. *Trop. Conserv. Sci.* **2017**, *10*. [CrossRef]

81. Sánchez-Montes, G.; Ariño, A.; Vizmanos, J.; Wang, J.; Martínez-Solano, Í. Effects of sample size and full sibs on genetic diversity characterization: A case study of three syntopic Iberian pond-breeding amphibians. *J. Hered.* **2017**, *108*, 535–543. [CrossRef] [PubMed]

82. Iacchei, M.; Butcher, E.; Portner, E.; Goetze, E. It's about time: Insights into temporal genetic patterns in oceanic zooplankton from biodiversity indices. *Limnol. Oceanogr.* **2017**, *62*, 1836–1852. [CrossRef]

83. Colwell, R. *The Princeton Guide to Ecology*; Chapter Biodiversity: Concepts, Patterns, and Measurement; Princeton University Press: Princeton, NJ, USA, 2009; pp. 257–263.

84. Whittaker, R. Evolution and measurement of species diversity. *Taxon* **1972**, *21*, 213–251. [CrossRef]

85. Veech, J.; Crist, T. Toward a unified view of diversity partitioning. *Ecology* **2010**, *91*, 1988–1992. [CrossRef] [PubMed]

86. Chao, A.; Chiu, C.H.; Jost, L. Unifying species diversity, phylogenetic diversity, functional diversity, and related similarity and differentiation measures through Hill numbers. *Annu. Rev. Ecol. Evol. Syst.* **2014**, *45*, 297–324. [CrossRef]

87. Anderson, M.; Crist, T.; Chase, J.; Vellend, M.; Inouye, B.; Freestone, A.; Sanders, N.; Cornell, H.; Comita, L.; Davies, K.; et al. Navigating the multiple meanings of β diversity: A roadmap for the practicing ecologist. *Ecol. Lett.* **2011**, *14*, 19–28. [CrossRef] [PubMed]

88. Tuomisto, H. A diversity of beta diversities: Straightening up a concept gone awry. Part 1. Defining beta diversity as a function of alpha and gamma diversity. *Ecography* **2010**, *33*, 2–22. [CrossRef]

89. Chao, A.; Chiu, C.H.; Hsieh, T. Proposing a resolution to debates on diversity partitioning. *Ecology* **2012**, *93*, 2037–2051. [CrossRef] [PubMed]

90. Reeve, R.; Leinster, T.; Cobbold, C.; Thompson, J.; Brummitt, N.; Mitchell, S.; Matthews, L. How to partition diversity. *arXiv* **2014**, arXiv:1404.6520.

91. Jost, L. Partitioning diversity into independent alpha and beta components. *Ecology* **2007**, *88*, 2427–2439. [CrossRef] [PubMed]

92. De Bello, F.; Lavergne, S.; Meynard, C.; Lepš, J.; Thuiller, W. The partitioning of diversity: Showing Theseus a way out of the labyrinth. *J. Veg. Sci.* **2010**, *21*, 992–1000. [CrossRef]

93. Botta-Dukát, Z. The generalized replication principle and the partitioning of functional diversity into independent alpha and beta components. *Ecography* **2018**, *41*, 40–50. [CrossRef]

94. Butturi-Gomes, D.; Petrere, M., Jr.; Giacomini, H.; Zocchi, S. Statistical performance of a multicomparison method for generalized species diversity indices under realistic empirical scenarios. *Ecol. Indic.* **2017**, *72*, 545–552. [CrossRef]

95. Butturi-Gomes, D.; Junior, M.; Giacomini, H.; Junior, P. Computer intensive methods for controlling bias in a generalized species diversity index. *Ecol. Indic.* **2014**, *37*, 90–98. [CrossRef]

96. Grabchak, M.; Marcon, E.; Lang, G.; Zhang, Z. The generalized Simpson's entropy is a measure of biodiversity. *PLoS ONE* **2017**, *12*, e0173305. [CrossRef] [PubMed]

97. Marcon, E. *Practical Estimation of Diversity from Abundance Data*; CCSD: Villeurbanne, France, 2015; pp. 1–29.

98. Chao, A. Nonparametric estimation of the number of classes in a population. *Scand. J. Stat.* **1984**, *11*, 265–270.

99. Chao, A. Estimating the population size for capture-recapture data with unequal catchability. *Biometrics* **1987**, *43*, 783–791. [CrossRef] [PubMed]

100. Brose, U.; Martinez, N.; Williams, R. Estimating species richness: Sensitivity to sample coverage and insensitivity to spatial patterns. *Ecology* **2003**, *84*, 2364–2377. [CrossRef]

101. Gotelli, N.; Colwell, R. *Biological Diversity: Frontiers in Measurement and Assessment*; Chapter Estimating Species Richness; Oxford University Press: Oxford, UK, 2011; pp. 39–54.

102. Chao, A.; Shen, T.J. Nonparametric estimation of Shannon's index of diversity when there are unseen species in sample. *Environ. Ecol. Stat.* **2003**, *10*, 429–443. [CrossRef]

103. Chao, A.; Jost, L. Estimating diversity and entropy profiles via discovery rates of new species. *Methods Ecol. Evol.* **2015**, *6*, 873–882. [CrossRef]

104. Haegeman, B.; Hamelin, J.; Moriarty, J.; Neal, P.; Dushoff, J.; Weitz, J. Robust estimation of microbial diversity in theory and in practice. *ISME J.* **2013**, *7*, 1092. [CrossRef] [PubMed]

105. Marcon, E.; Zhang, Z.; Hérault, B. *The Decomposition of Similarity-Based Diversity and Its Bias Correction*; CCSD: Villeurbanne, France, 2014; pp. 1–12.

106. Heip, C.; Herman, P.; Soetaert, K. Indices of diversity and evenness. *Oceanis* **1998**, *24*, 61–88.

107. Chao, A.; Chiu, C.H.; Jost, L. Phylogenetic diversity measures based on Hill numbers. *Philos. Trans. R. Soc. Lond. B Biol. Sci.* **2010**, *365*, 3599–3609. [CrossRef] [PubMed]

108. Chiu, C.H.; Chao, A. Distance-based functional diversity measures and their decomposition: A framework based on Hill numbers. *PLoS ONE* **2014**, *9*, e100014. [CrossRef] [PubMed]

109. Schleuter, D.; Daufresne, M.; Massol, F.; Argillier, C. A user's guide to functional diversity indices. *Ecol. Monogr.* **2010**, *80*, 469–484. [CrossRef]

MDPI

St. Alban-Anlage 66

4052 Basel

Switzerland

Tel. +41 61 683 77 34

Fax +41 61 302 89 18

www.mdpi.com

Mathematics Editorial Office

E-mail: mathematics@mdpi.com

www.mdpi.com/journal/mathematics

www.ingramcontent.com/pod-product-compliance
Lightning Source LLC
Chambersburg PA
CBHW051846210326
41597CB00033B/5798